Structural features	of [illegible] [illegible] [illegible] [illegible] lines [illegible] [illegible] definite direction and a definite system and distributed successively in rows. Contrast of differential movement large. Structural lines conspicuous.	[illegible] generally of [illegible] [illegible] structures short [illegible] different directions and [illegible] and distributed in [illegible]. Contrast of differential movement small. Structural lines not conspicuous.	occurring as [illegible] lines. Block faulting well developed, sometimes [illegible] the [illegible]. Thrust or overthrust also occur. Structures arranged in a definite direction and a definite system and distributed successively in rows. Contrast of differential movement sharp and reflected in the Moho discontinuity. Structural lines conspicuous.
Paleogeography	I^1: Trough and island chain lying as a long line. I^{2-3}: Fold mountain, foredeep and intermontane depression (continental facies and red-bed climate in I^{2-3}). Paleotopographic contrast large.	[illegible] sea basin (or island sea) and continent. Paleotopographic contrast small.	Continental (rarely marine) basin lying as a short belt. Fault-fold mountain. Coal-generating (often in II_1) or red-bed climate (often in III^{2-3}). Paleotopographic contrast large.
Neotectonic movements	Severe. Seismicity high (the largest earthquake has M≥8). Volcanoes. Quaternary folds, faults and tilting. Tectonic contrast large. Differential movement strong.	Relatively weak. Seismicity comparatively low (earthquakes generally have M≤5 1/2). Uplift and subsidence affecting a large area. Tectonic contrast small. Differential movement weak.	Severe. Seismicity high (the largest earthquake has M≥8). Volcanoes. Quaternary block-faulting, folds and tilting. Arcogenesis. Tectonic contrast large. Differential movement strong.
Landforms (present-day)	Topographic contrast large. Trough alternating with island chain. Topographic youth. High mountain and deep [illegible]	Contrast small. Peneplain. Topographic old age.	Contrast large. Mainly topographic youth. Plateau and depression (low to middle mountain in the later period). Mountain range lying as a short or long

TECTONICS OF CHINA

CHEN GUODA
Academician, Professor of Geotectonics
Honorary Director of Changsha Institute
of Geotectonics, Academia Sinica

IAP

INTERNATIONAL ACADEMIC PUBLISHERS
A Pergamon-CNPIEC Joint Venture

PERGAMON PRESS
OXFORD · NEW YORK · BEIJING · FRANKFURT
SÃO PAULO · SYDNEY · TOKYO · TORONTO

DISTRIBUTORS

U.K.	Pergamon Press plc, Headington Hill Hall, Oxford OX3 0BW, England
U.S.A.	Pergamon Press, Inc., Maxwell House, Fairview Park, Elmsford, New York 10523, U.S.A.
PEOPLE'S REPUBLIC OF CHINA	International Academic Publishers, Xizhimenwai Dajie, Beijing Exhibition Centre, Beijing 100044, People's Republic of China
FEDERAL REPUBLIC OF GERMANY	Pergamon Press GmbH, Hammerweg 6, D-6242 Kronberg, Federal Republic of Germany
BRAZIL	Pergamon Editora Ltda, Rua Eça de Queiros, 346, CEP 04011, Paraiso, São Paulo, Brazil
AUSTRALIA	Pergamon Press Australia Pty Ltd., P.O. Box 544, Potts Point, N.S.W. 2011, Australia
JAPAN	Pergamon Press, 5th Floor, Matsuoka Central Building, 1-7-1 Nishishinjuku, Shinjuku-ku, Tokyo 160, Japan
CANADA	Pergamon Press Canada Ltd., Suite No. 271, 253 College Street, Toronto, Ontario, Canada M5T 1R5

First edition 1989

Library of Congress Cataloging in Publication Data

Ch'en, Kuo-ta, 1912–
Tectonics of China.
Translated from the Chinese.
"A Pergamon-CNPIEC Joint Venture".
Bibliography: p.
1. Geology, Structural. 2. Geology—China. I. Title.
QE601.C433 1988 551.8'0951 89–2966

British Library Cataloguing in Publication Data

Chen, Guoda
Tectonics of China
1. Tectonics
I. Title
551.1'36

ISBN 0-08-037033-0

Printed in the People's Republic of China

PREFACE

Geologic cause has attained a rapid development since the founding of the People's Republic of China. In recent years research work on geologic mapping, prospecting and exploration of mineral deposits and earthquake prediction has yielded important results, and a wealth of data has been accumulated. This has enabled geologists, on the basis of the achievements gained by predecessors, to comprehensively study the tectonics of China and compile new tectonic maps. This book is an explanatory text to accompany the Tectonic Map of China on a scale of 1 : 4,000,000 [37], compiled from the viewpoint of diwa (geodepression)[1] and based on numerous geological, seismological, geophysical and geochemical data. For the convenience of our readers, the map is reduced to 1 : 40,000,000 and included here in this book (Fig. 3.8.).

The diwa theory, advanced by the author in the 1950s, has theoretically broken through the geosynclinal concept founded by the American geologists J. Hall and J. D. Dana, and singled out the third main tectonic element of the earth's crust named "diwa" (1959) or "activated" (1956) region. Practically it probes into metallogenic regularities from a new point of view and thus expands the clue and domain of the search for mineral deposits.

The crust of China mentioned in this book is divided into 23 tectonic regions by using a synthetic analysis method. The analysis of geological history has been integrated with the consideration of tectonic stresses.

China has its specific geologic features and complex crustal evolution. The results concerning its tectonic character, development history, metallogenic and seismological characteristics are of basic importance for the understanding of origin and evolution of global tectonic features, and metallogenic and seismological characteristics of other parts of the world.

---

1. The term "diwa" is derived from Chinese words meaning ge + depression, being a special kind of intermontane structural basins occurring mostly on a platform region, filled chiefly with mollasoid sediments. A diwa region is marked with high relief by basins diwa as negative units intercalated with ranges as positive units called "geodome". It is a post-platform mobile region, the third main tectonic element of the earth's crust besides geosynclinal and platform regions, proposed by the author in 1959 [12, 19, 30, 38]. Its synonym is "activated" region proposed by the author in 1956, because it is formed by the activation of platform [9, 15].

I am indebted to many geological and seismological organizations and individuals for their help. Space does not permit a complete listing but I would like to acknowledge to the following my appreciation for material aid in various places of the book: geological bureaus of all provinces, municipalities and autonomous regions in China, the Chinese Academy of Geological Sciences, the National Bureau of Seismology and geological or seismological teams (institutes) under them. I wish to express my thanks for the use of their material. I am particularly grateful to Huang Su, Guan Yinwen, Xue Jiamou, Wei Bailin, Chen Jiachao, Liu Yixuan, Deng Jing, Wen Shanji, Wei Zhouling, Hu Huoyan and Huang Ruihua for their assistance in my writing. Special appreciation goes to Wen Changen, Xu Jianqing, Tan Pixian, Song Dayun, Cai Shuzhi, Huang Zhenying and Ye Dengzi, who have prepared all of the drawings.

The English version of the book is revised and some new data are included. I hope that this book will become a most useful reference book to those who are interested in the tectonics, metallogeny, neotectonics, seismotectonics, plate tectonics and geomorphology of China.

Chen Guoda

Changsha Institute of Geotectonics

Academia Sinica

June, 1988

About the Author

Academician, Chinese Academy of Sciences; Vice-Chairman, Commission on Tectonics of Ore Deposits, IAGOD, IUGS since 1980; Vice-Chairman, Geological Society of China, 1979-1983; Honorary Director, Changsha Institute of Geotectonics, Academia Sinica; Academic Advisor and Professor, Central-South University of Technology; Winner of the National Prize for Natural Sciences, 1982.

CONTENTS

magmatic formations, predominantly granitic, were intruded in the eastern Kunlun Mountains. In Carboniferous time the geosynclines continued their development. Carboniferous strata in the northern part of the western Kunlun Mountains are composed of limestones, shales and intercalated intermediate-composition effusive rocks with a thickness of 4,300 m. They constitute a limestone subformation and a shale subformation, respectively. Eastwards, the Carboniferous is a shale formation interbedded with sandstones and limestones; the total thickness is more than 2,200 m. Southwards, in the northern part of the western Kunlun Mountains, Lower Carboniferous strata consist of sandstones, phyllites, limestones and intermediate-composition volcanic rocks and appear to be a terrigenous volcanic subformation. The Middle and Upper Carboniferous strata consist chiefly of limestones and rarely of basalts. In the western segment of the eastern Kunlun Mountains, the Carboniferous strata consist primarily of limestones with many gravels and sandy-argillaceous rocks, attaining a thickness of 7,000 m. It grades into a terrigenous-carbonate formation again in [illegible] Mountain.

In the western segment of the northern part of the western Kunlun Mountains, the Lower Permian consists of intermediate-composition volcanics and tuff clastics, more than 3,000 m thick. In the northern zone of the southern part of the central Kunlun Mountains, it consists chiefly of sandstones, argillaceous limestones and intermediate volcanic rocks, more than 2,000 m thick, constituting a pyroclastic formation. But in the southern zone it consists mainly of [illegible] and sandy-argillaceous deposits, more than 1,050 m thick. Observed in the eastern Kunlun Mountains is chiefly the Lower Permian. Its lower part consists [illegible] conglomerates, graywackes and slates, 1,616-2,600 m thick; its middle part consists of limestones, 1,000-1,402 m thick; and its upper part is composed of [illegible] sandstones intercalated with black slates and crystalline limestones, 3,155-6,000 m thick. They group into a thick terrigenous-carbonate formation.

As concerns the sequence of occurrence of the geosynclines, the central Kunlun Mountains rose first and then the geosynclines moved southwards and northwards. At the end of the Permian the geosynclines closed first in the western Kunlun Mountains and then in the eastern where the geosynclinal regime existed probably till the end of the Triassic and where a series of linear [illegible] Magmatic intrusion of Hercynian age spread over the whole region [illegible] granites. After that the region entered into a [illegible] period. The Triassic-Quaternary section is composed of a molasse formation which reaches a thickness of 1,875 m on the south slope of the western Kunlun Mountains. In addition, there were also polyphasic volcanic extrusions of [illegible] composition. Neotectonic movement is prominent. [illegible] volcanic explosion still occurred at the turn of an arcuate structure [illegible] toward the south in the central part of the region. Quaternary lavas have been [illegible] along major fractures in the southern part of Aqikekuli Lake (Fig. 2.1[illegible]).

Generally speaking, the region still shows tectonic activities in the Himalayan orogeny: block-faulting and upward movement on the basis of the [illegible] fold belt and indications of reactivation of old structures in the new period. On account of a scarcity of geological data, it is tentatively assigned to a [illegible] fold region.

CHAPTER 1
GENERAL CONSIDERATIONS

Introduction

This bock is written frcm the viewpoint of diwa theory [12, 15, 17, 18, 19, 30, 38, 161] and its extension----theory of progression with transformation between mobile and "stable" regions[13,14]. It attempts to approach the tectonic characteristics and crustal evolution of China as well as the spatial and temporal relationships between them and mineral deposits and earthquakes in this country.

Regions or periods of different tectonic character differ from each other in magmatism, tectonism, sedimentation, metamorphism, and geochemical processes, thus having different metallogenic conditions and yielding different minerals or deposit types. Therefore, formation and localization of any mineral deposits are controlled by definite regional geologic features and tectonic environment. Endogenic ore deposits are related mainly to magmatic formations, structural features and country rocks existing in a given tectonic region during a corresponding metallogenic epoch. Exogenic ore deposits are controlled by sedimentary formations, paleogeographic (geomorphologic and climatic) and paleogeologic conditions prevailing in a given tectonic region during a corresponding metallogenic epoch. Therefore, we can, by way of understanding the tectonics of China, apply these relationships to the study of the metallogeny of China and to the analysis of the age and location of various mineral deposits in order to search for ore potential areas on a broader scale.

Next, different tectonic regions often possess different features of crustal movements and display neotectonic movements of varying intensity and form, thus yielding different seismogenic structures and producing or being pregnant with earthquakes of different magnitude. Therefore, we can, by way of understanding the tectonics of China, apply these relationships to the study of the peculiarities and spatial and temporal regularities in earthquake occurrence as a basis for seismological studies and long-term earthquake prediction.

I. Tectonic division of China

1. Main tectonic elements of the earth's crust in China

Tectonic elements of the earth's crust are basic tectonic units. Each tectonic element possesses specific characteristics of its own reflecting the definite features of geologic development.

Any part of the earth's crust moves, changes and develops unceasingly. But the markedness of its movement varies in different portions of the earth's crust, and the form of its expression is also different. Therefore, according to the intensity of tectonic activity and the mode of its expression, main tectonic elements of the earth's crust fall into two groups: (strongly) mobile and (relatively) stable regions. The former possess a high mobility and a strong tectonic activity, while the latter have a low mobility and a weak tectonic activity.

Both the mobile and stable regions exist in China. On the basis of their principal features and construction[28], they may be divided into several types. So far as the well-studied tectonic elements are concerned, mobile regions include geosynclinal and diwa regions; while stable ones, platform regions. Among them geosynclinal and platform regions are the only two tectonic elements by geosynclinal-platform theory. The diwa region is a new type post-platform mobile region, which has been put forward and expounded on the basis of summing-up of production practice in China[9, 12, 15, 19].

2. Criteria for recognition of tectonic elements in China

In this book tectonic elements are recognized by using a synthetic analysis method, that is, the analysis of geologic history has been integrated with the consideration of tectonic stresses.

The criteria for recognition of tectonic elements of different character and types by using the synthetic analysis method may be summarized as follows:

(1) Tectonic contrast

By tectonic contrast (C) we mean the tectonic relief of a given area during a given period[31]. It is a function of the density of tectonic relief (d), the amplitude of tectonic relief (h) and the velocity of tectonic relief (v). The relationship is:

$$C = f(d, h, v)$$

where d is the number of tectonic relief at a unit distance in a cross section, h is the greatest amplitude of tectonic relief, and v is the height difference of tectonic relief in a unit time.

At any place and during any period, tectonic contrast will inevitably express itself in the following respects: sedimentary formations and structural features. They are the principal indicators and the former is more important for the understanding of tectonic contrast in geologic history. In addition, paleotopographic contrast, estimated from an analysis of sedimentary formations, can also be used, to some extent, as a supplementary indicator of paleotectonic contrast. As for the modern tectonic epoch, tectonic contrast is reflected in Quaternary sedimentary formations, neotectonic features and topographic contrast.

By sedimentary formations we mean assemblages of sedimentary rocks genetically associated with each other in lithology and facies and deposited in a tectonic region with definite character and type, during a definite development stage and under a definite geographical and climatic condition. By structural features we mean structural forms (folds, faults, etc.), their types and dimensions as well as intensity of structural disturbances[31].

(2) Magmatism

This includes rock types and attitude (intrusion or extrusion) of magmatic formations as well as intensity, scale, frequency, age and sequence of magmatic activities.

(3) Metamorphism

This includes rock types of metamorphic formations, grade and cause of metamorphism, and scale and sphere of its influence.

(4) Types of crustal movements

By these we mean horizontal and vertical movements. They are recognized chiefly from an analysis of the above-mentioned three formations and one reformation. As regards horizontal movement, one can infer and reconstruct the stress field in a given region by means of analysing the directions of structural lines and tectonic systems composed of them, using the strain ellipsoid theory and applying corresponding modelling. Thus, it is possible to solve the problem of mechanism as related to the orientation and shape of tectonic elements (especially geosynclinal and diwa regions, including their internal positive and negative units) controlled by these tectonic systems.

(5) Construction of a tectonic region

By construction of a tectonic region we mean the number of structural layers within it[28]. A structural layer is a stratigraphic assemblage having a definite type of sedimentary formations deposited in a definite tectonic environment during a certain development stage and accompanied by a corresponding type of structural features and sometimes magmatic and metamorphic formations. The structural layer may be divided into three classes, large, medium, and small, according to the duration of the representative development stage.

1) Basic structural layer (hereafter called "structural layer"). It represents a basic stage of crustal development, e.g. pre-geosynclinal, geosynclinal, platform and diwa structural layers (symbolized by X, I, II, and III in succession).

2) Structural sublayer. It represents a minor stage within a certain basic development stage, e.g. structural sublayers of the early and late periods of a geosynclinal stage (symbolized by I^1 and I^3) and of the early, middle and late periods of a platform and a diwa stage (symbolized by II^1, II^2, II^3, and III^1, III^2, III^3 respectively).

3) Small structural sublayer. It represents a still smaller stage within a minor development stage, e.g., I^{1a}, I^{1b} and other small structural sublayers within an early geosynclinal structural sublayer.

Each tectonic element has a representative structural layer of its own, the newest structural layer formed during the whole time of its existence. The representative structural layer of a geosynclinal region is the geosynclinal structural layer. Where a geosyncline still exists, the geosynclinal structural layer is the newest one there.

Next, any tectonic element may inherit representative structural layers of its predecessors, on which it has developed. For example, a geosynclinal region is developed on a pre-geosynclinal tectonic element. Consequently, under a geosynclinal structural layer generally exists a pre-geosynclinal one (crystalline basement). A platform region is developed on a geosynclinal region or a pre-geosynclinal tectonic element. Consequently, under a platform structural layer often exists a geosynclinal or pre-geosynclinal one.

Finally, for the two above-mentioned reasons, tectonic elements formed at different stages of crustal development vary in number of structural layers, i.e., in construction. Tectonic element formed at an earlier stage has a simple construction, while that formed at a later stage possesses a complex one. As calculated from the geosynclinal structural layer, the geosynclinal region has only one structural layer, the platform region a two-storied construction, whereas the diwa region a three-storied construction (Fig. 1.1).

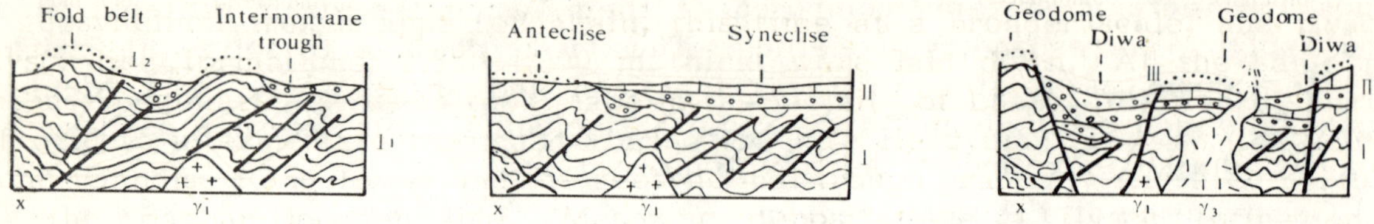

Fig. 1.1. Difference in construction between diwa (right), platform (center) and geosynclinal (left) regions. Structural layer: X, Pre-geosynclinal; I, Geosynclinal; II, Platform; III, Diwa. Magmatic rock: γ_{I}, Geosynclinal; γ_{III}, Diwa. Dotted line represents the eroded part of a preceding structural layer.

It is worth noting that in any tectonic element some structural layers may be absent or partly absent (for instance, in an uplifted area).

3. Principal characteristic features of some tectonic elements in China

The principal characteristic features of some well-defined tectonic elements in China, i.e., geosynclinal, platform and diwa regions, may be summarized as follows.

(1) Characteristic features in the process of development

1) Geosynclinal region. It is developed on a pre-geosynclinal tectonic element. Therefore, a pre-geosynclinal stage was present there. At the boundary between the geosynclinal and the pre-geosynclinal stage there generally occurred strong crustal movement leading to an unconformity, which separates the geosynclinal structural layer from the crystalline basement.

Three periods can be distinguished in the history of the tectonic development of a geosynclinal region in the light of the degree of mobility (Fig. 1.2). According to the regularities in the development of mobile regions, the early period of development of a geosynclinal region is the initial-mobility period. During this period, folding and faulting produce tectonic and topographic relief with a gradually increasing contrast, and geanticlines (positive units) and intervening geosynclines (negative units) take shape. In geosynclinal troughs clastic and other geosynclinal deposits derived from geanticlines and neighbouring uplifts accumulate. Meanwhile, magmatic activity (volcanic extrusion) occurs. The middle period is the maximum-mobility period. During this period horizontal

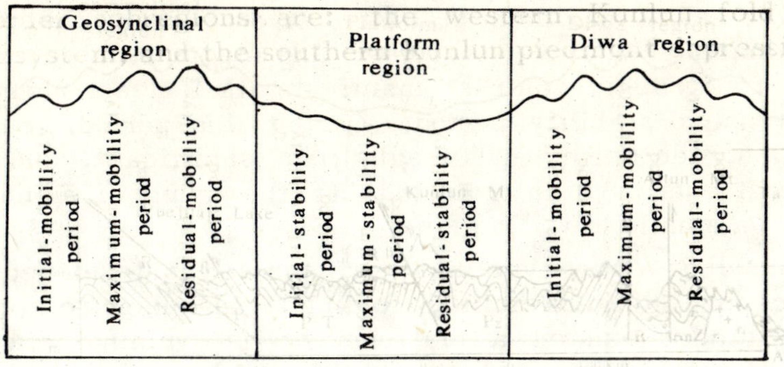

Fig. 1.2. Process of development of geosynclinal, platform and diwa regions. The curve designates intensity of tectonic mobility.

movement predominates as a whole; tectonic and topographic contrast manifests itself most sharply; folding, faulting, magmatism and metamorphism occur strongly. Geosynclinal deposits experience folding and uplifting, and eventually give rise to fold belts (positive units) and intermontane or piedmont depressions (negative units). The final period is the residual-mobility period. During this period tectonism and magmatism become weaker; tectonic and topographic contrast manifests itself less sharply; fold belts are subjected to erosion, and sedimentary processes take place in intermontane (and piedmont) depressions. As a result of tectonic evolution in this period, the geosynclinal region dies out gradually, giving place to a stable region, that is the platform.

The development of a geosynclinal region can also be divided into two periods in the light of its principal features. The earlier period is characterized by the appearance of geosynclines and the accumulation of sediments. It is named geosynclinal period corresponding to the initial-mobility and the early maximum-mobility period. The later period is marked by the appearance of a fold belt and its erosion. It is called fold belt period (or fold period for short) and corresponds to the late maximum-mobility and the residual-mobility periods.

When a geosynclinal region enters into the fold period, it is also called geosynclinal fold region.

2) Platform region (or platform for short). It is developed on a geosynclinal region (sometimes directly on an ancient pre-geosynclinal tectonic element), so that a geosynclinal stage must have been present there. The boundary between the platform and the geosynclinal stage is marked by the general peneplanation of a geosynclinal fold belt. Therefore, the platform structural layer has a planar distribution and is separated from the underlying geosynclinal structural layer by a regional unconformity representing a long-term erosion.

Three periods can be outlined in the history of the tectonic development of a platform region in the light of the degree of stability (Fig. 1.2). The early period is called initial-stability period. During this period, the region has just been transformed from a geosyncline into a platform, so tectonic activity is still high, magmatism significant and tectonic and topographic contrast sharp.

The middle period is called maximum-stability period. During this period, vertical movement predominates as a whole, tectonism and magmatism occur most weakly, and tectonic and topographic contrast manifests itself least sharply. The late period is called residual-stability period. During this period, tectonic activity becomes high again, tectonism and magmatism significant, and tectonic and topographic contrast sharp. It is a prelude to the dying of a platform and its transformation into a mobile (diwa) region.

3) Diwa region. It is a new-type mobile region and is developed on a platform or a region of completed geosynclinal folding (and older tectonic elements). Therefore, mostly there existed a platform or geosynclinal stage. The diwa structural layer is separated from the platform one by intense folding and faulting marked generally by an apparent unconformity.

Three periods, initial-, maximum- and residual-mobility periods, can also be distinguished in the history of the tectonic development of a diwa region in the light of the degree of mobility and according to the regularities in the development of mobile regions. During the initial-mobility period, older structural layers, upon which the diwa region has been developed, were intensely reworked by arcogenesis, folding and faulting. Tectonic and topographic relief with an increasingly sharpening contrast appears in the region. Geodome uplifts (positive units) and intervening diwa basins (negative units) take shape. The latter begin to receive diwa deposits composed mostly of continental clastics and transported from geodomes or adjacent uplifts. During the maximum-mobility period, horizontal crustal movement predominates, and folding and especially block-faulting occur most intensely. Diwa deposits accumulated in this period and earlier are disturbed together with the underlying platform, geosynclinal and pre-geosynclinal structural layers. They form tectonically and topographically more contrasting geodome uplifts (fold-fault or block-faulting belts) and intervening new-built diwa basins (intermontane depressions). During the residual-mobility period, folding, faulting and topographic contrast become less sharp, fold-fault (block-faulting) belts are eroded and reduced to a lower level, and late diwa deposits continue to accumulate in intermontane depressions.

The development of a diwa region can also be divided into two periods in the light of its principal features. The earlier period is featured by the appearance of earlier diwa basins and is called diwa period corresponding to the initial-mobility and the early maximum-mobility period. The later period is marked by the appearance of fold-fault belts (or block-faulting belts corresponding to later geodomes) and is called fold-fault belt period (fold-fault period for short) corresponding to the late maximum-mobility and the residual-mobility periods.

After entering into the fold-fault (belt) period, the diwa region may also be called diwa fold-fault region or fold-fault region for short.

(2) Other features

For the sake of comparison, other features and recognition criteria of several well-defined tectonic elements are summarized in Table 1.1.

Table 1.1. Comparison of the principal features of the main tectonic elements in China

Principal features[1)]			Geosynclinal region	Platform region	Diwa region
Construction[2)]			One-storied (X + I)	Two-storied (X + I + II)	Three-storied (X + I + II + III)
Tectonic contrast			Large	Small	Large
Sedimentary formations	Peculiarities of distribution		Arranged in a long belt, distributed in a definite direction and grouped in a system. I^1 is controlled by geosynclines and geanticlines, I^{2-3} by piedmont and intermontane depresssions. lying on X.	Arranged in a planar pattern. Covering I and / or X.	Arranged in a short belt, distributed in a definite direction and grouped in a system. Controlled by diwa basins and lying in small pieces on II, I and / or X[3)].
	Sedimentary facies	Rocks	I^1: Chiefly sandstone and shale, intercalated with pyroclastic rocks and sometimes with limestone (generally rich in argillaceous strips). I^3 is rich in conglomerate.	Sandstone and shale; limestone common; conglomerate rare (generally in $II^{1,3}$)	Conglomerate (especially fanglomerate) abundant; sandstone and shale; limestone rare. Sometimes pyroclastic rocks abundant.
		fossils	I^1: Chiefly marine. Plankton fossils common, but benthos rare.	$II^{1,3}$: Continental or marine. II^2: Chiefly marine. Sometimes alternating marine and	Chiefly continental (piedmont proluvial, intermontane alluvial and limnetic), occasionally

Sedimentary formation	Sedimentary facies	Facies &	I^{2-3}: transitional into continental in the upper part.	continental, benthos fossils abundant in marine deposits.	intercalated with marine and lagoon facies or marine-continental transitional facies (in $III^{1,3}$)
		Sorting	Composition, grain size and roundness all poorly sorted, especially in I^{2-3}. Polymictic sandstone abundant.	Well-sorted, especially in II^2. Monomictic sandstone abundant. Roundness grades frequently III-IV; sorting coefficient S_o generally < 2.15	Often poorly sorted, especially in III^3. Polymictic sandstone well-developed. Roundness grades frequently 0-I; S_o generally > 2.15, in some cases 7-8, or even 10.
		Variability	Lithology, facies and thickness highly variable, especially across geosynclines, piedmont and intermontane depressions. I^1 thick in geosynclines (usually more than 10,000 m), but thin or absent on geanticlines. I^{2-3} present only in piedmont and intermontane depressions, generally thick.	Less variable, especially for fine-grained and precipitated rocks in II^2. Thick in syneclises (individual bed measured in hundreds of meters, total thickness in thousands of meters), but thin or absent on anteclises.	Highly variable, especially for coarse-grained rocks. Thick or widespread in diwa basins (measured generally in several thousand, sometimes up to ten thousand meters), but thin, limited or absent on geodomes.
		Sediment rhythm	I^1: Marine flysch or flysch-like. Frequently of discontinuity type. Elements more. Individual element thin (measured in centimeters).	Also present. Frequently of continuity type. Elements generally less. Individual element thick.	Continental flysch. Frequently of discontinuity type. Elements more. Individual element thick (measured in one or ten meters).
	Principal formation types		I^1: Flysch and graywacke formations distributed widely. Sometimes spilite-	A large number of carbonate, oolitic-textured and quartzose sandstone formations.	Cathysian (molasse-like) -type formation, subdivided into Pingxiang-type (coal-bearing), Nanxiong-type (red beds), Jiande-

Sedimentary formation	Principal formation types	I^{2-3} : Molasse formation.		type (red beds, containing volcanic material) and Danxia-type (red beds, unusually rich in coarse grains) subformations.
	*	Unconformities common, but I^1 can graduate into I^{2-3}.	Stratigraphic breaks frequent, marked generally by disconformities.	Unconformities common.
	* *	Unconformity	Unconformity	Usually unconformity.
Magmatism		I^{1-2}: Strong, large-scale. Rocks complicated. General order: Basic-ultrabasic → acid-intermediate; extrusion→intrusion[4)]. Extrusion occurs generally in I^1 under marine conditions. Acid-intermediate intrusion in a great quantity takes place in I^2, but relatively weakly in I^3. Petrochemistry: K and Na contents comparatively low, Fe and Mg contents high.	Relatively weak, small-scale, and simple. Predominantly basic rocks, frequently seen in $II^{1,3}$.	Strong, large-scale. Rocks complicated. General order: acid-intermediate and alkaline → basic-ultrabasic; intrusion → extrusion. Extrusion occurs under continental conditions. Acid-intermediate intrusion in a great quantity takes place frequently in III^{1-2}, but basic extrusion in III^3. Petrochemistry[5)]: K and Na contents comparatively high, Fe and Mg contents low.
Metamorphism		I^2: Regional metamorphism, low-grade (or high-grade); I^3: occasionally present, but weak.	Usually absent.	III^{1-2}: Dynamic and contact, especially dislocation metamorphism, low-to-high-grade; III^3: low-grade.
Structural features		Deformation intense, on a large scale and with many phases. Fold often	Deformation weak. Folds often of discontinuous type (or broad type on the bor-	Deformation intense, on a large scale and with many phases. Fold often of broad or tight type,

* Relations between sedimentary sequences.

** Relation with base.

Structural features	of tight type (or broad type for I^3), occurring as a long line. Large-scale thrusts. Structures arranged in a definite direction and a definite system and distributed successively in rows. Contrast of differential movement large. Structural lines conspicuous.	ders). Faults generally of a high angle. Structures short of any definite directions and systems and distributed in isolation. Contrast of differential movement small. Structural lines not conspicuous.	occurring as a short line. Block-faulting well-developed, sometimes penetrating the crust. Thrust or even nappe also occur. Structures arranged in a definite direction and a definite system and distributed successively in rows. Contrast of differential movement sharp and reflected in the M- discontinuity. Structural lines conspicuous.
Paleogeography	I^1:Trough and island chain lying as a long line. I^{2-3}: Fold mountain, foredeep and intermontane depression (continental facies and red-bed climate in I^{2-3}). Paleotopographic contrast large.	Broad sea basin (or island sea) and continent. Paleotopographic contrast small.	Continental (rarely marine)basin lying as a short belt. Fault-fold mountain. Coal-generating (often in III^1) or red-bed climate (often in III^{2-3}). Paleotopographic contrast large.
Neotectonic movements	Severe. Seismicity high (the largest earthquake has $M \geqslant 8$). Volcanoes. Quaternary folds, faults and tilting. Tectonic contrast large. Differential movement strong.	Relatively weak. Seismicity comparatively low (earthquakes generally have $M \leqslant 5\ 1/2$). Uplift and subsidence affecting a large area. Tectonic contrast small. Differential movement weak.	Severe. Seismicity high (the largest earthquake has $M \geqslant 8$). Volcanoes. Quaternary block-faulting, folds and tilting. Arcogenesis. Tectonic contrast large. Differential movement strong.
Landforms (present-day)	Topographic contrast large. I^1:Trough alternates with island chain. I^2: Topographic youth; high mountain and deep	Contrast small. Peneplain. Topographic old age.	Contrast large. Mainly topographic youth. Plateau and depression (low to middle mountain in the later period). Mountain range lying as a short or long

Landforms (present-day)	valley. I^3: Low to middle mountain. Mountain range lying as a long line.		line. Block-faulting control obvious.
Crustal movement	Mainly horizontal. Folding strong in I^2, but gradually weak in I^3.	Mainly vertical. Oscillatory movement predominant.	Mainly horizontal for the most part of the stage. Compression predominant in III^{1-2} while tension in III^3. Faulting strong, especially in III^3.
Geophysical characteristics	Gravity anomalies arranged in a belt, distributed in a definite direction and grouped in a system. I^1: The positive alternates with the negative or predominant[6)]; I^3: The negative predominant[7)]. Heat flow values high[10)], especially in I^{1-2}. P_n 7.3-8.0[13)].	Gravity anomalies well-distributed and short of a a clear linear arrangement[8)]. Heat flow values low[11)]. $P_n > 8.2$[13)].	Gravity anomalies arranged in a belt, distributed in a definite direction and grouped in a system. The negative predominant. The more a region uplifts, the more a Bouguer anomaly decreases (especially in III^2)[9)]. Heat flow values high[12)], especially in III^2. Low velocity layer may be shallow. P_n 7.3-8.1[13)].
Ore formation and its existence	I^1: Endogenic mineral deposits particularly abundant and associated with basic and ultrabasic as well as intermediate and acid rocks. Volcanic mineral deposits generally seen in I^1 and formed under marine conditions. Intrusive mineral deposits formed in I^2 and found in I and X. Exogenic mineral deposits plentiful in I^1 and I^3. Mineral deposits inherited from X are controlled by the original tectonic regime and reworked by geosynclinal process.	Exogenic mineral deposits particularly rich, formed under various marine and continental conditions and found in II. Endogenic mineral deposits associated mostly with basic rocks and found in II, I and X. Mineral deposits inherited from X and I are controlled by the original tectonic regimes respectively and reworked occasionally by platform process.	Endogenic mineral deposits particularly abundant and associated with acid-intermediate and alkaline as well as basic and ultrabasic rocks. Volcanic ores formed mostly in III^{2-3} under continental conditions. Intrusive ores found in all structural layers. Exogenic mineral deposits chiefly continental in origin and found in III (but oil and gas can migrate). Mineral deposits inherited from X, I and II are controlled by the original tectonic regimes respectively and reworked by diwa process.

1) All the principal features are indicative of the representative structural layer or development stage, with the exception of the "construction". They are most typical of the middle period, except the indicated cases.

2) X indicates the pre-geosynclinal structural layers. I, II and III represent the geosynclinal, platform and diwa structural layers, respectively. Anyone of these structural layers may be absent. The main structural layer can be subdivided into structural sublayers symbolized as $I^{1, 2, 3,}$, $II^{1, 2, 3,}$ and $III^{1,2, 3,}$.

3) II in the diwa region may be affected by diwa tectonism and no longer kept intact as in the platform region.

4) In eugeosynclines and eudiwa, magmatic activity occurs relatively strong and complete, whereas in miogeosynclines and miodiwa----relatively weak, and volcanic activity takes place seldom.

5) To take southeast China as an example, the SiO_2 content in diwa granite is relatively high, generally 74-78%, and even 86% (acid-ultraacid rocks); $K_2O + Na_2O$ generally > 8-8.5%, and in most cases $K_2O : Na_2O > 1$. In diwa basalt $K_2O + Na_2O > 4\%$.

6) Generally from -50 to +100 mGal.

7) Generally from -200 to -300 mGal.

8) Generally from -10 to -30 (or -40) mGal.

9) Generally from -200 to -300 mGal (still lower in the maximum-mobility period, e.g. -550 mGal in the Qinghai-Xizang plateau).

10) Generally from 1.7 to 2 HFU (for example, in the Alps), but they decrease in fold belts of old age, where they are generally from 0.7 to 1 HFU.

11) Generally from 0.7 to 0.8 (sometimes about 1) HFU.

12) Generally from 1.7 to 2.5 (for example, in the Rocky Mountains, U.S.A.) or 2.6 HFU (in Baikal, U.S.S.R); a little lower in the residual-mobility period (generally from 1.2 to 1.5, rarely 1.95 HFU in north China).

13) Velocity of P_n at the crust-mantle boundary.

4. Types of China's geosynclinal, platform and diwa regions

(1) Classification on the basis of the age of formation

According to the age of closure of geosynclines (transformation of a geosynclinal into a fold belt period), geosynclinal regions in China, both active and dead, are of Luliang, Jinning, Caledonian, Hercynian, Pacific (subdividing into Indosinian and Yanshanian), and Himalayan ages. Platform regions in China, on the basis of the age of their onset, fall into two groups: old (transformed from Precambrian geosynclinal fold regions) and young (from post-Precambrian geosynclinal fold regions). To the former belong post-Luliang and post-Jinning platforms, while to the latter post-Hercynian platforms. Diwa regions in China, according to the age or the maximum-mobility period, are divisible into Cathysian ones, the maximum-mobility period of which is within the Mesozoic Pacific orogeny, and Central Asia ones, the maximum-mobility period of which is within the Cenozoic Himalayan orogeny[15, 19].

(2) Classification on the basis of principal features

Mobile regions can be modified by a prefix eu or mio depending upon the presence or absence of volcanism or according to its intensity. For geosynclines, those with strong volcanism are called eugeosynclines and those with weak or no volcanism miogeosynclines. The same is true of diwa regions, and eudiwa (region) and miodiwa (region) are named.

Diwa regions, on the basis of their principal genetic features, fall into two types, arcogenesis-faulting type and block-faulting type. Geodomes and diwa of the first type owe their origin to arcogenesis and faulting, while those of the second type chiefly to block-faulting [15, 19].

Among diwa regions of arcogenesis-faulting type only Cathysian subtype has been observed in China.[1] It is featured by development of small diwa basins and very strong magmatism and is seen mostly in east China.It may be subdivided into the following three styles.

1) Fujian-Zhejiang style: It is characterized by the most intense magmatism, thickly dotted intrusions and numerous extrusions and is most typical of Fujian and Zhejiang Provinces.

2) Hubei style: It is characterized by moderate magmatism, less intrusions and unnoticeable volcanism, and has been recognized in east Hubei, Hunan and Jiangxi Provinces.

3) Guizhou-Guangxi style: It is characterized by relatively weak magmatism and lack of volcanism and exists in Guizhou, northwest Hunan and west Hubei Provinces.

---

1. The other one is Donetz subtype, which is characterized chiefly by the development of large diwa basins and weak magmatism, e.g., Donetz basin, USSR, and Wichita, USA. This subtype has not been recognized in China[15, 19].

Among diwa regions of block-faulting type only Central Asia subtype has been observed in China.[1] It is characterized chiefly by the development of large diwa basins and relatively weak magmatism and is recognized in west China and in other parts of Central Asia.

II. Tectonic regionalization of China

1. Definition

The term tectonic region is defined here as a region or part of the region (subregion) having, during a certain period and within a definite domain, a certain tectonic character (mobile and stable) and type (geosyncline, platform and diwa). It reflects the character and type of crustal movement in the discussed region during a certain period and its whole geologic history before the above-mentioned period--the tectonic character and types of various generations and the process of their evolution. Its position in space and geographical limits suggest the sphere of influence of a certain type crustal movement during a given period, and the duration of its existance indicates the time of action of this type crustal movement within the limits.

2. Steps in tectonic regionalization

In this book the tectonic regionalization of China takes the following steps.

(1) Synthesizing data on the same development stage and singling out a structural layer

The above-mentioned data should be synthetically analysed and summarized as indicated in Table 1.1, and those belonging to the same development stage be placed together. Then we can determine the tectonic character and type of the discussed region at a given time and single out its representative structural layer.

Having the same type of sedimentary formations, strata deposited during the same major epoch (basic development stage) and accompanied by a corresponding type of structural features (as well as magmatic and metamorphic formations) may be regarded as a major (basic) structural layer, e.g. pre-geosynclinal (X), geosynclinal (I), platform (II) and diwa (III) structural layers.

By analogy, we can single out structural sublayers representing minor development stages. These are: the (structural) sublayers of the proper geosynclinal (I^1) and the fold belt period (I^{2-3}) within a geosynclinal structural layer; the sublayers of the initial- (II^1), maximum- (II^2) and residual-stability periods (II^3) within a platform structural layer; and the sublayers of the initial- (III^1), maximum- (III^2) and residual-mobility periods (III^3) within a diwa structural layer.

In the case of necessity (e.g. need of the search for minerals) and possibility (if scale permits), a structural sublayer can further be divisible into still smaller structural layers, e.g. some minor structural layers (I^{1a} and I^{1b}) within a structural sublayer of the proper geosynclinal period.

---

1. The other one is Eastern African subtype, which is characterized chiefly by the development of small diwa basins and intense volcanism. This subtype has been recognized in Eastern Africa and Arabia but not observed in China [15, 19].

(2) Reconstructing the whole process of geologic development according to the structural layers in their chronological order

One should, based on the above-mentioned structural layers, determine the number, tectonic character, type and chronological order of the basic structural layers in the study region, clarify its construction and then reconstruct the whole history of geologic development in order to know the whole process of crustal evolution in the discussed region and the tectonic situation of various development stages.

(3) Identifying tectonic regions on the basis of the present development stage

In identifying tectonic regions we, according to the progression theory[13], take the tectonic character and type of the present structural layer as a criterion. Such a tectonic regionalization based on the present development stage can reflect the whole tectonic history of the discussed region.

3. Subdivision systems of the main tectonic elements

(1) Subdivision system of a geosynclinal region

1	2	3
Geosynclinal (fold) region	Fold system	Fold belt
	intermontane (piedmont) depression system	intermontane (piedmont) depression

(2) Subdivision system of a platform region

1	2	3	4
Platform region	Shield	Anteclise	Swell
	Table	Syneclise	Basin

(3) Subdivision system of a diwa region

1	2	3	4
Diwa region	Geodome system	Geodome series	Geodom
	Diwa system	Diwa series	Diwa

In compiling tectonic maps, because of small scale and paucity of material, not all subregions can be singled out in every region. In addition, platform regions in China are all remnants of the Chinese paleoplatform disintegrated in mid-Mesozoic time. They are limited in areal extent and divisible only into third-order subregions.

4. Principles of nomenclature

(1) Subdivision of a geosynclinal region and basis for its nomenclature

1) Regions of geosynclinal character within a definite area are designated as the same first-order unit, if their maximum-mobility period is of the same age and their structural lines are intimately related to each other.

2) Different parts of the first-order tectonic region, differentiated from each other in certain aspects of geosynclinal process or in the history of their development, are classified, according to their extent, into various subregions and given subregion names of a corresponding order.

3) The degree of the development of the upper (fold belt period) structural sublayer and the extent of its distribution can be used as a criterion for the distinction between negative (where the sublayer is thick) and positive (where it is thin or absent) units.

(2) Subdivision of a platform region and basis for its nomenclature

1) Regions of platform character within a definite area are designated as the same first-order unit, if they are set up in the same age.

2) Different parts of the first-order tectonic region are classified, according to the specific features of platform process or the history of their development and the extent of their distribution, into various subregions of a smaller order and given subregion names of a corresponding order.

3) The difference in the thickness of the platform structural layer can be used as a criterion for the distinction between negative (where the layer is thick) and positive (where it is thin or absent) units.

(3) Subdivision of a diwa region and basis for its nomenclature

1) Regions of diwa character within a definite area are designated as the same first-order unit, if their maximum-mobility period is of the same age, their structural lines are intimately related and their geologic history is similar to each other. But they are divided into two or more first-order units, if they are overextended in area and markedly distinct in their geologic history and structural lines.

2) Different parts of the first-order tectonic region, differentiated from each other in certain aspects of diwa process or in the history of their development, are classified, according to their extent, into various subregions and given subregion names of a corresponding order.

3) The degree of the development of the diwa structural layer and the extent of its distribution can be used as a criterion for distinction between negative and positive units. In the former case, the diwa structural layer spreads widely, while in the latter it has only a limited distribution or is absent. If an area is wholly covered by the diwa structural layer, the portions with its relatively greater thickness are negative units, but those with a smaller thickness are positive ones.

CHAPTER 2
BRIEF DESCRIPTION OF TECTONIC REGIONS IN CHINA

According to the tectonic character of the present development stage, 23 tectonic regions can be singled out in China.

I. Geosynclinal regions

There are five active geosynclinal regions in China.

1. Geosynclinal (fold) regions of Hercynian age

The Kunlun geosynclinal region[1]

It is situated between Xinjiang, Qinghai and Xizang (Tibet), stretches E-W in a narrow belt, and coincides approximately with the Kunlun Mountains[2] (see Fig. 3.8)[56, 86, 87, 99, 100].

The Precambrian was a pre-geosynclinal stage. Its representative structural layer is composed of deeply metamorphosed rocks: gneiss, schist, marble and quartzite, several thousand meters thick, in the western and central Kunlun Mountains, gneiss, schist (Jinshui Formation), dolomite, marble and siliceous banded crystalline limestone (Binggou Formation) in the eastern Kunlun Mountains. The tectonic regime at this stage has not yet been clarified.

In the Early Cambrian the region entered into a geosynclinal stage. At this time a geanticline, the central Kunlun uplift, appeared in the central part. To the north and south lay geosynclinal troughs. Their representative structural layer is composed of Cambrian and overlying strata. During the Cambrian-Devonian interval, crustal mobility gradually increased and varied in intensity and mode in different parts of the region. In a geosyncline north of the western Kunlun Mountains there is a carbonate formation, 800-2,000 m thick, composed of Cambrian-Ordovician limestone and dolomite, and a sandstone-argillite formation, 400-1,400 m thick, composed of Silurian-Devonian clastic rocks. Cambrian-Ordovician rocks in a geosyncline south of the western Kunlun Mountains have been mostly metamorphosed into schist and gneiss with a thickness of more than several thousand meters. The overlying Silurian-Devonian rocks have been weakly metamorphosed into phyllite, slate and metasandstone; sometimes the upper part is composed chiefly of limestone with a total thickness of 1,830 m. Eastwards, in Burhan Budai Mountain, because of frequent volcanism, the Lower Paleozoic Nachitai Group is composed of green schist, gneiss and metavolcanics with a thickness of 2,400 m and appears as a lower terrigenous-volcanic formation. In the Caledonian orogeny, magmatic activity occurred relatively intensely and

---

1. Chen Jiachao.
2. The Precambrian rocks in the eastern part of the northern slope of the western Kunlun Mountains are represented by a spilite-keratophyre formation, more than 3,500 m thick; in the western part by greenish phyllite, sandstone and acid-basic effusives, more than 3,900 m thick. These rocks, different from the above-mentioned Precambrian, are of geosynclinal type.

magmatic formations, predominantly granitic, were intruded in the eastern Kunlun Mountains. In Carboniferous time the geosynclines continued their development. Carboniferous strata in the northern part of the western Kunlun Mountains are composed of limestones, shales and intercalated intermediate-composition effusive rocks with a thickness of 4,300 m. They constitute a limestone subformation and a shale subformation, respectively. Eastwards, the Carboniferous is a shale formation interbedded with sandstones and limestones; the total thickness is more than 2,200 m. Southwards, in the northern part of the western Kunlun Mountains, Lower Carboniferous strata consist of sandstones, phyllites, limestones and intermediate-composition volcanic rocks and appear to be a terrigenous volcanic subformation. The Middle and Upper Carboniferous strata consist chiefly of limestones and rarely of basalts. In the western segment of the eastern Kunlun Mountains, the Carboniferous strata consist primarily of limestones with many gravels and sandy-argillaceous rocks, attaining a thickness of 2,000 m. It grades into a terrigenous-carbonate formation again in Burhan Budai Mountain.

In the western segment of the northern part of the western Kunlun Mountains, the Lower Permian consists of intermediate-composition volcanics and tuff clastics, more than 3,000 m thick. In the northern zone of the southern part of the central Kunlun Mountains, it consists chiefly of sandstones, argillaceous limestones, acid and intermediate volcanic rocks, more than 2,000 m thick, constituting a pyroclastic formation. But in the southern zone it consists mainly of calcareous and sandy-argillaceous deposits, more than 1,050 m thick. Observed in the eastern Kunlun Mountains is chiefly the Lower Permian. Its lower part comprises principally conglomerates, graywackes and slates, 1,616-2,600 m thick; its middle part consists of limestones, 1,000-1,402 m thick; and its upper part is composed of polymineralic sandstones intercalated with black slates and crystalline limestones, 3,255-6,000 m thick. They group into a thick terrigenous-carbonate formation.

As concerns the sequence of occurrence of the geosynclines, the central Kunlun Mountains rose first and then the geosynclines moved southwards and northwards. At the end of the Permian the geosynclines closed first in the western Kunlun Mountains and then in the eastern where the geosynclinal regime existed probably till the end of the Triassic and where a series of linear tight folds were formed. Magmatic intrusion of Hercynian age spread over the whole region and yielded granites. After that the region entered into a fold belt period. The Triassic-Quaternary section is composed of a molasse formation which reaches a thickness of 1,875 m on the south slope of the western Kunlun Mountains. In addition, there were also polyphasic volcanic extrusions of acid and basic composition. Neotectonic movement is prominent. Even in 1951 a volcanic explosion still occurred at the turn of an arcuate structure convex toward the south in the central part of the region. Quaternary lavas have been found along major fractures in the southern part of Aqikekuli Lake (Fig. 2.1) [99].

Generally speaking, the region still shows tectonic activities in the Himalayan orogeny: block-faulting and upward movement on the basis of the Hercynian fold belt and indications of reactivation of old structures in the new period. On account of a scarcity of geological data, it is tentatively assigned to a geosynclinal (fold) region.

Second-order subregions are: the western Kunlun fold system, the eastern Kunlun fold system, and the southern Kunlun piedmont depression system.

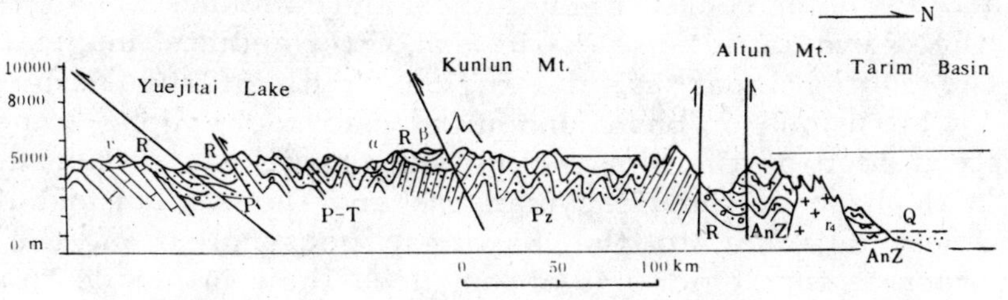

Fig. 2.1. Diagrammatic cross section through the Kunlun Mountains (after Jiang Zhongti, 1979). The area from the overthrust on the north bank of Yuejitai Lake to the fault on the south slope of Altun Mountain is the Kunlun geosynclinal fold region.

2. Geosynclinal (fold) regions of Pacific age

(1) The Bayanhar geosynclinal region[1]

It is an early-Pacific (Indosinian)geosynclinal fold region at the junction of Qinghai, Xizang (Tibet), Sichuan, Yunnan, and Shaanxi, and occupies the Bayanhar Mountains, Hoh Xil Range, part of the Hengduan Mountains, the western segment of the Qinling Mountains, and the Nanshan region of Qinghai Province [56, 87, 99, 100].

The Dege and Danba Groups, which were developed in the south and east of the geosynclinal region, consist of flysch deposits, carbonate rocks and probably intermediate-to-basic volcanics. The Dege Group, more than 11,600 m thick, is an alternation of quartz-mica schist with green schist. The Danba Group, more than 13,000 m thick, is composed of quartzite, schist and marble.[2] This suggests that the region (at least its southeastern part) was at a geosynclinal stage in early Paleozoic time. At the end of the Caledonian orogeny, the geosynclinal regime died out along the Jinsha River (and probably in the Danba area), accompanied by magmatic intrusions of acid and intermediate composition. In the larger part of the region, however, geosynclines continued their development. During late Paleozoic time geosynclinal troughs extended into the western Qinling Mountains. In the southern part of the region the representative deposits (Devonian-Permian) are composed chiefly of a carbonate formation intercalated with clastic and intermediate-to-basic volcanic rocks (green schist in the Baiyu-Yidun area). In the western Qinling Mountains both clastic (Devonian) and carbonate (Permo-Carboniferous) rocks are well-developed. The former includes the Xihanshui

1. Xue Jiamou.
2. The lower part of the Danba Group and the metamorphic sequence along the Jinsha River has been deeply metamorphosed and has a complex lithologic composition. It is probably a product of a pre-geosynclinal stage.

Formation of the northern part of the western Qinling Mountains and the Daicaotan Formation of west Gansu Province, having a total thickness of 5,000-6,000 m. The latter embraces the Muzhailing Formation north of Min County and other Carboniferous and Permian rocks, their facies and thickness changes being very rapid. During the Permian intense basic and intermediate magmatic extrusion took place in the southern part of the region (south of the Yushu-Garze-Dawu line), followed by intrusions of basic and ultrabasic rocks along a major fracture zone at the eastern edge of the geosyncline and granitic intrustion at its western edge. At the final phase of the Hercynian orogeny the Kunlun and North Xizang (Tibet) geosynclines, adjacent to the Bayanhar geosyncline on the west, were uplifted. The border of the region has experienced their influence to some extent. As a result, an angular unconformity appeared between the Paleozoic and the overlying Triassic along the Jinsha River and in the northwestern part of the region adjoining the Kunlun geosynclinal region, but the geosynclinal process was not completed. During the Triassic the whole region again subsided deeply and was covered extensively by the sea which extended into the Nanshan Mountains, Qinghai Province. The center of subsidence was in the southern zone of the region. Here flysch or flysch-like formations (Bayanhar, Caoba and Xikang Groups[1]) accumulated and liparite poured out at the west edge (in the Yushu area). The Triassic section in the northern zone (in the western Qinling Mountains) is composed of sandstone, shale and conglomerate, which are similar to the contemporaneous rocks of the southern zone in sedimentary features, but far inferior in its thickness amounting to less than 1,000 m. At the end of the Triassic the sea retreated and the whole region was folded and uplifted, forming a fold belt with the accompanying intrusion of acid-to-intermediate magmas and slight regional metamorphism. The formed structural lines trend in a NW-SE direction, but in the eastern segment, as a result of horizontal differential movement, part of the structural lines form arc-shaped curves convex towards the northeast and running at the east end parallel approximately to the N-S structural lines in the N-S diwa region. Beginning in the Jurassic, a molasse formation appeared. However, except appreciable amounts of deposits in the Hoh Xil piedmont depression, only fragmentary deposits of intermontane depressions can be found in other areas. At present the whole region is intensely rising and earthquakes occur frequently, especially in the south and east and in the Garze-Luhuo area. The Luhuo earthquake of February 6, 1973 brought about a fracture extending off and on more than 80 km along the Xianshui River. All signs indicate that the region is still in its geosynclinal fold belt period. Now it is a rising plateau, 4,000 m high above sea level. The Kali River near Xingsuhai is the source of the Huanghe (Yellow) River (Fig. 2.2).

Second-order subregions are: the Jinsha River fold system, the Bayanhar fold system, the western Qinling fold system, and the Hoh Xil piedmont depression system.

---

1. Recent work has revealed that the original Xikang Group includes part of the Upper Paleozoic as well as the Triassic. Here we denote only its Triassic part.

Fig. 2.2. Kali River, source of Huanghe River (after Jia Yujiang et al., 1979).

(2) The Gandise geosynclinal region[1]

It is a geosynclinal fold region of late Pacific (Yanshanian) age, located between the North Xizang (Tibet) diwa region and the Yarlung Zangbo River-Himalayan geosynclinal region. The western segment of its northern limit is not quite clear, while the eastern segment is the Kangtog-Mainling deep fracture, adjoining the Subanxili geodome system at the east end of the southern Himalayan diwa region. The southern limit is defined by the Yarlung Zangbo River deep fracture. It includes the Gandise and Nyainqentanglha Ranges and the area near Lhasa [7, 56, 85, 86, 99, 100, 152].

The Precambrian ultrametamorphic rocks exposed in the Nyainqentanglha Range and in the vicinity of the Shiquan River and composed of schist and gneiss probably belong to the pre-geosynclinal structural layer. In the early Paleozoic a geosynclinal trough was formed in the Nyainqentanglha Range and the Nyainqentanglha Group[2] accumulated. During the late Caledonian orogeny the geosyncline partly closed and the core of the Nyainqentanglha Range was folded

1. Xue Jiamou.
2. The lower part of the Nyainqentanglha Group has been deeply metamorphosed and is composed of augen gneiss and schist. It probably partly belongs to the pre-geosynclinal structural layer. Owing to the uncertainty as to its contact with the upper part consisting of slightly metamorphosed rocks, it is tentatively assigned to the geosynclinal structural layer.

and uplifted to form a geanticline. At the same time, magmas were injected and strata experienced regional metamorphism. Devonian (mostly Upper Devonian) deposits are exposed in a small amount only in the southeast and at the edge of the Nyainqentanglha Range. This suggests that in Early Devonian time the greater part of the region was in an uplifted manner. In the Carboniferous and Permian a geosynclinal trough appeared again, this time at a broader scale, and flysch or flysch-like formations, 2,500-4,000 m thick, were laid down. At the same time, intense magmatic extrusion took place in the vicinity of Lhasa, followed by intrusion of acid-to-intermediate magmas. During the late Hercynian orogeny, the southern zone (Gandise Range) rose and the geosynclinal trough gradually migrated northward to the Banbar-Dengqen line. Mesozoic deposits are a flysch formation and represented by the Shading Slate in Dengqen. In the Cretaceous, during the Yanshanian movement, the geosyncline was wholly closed and folded, with accompanying intense acid and intermediate magmatic activity and slight regional metamorphism (Fig. 2.3) [99]. According to Zhou Yunsheng et al. (1981) [80], the most intense acid and intermediate magmatic activity is dated at 40-80 million years, i.e., Late Cretaceous to Eocene (Figs. 2.4 and 2.5). The magma in the south of the batholith zone has a high temperature and high contents of iron, magnesium and calcium, and may have been derived from the lower crust. Structural lines trend approximately in a WNW-ESE direction to form a S-shape. In accompanying the folding and rising of the principal part of the geosynclinal region, a piedmont depression formed at the north edge and was filled with Cretaceous-Tertiary acid and intermediate volcanic rocks. At present this region is rapidly rising with the Qinghai-Xizang Plateau. Earthquakes occur frequently and intensely, particularly along a major fracture at the northwest edge of the Nyainqentanglha Range. There is a belt of hot springs extending along the south foot of the Gandise Range, where the famous Yangbajing geothermal field lying 90 km northwest of Lhasa is located (Fig. 2.6). The above-mentioned facts indicate that since the emergence of the geosyncline in early Paleozoic time the centers of sedimentation have migrated several times, and folding and rising also repeatedly occurred. However, the geosyncline did not lose its high mobility until the Yanshanian movement. Gravimetric, seismological, and magnetotelluric investigations indicate that the earth's crust is 70 km thick near Lhasa.

Second-order subregions are: the Gandise fold system, the Nyainqentanglha fold system, and the northern Gandise piedmont depression system.

3. Geosynclinal (fold) regions of Himalayan age

(1) The Yarlung Zangbo River-Himalayan geosynclinal region

It is situated in the southwestern part of China, in close vicinity to the boundary. Its northern limit is defined by the Yarlung Zangbo River deep fracture and adjoins the Gandise geosynclinal region, while its southern limit extends from west of Zanda through south of Zhongba and Kangmar to Moindawang, and adjoins the Southern Himalayan diwa region. The eastern end is separated from the latter by the Subanxili fracture. The region includes the Northern Himalaya and the Yarlung Zangbo River valley, corresponding to the Northern Himalayan deformed flysch zone and the Yarlung Zangbo River-Ayila Mountain "suture" zone, as named by Zhou Yunsheng et al.[155], or to the Yarlung Zangbo River-Himalayan stratigraphic subprovince suggested by Gu Qingge et al.[65] in their stratigraphic division of Xizang [8, 34, 56, 85, 86, 99, 100,152].

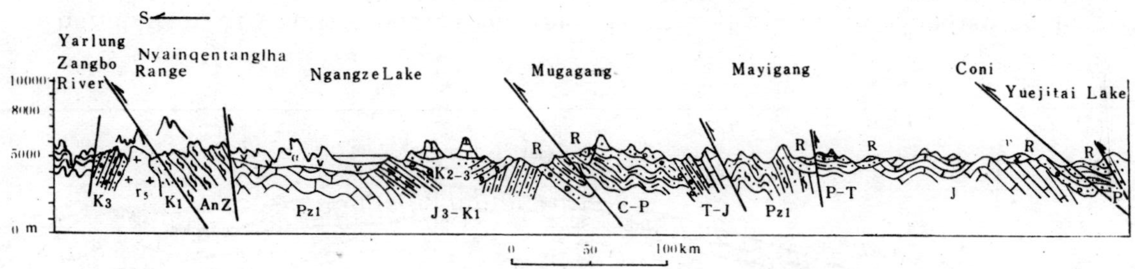

Fig. 2.3. Diagrammatic cross section through the Qinghai-Xizang Plateau from the Yarlung Zangbo River to Yuejitai Lake (after Jiang Zhongti,1979). The area from the reverse fault on the north bank of the Yarlung Zangbo River to the fault at the north foot of the Nyainqentanglha Range is the Gandise geosynclinal fold region. The area from the middle and right end to the overthrust on the north bank of Yuejitai Lake is the North Xizang (Tibet) diwa region.

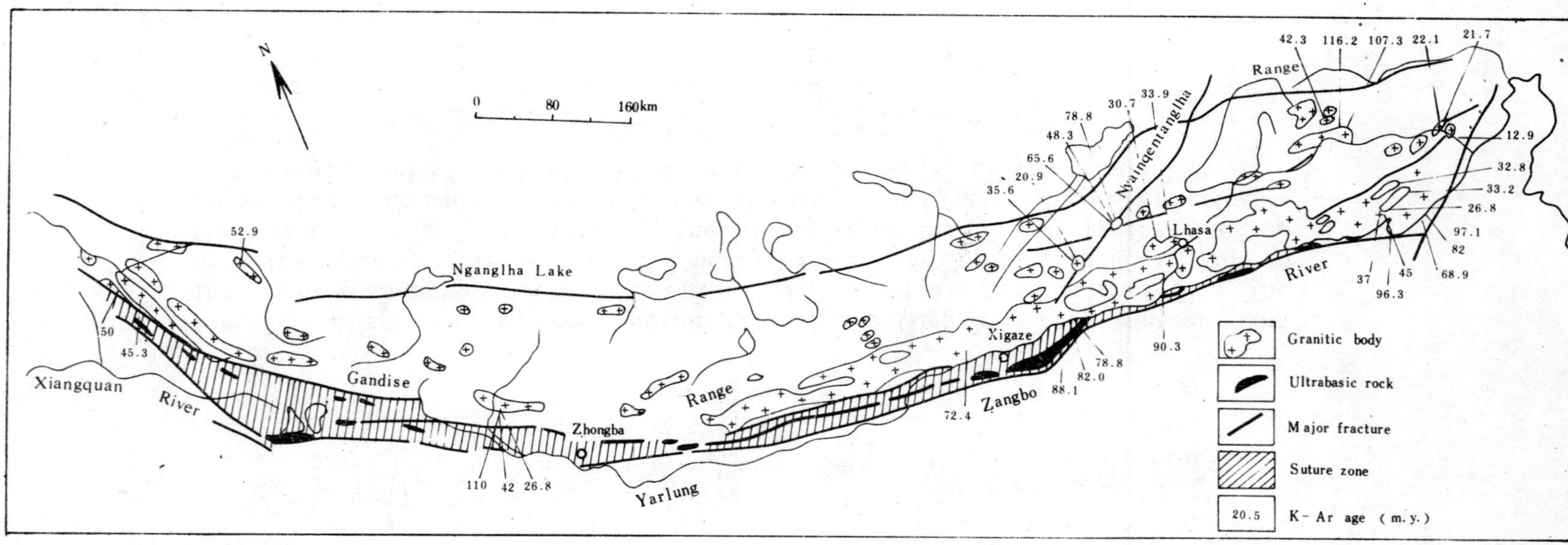

Fig. 2.4. Distribution of granitic bodies and their isotopic ages in the Gandise acid-to-intermediate intrusive rock belt (after Zhou Yunsheng et al., 1979). The suture zone is the Yarlung Zangbo River-Ayila Mountain ultrabasic rock belt belonging to the northern zone of the Yarlung Zangbo River-Himalayan geosynclinal fold region.

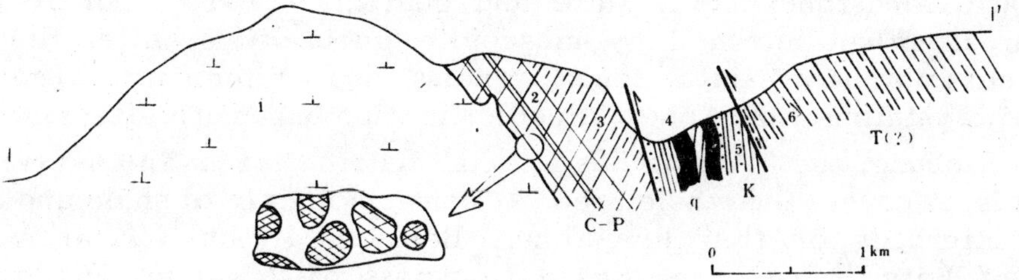

Fig. 2.5. Contact of a quartz diorite body with the country rock in the Gandise acid-to-intermediate intrusive rock belt (after Zhou Yunsheng et al., 1981). 1, Quartz diorite; 2, Hornblende; 3, Schist; 4, Ultrabasic rock; 5, Sandstone and shale; 6, Slate and phyllite.

Fig. 2.6. Yangbajing geothermal field (after China Pictorial, 1980).

History of crustal evolution.

We know very little of the history of crustal development at the pre-geosynclinal stage in this region. The earliest known record of the geosynclinal stage began in the late Paleozoic. Stratigraphic studies in Xizang, however, have led Gu Qingge [65] to the conclusion that the region is an eugeosyncline developing for a long time and having formations of geosynclinal type. It is a segment of Ancestral Tethys.

Permo-Carboniferous geosynclinal deposits are observed in the southern zone of the region, mainly in the cores of arch-anticlines at low drainage divides in south Xizang. They are represented by slightly metamorphosed marine clastics (conglomeratic sandstone, sandy slate and quartzite) and carbonate rocks in the Kangmar area. When intruded by muscovite gneissose granite, they have been intensely metamorphosed into schist, gneiss and migmatite. The Permian is composed of crystalline limestone with silicification and marbleization.

West of Rinbung, the Triassic has a zonal distribution in an east-west direction, but eastwards, it covers the whole area. It consists mainly of shale and sandy-clayey rock with interbeds of thin limestone. Being a flysch formation, the Jilong Formation of Late Triassic age has a thickness of 2,500 m, and contains many beds of volcanic rocks. Above lies the Langjiexue Formation, 1,444-9,750 m thick, consisting of an alternation of sandstone, shale and slate to form a flysch series, sometimes with interbedded siliceous slate. In Lakang of Lhozhag County and in Cona County, the Gabo Formation of Middle and Late Triassic age greatly varies in lithology. In the Snake area, the lower part of the metamorphic sequence was intruded by granite and, therefore, was turned into migmatite. The middle and upper parts consist of marble and slate intercalated with kyanite schist, with a visible thickness of 3,089 m. The Mendanggou Marl Member, containing quartz schist and basic volcanic rocks at the top, attains a thickness of 6,794 m. The Wuyula Slate Member, 1,050 m thick, consists of sandy slate interbedded with thin limestone and basic volcanic rocks.

The Jurassic is also widespread. In Lhunze and Gyangze, it belongs to alternating beds of sandstone and shale, containing many siliceous and argillaceous nodules, with interbeds of thin conglomerate, conglomeratic sandstone and limestone. The Middle Jurassic in Longzi attains a thickness of 6,000 m and contains 130 m thick basalt and volcanic breccia at the bottom.

The Cretaceous is distributed on the north and south sides of the belt of ultrabasic rocks, in the area from west of Lhaze to Lanzhuohong Lake, Gyangze County. It has a great thickness, complex facies and many sediment types. The Bugajiala Formation of Early Cretaceous age, 1,670 m thick and rich in ammonite, consists of calcareous and siliceous shales and siltstone, with interbeds of chert and conglomerate. The Zongzhuo Formation of Late Cretaceous age, 800 m thick, is composed of siliceous flysch sediments. The Lhaze Group, 1,000 m thick, consists of chert, siliceous shale, limestone, and andesite. The above-mentioned rocks are all the products of the early geosynclinal stage of the region.

The geosyncline began to close in Late Cretaceous time, when local marine regression and molasse appeared. The molasse is represented by the Xigaze Group. Its lower part is the Qiuwu Formation which is composed of thick conglomerate at the bottom. At first its boulders, having a diameter greater than 50 centimeters, were derived almost all from the underlying rock. The upper part, 300 m thick, is alternating beds of mudstone, siltstone, sandstone and conglomerate intercalated with thin coal beds. In the middle and upper parts, however, sometimes flyshoid marine sediments with ammonite can be seen (Ngamring Formation), reflecting the sedimentary features of a transitional period from sea to land in the process of geosynclinal inversion. These sediments were also strongly folded.

In the Tertiary the region entered into the fold belt period. During this time topographic contrast increased. With the exception of a local residual sea at the western extremity in the early Tertiary, there were mainly intermontane depressions receiving continental sediments. This reflects a general uplift, with a sharp tectono-geomorphic relief of high mountains and deep valleys. Such a relief can be observed in Gar County of the western and in Nedong County of the eastern segment. The sediments attain a thickness of 804 m, being laid down in intermontane depressions formed after the general inversion in the late geosynclinal stage of the region.

Neotectonic movement in the region is very severe. The region shows high and frequent seismicity and, therefore, is an intense seismic region. The greatest earthquake recorded has a magnitude of 8.5. It can be concluded, therefore, that the region is still in the maximum-mobility period of the geosynclinal development.

Magmatic activity took place intensely in the region. All kinds of igneous rocks, including acid, intermediate, basic and ultrabasic, can be seen. They have a zonal distribution in an east-west direction (Fig. 2.7). There are two main belts of rocks: the Yarlung Zangbo belt of ultrabasic rocks in the north and the Lhaguigangri belt of granite in the south. The former was emplaced in Late Cretaceous to Eocene time, and also intruded into the Xigaze Group. This is one of the largest belts of ultrabasic rocks in China and consists chiefly of dunite and peridotite. The latter is characterized by a wide distribution of medium- and small - size granitic bodies and related rocks, which yielded a K-Ar age of 20-40 million years (Eocene to Miocene). The granitic bodies are controlled by the axis of folds and the interface of thrusts. The rock has relatively high contents of CaO, MgO and iron and is rich in Sr, Ba and Y. This is the northern one of the two belts of granite in the Himalayas, the "Lhaguigangri belt of gneissoid granite" suggested by Zhou Yunsheng et al. (1981)[155].

Subregions and their principal features.

According to the history of the geosynclinal development and the differences in its peculiarities, the geosyncline in the region can be divided into two zones, the northern and the southern. They were folded and uplifted to become fold belts one after another in the Himalayan orogeny (Fig. 2.7, the right end). In the northern zone there are also well-developed intermontane depressions. Accordingly, the region can be divided into the following two second-order subregions (Fig. 2.7).

1) The northern Himalayan miogeosynclinal fold system. It is the southern zone of the geosyncline, which includes the southern slope of Ayila Mountain in the western segment and the south Xizang (Tibet) low water drainage(the Lhaguigangri water drainage between the Yarlung Zangbo River water system and the Pengqu River, Baxi River and Xiaqu River water system) in the eastern segment. Strata exposed in the zone are represented, apart from the Carboniferous and Permian observed locally, mainly by a flysch series of Triassic age ranging from several to ten thousand meters in thickness. It is accompanied by sandstone,

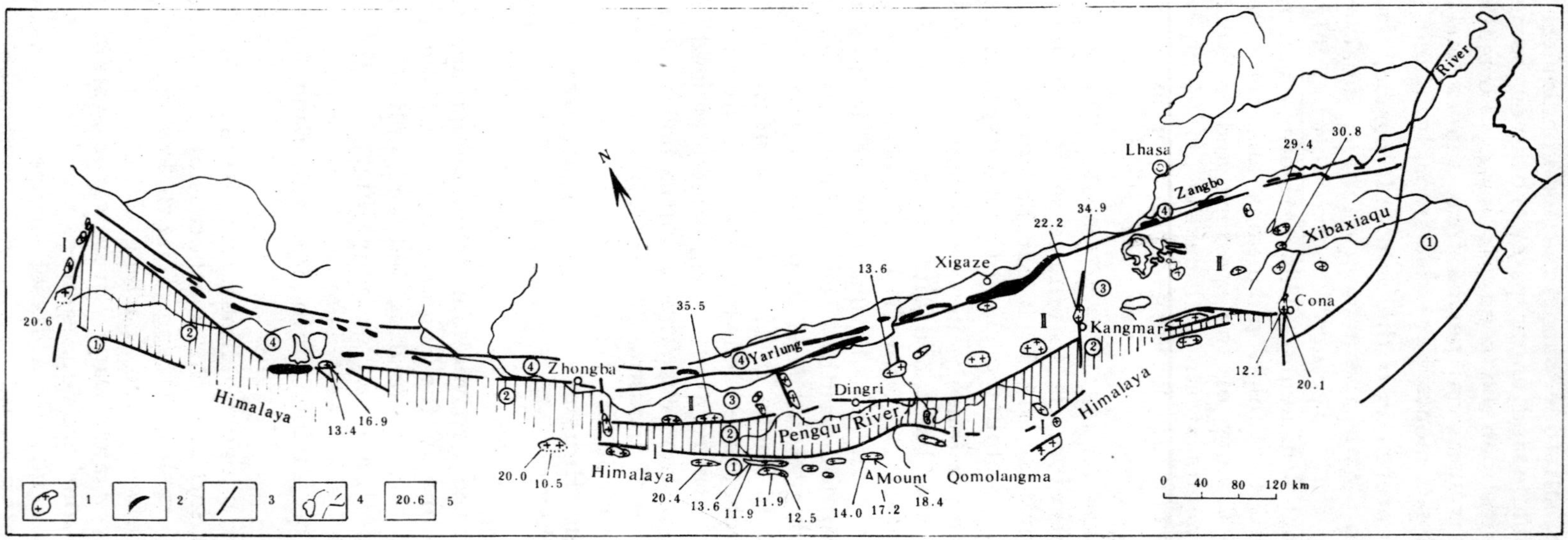

Fig.2.7. Distribution of granitic bodies and their K-Ar ages in the Himalayas in Chinese territory, supplemented by the Yarlung Zangbo belt of ultrabasic rocks (after Zhou Yunsheng, 1981). 1, Granitic body; 2, Ultrabasic rock; 3, Main fracture; 4, River and lake; 5, K-Ar age value (million years). Granite belt: I-Higher Himalayan tourmaline granite belt; II-Lhaguigangri gneissoid two-mica granite belt. Tectonic division: 1 and 2, Southern Himalayan diwa region; 3 and 4, Yarlung Zangbo River-Himalayan geosynclinal fold region (including: 3, Northern Himalayan miogeosynclinal fold system, an[illegible] Zangbo eugeosynclinal fold belt-intermontane depression system).

thin limestone and intermediate-basic volcanic rocks, and poor in fossils. Probably the zone was rising in Late Cretaceous time, and the strata underwent intense folding and slight metamorphism everywhere. In the Eocene, fracturing took place on a large scale. Along the main ridge of Lhaguigangri, granitic intrusion and migmatization occurred. This place is the site of the above-mentioned granite belt of this geosynclinal region.

2) The Yarlung Zangbo eugeosynclinal fold belt-intermontane depression system. It is the northern zone of the geosyncline, which lies as a narrow belt stretching approximately along the southern bank of the Yarlung Zangbo River and south of the main ridge of Ayila Mountain. According to Zhou Yunsheng et al. (1981)[155], here Triassic volcanic and siliceous rocks as well as Tertiary flysch deposits are exposed on the surface. In addition, there are also Tertiary molasse deposits accumulating in fault-bounded intermontane depressions. As the geosynclinal deposits are accompanied by extensive intermediate and basic volcanic rocks, the zone evidently has an eugeosynclinal character. The period of folding and rising set up in the Eocene when intense folding and reversal of strata took place and high-pressure metamorphism occurred. Ultrabasic rocks are widespread, constituting the main zone of distribution of ultrabasic rocks in this geosynclinal region, i.e., the above-mentioned famous Yarlung Zangbo belt of ultrabasic rocks (Fig. 2.7).

(2) The Taiwan geosynclinal region

It includes Taiwan Island proper, its surrounding islands and adjacent sea area. It is a developing geosynclinal region[56, 70, 88, 97, 99, 100].

The pre-geosynclinal stage is represented by a crystalline basement composed of deeply metamorphosed, structurally complicated Paleozoic and Mesozoic rocks (schist, crystalline limestone and basic volcanic rocks). During this stage the region experienced the Hercynian and Pacific orogenies which were accompanied by magmatic activity and have made the pre-geosynclinal structural layer intensely folded.

The region stepped into the geosynclinal stage in the Cretaceous. At this time a NNE-trending geosyncline appeared on the crystalline basement. The eastern part subsided earlier and was filled with a flysch formation of pre-Paleocene age (Bihou Group), 8,000 m thick, with intercalations of submarine volcanic rocks. During the Paleocene, the eastern part experienced uplifting which resulted in a stratigraphic break. The western part was in a uplifted state all the time. At the beginning of the Eocene, the whole region subsided and a sandy-argillaceous flysch formation represented by the Wulai Group, 3,000 m thick, was deposited in the western part. In the eastern part, besides the flysch formation, there are basic and intermediate volcanic rocks. At the end of the Oligocene, the geosyncline began to close partly, giving rise to NNE-striking linear tight folds forming Central Mountain and Mount Yushan of Taiwan; at the same time regional metamorphism took place. Afterwards, the geosyncline migrated eastwards and westwards. The eastern segment is an eugeosyncline, in which, from Miocene to Pliocene, preceded by an andesite formation, a flysch formation composed of conglomerates, sandstones, shales and limestones was deposited. The western segment is a miogeosyncline, in which a flysch formation and a coal- and oil-bearing formation of flysch type, 7,000 m thick, were laid down. At the end of the Tertiary and beginning of the Quaternary, the Taiwan movement corresponding to the third phase of the Himalayan orogeny affected the whole region, and both the eastern and western geosynclines came to an end forming a fold belt. Meanwhile, acid, intermediate and basic magmas were injected to broaden the fold belt. Mountains and a longitudinal valley on east and Ali Mountain on west Taiwan came into being during this period. The molasse formation represented by the Beinanshan Conglomerate in the east and the Toukeshan Sandstone in the west appeared to accompany the uplift and erosion of the fold belt. The east Taiwan grand fault and schuppen structures at Ali Mountain were also formed during this period (Fig. 2.8)[48]. From the Quaternary onwards, differential movements of Taiwan Island have still been prominent. Mountains in Taiwan are still rising and the bottom of the sea on both sides (particularly on the east side) is rapidly subsiding. Modern topographic contrast is also sharp. The highest peak of Mount Yushan is 3,950 m high, but the east Taiwan longitudinal valley is 50 m below sea level. Earthquakes are frequent and concentrated mainly in the eastern part.

Diaoyu Island consists of Tertiary rocks, chiefly sandstone and conglomerate, with intercalations of mudstone and thin limestone. The rocks have experienced tilting and were intruded by acid magma. Around the island there are coral reefs, 1 m thick. Chiwei Islet is composed of andesite agglomerate. On the continental shelf near Diaoyu Island, thick Neogene sediments were deposited and lie almost horizontally (Fig. 2.9). The topography of the continental shelf in the East China Sea is related to their bedding plane.

Separated by the Taizhong-Gaoxiong fracture, two second-order subregions, the Taiwan fold system in the east and the West Taiwan piedmont depression system in the west can be singled out.

II. Platform regions

There are only three surviving platform regions in China which are ascribed to three epochs. They are relics (residual platforms) of the broad Chinese paleoplatform, which existed in the Paleozoic, and was broken down and largely transformed into a series of diwa regions in mid-Mesozoic time. These platforms are influenced to some extent by diwa activity, and no longer preserve fully the peculiarities of the original platform.

1. Platform regions of post-Luliang age

The Yishan (Ordos) platform region[1]

It is the oldest residual platform in China, situated in a rectangular area south of the major Dalad Banner fracture, east of the major fracture at the east edge of Zhuozi Mountain, and west of the Fenhe-Weihe graben[56, 99, 100].

The Archean was a pre-geosynclinal stage. At this time gneiss and schist, moderately or deeply regionally metamorphosed and represented by the Jichekou and Luliang Groups, were formed. They constitute the crystalline basement intruded by ultrabasic and granitic rocks.

The Early Proterozoic was a geosynclinal stage. The geosyncline formed at this time was the western branch of the North China paleogeosyncline. It was filled with terrigenous rock in the lower part represented by the Lanhe Group,

1. Wei Bailin.

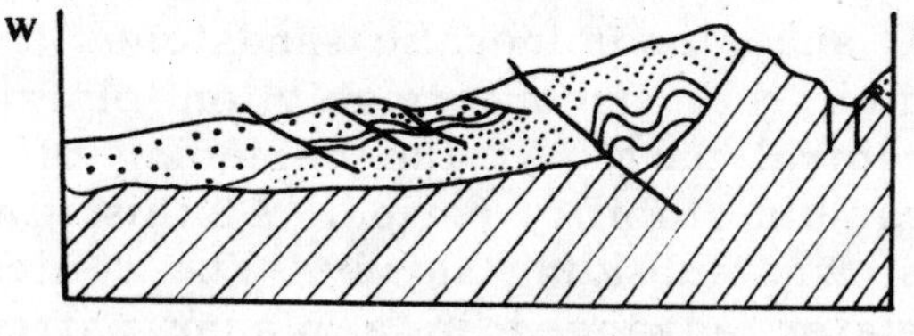

Fig.2.8. Pleistocene deformations in the Neogene geosyncline (after He Chunxun, 1967).

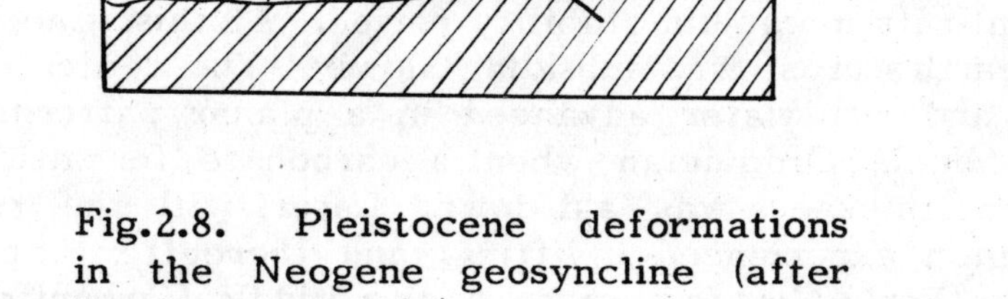

Fig. 2.9. Seismic section across the continental shelf near Diaoyu Island in the East China Sea (After Ministry of Petroleum Industry).

volcanic rock in the middle represented by the Yejishan Group, and terrigenous rock in the upper represented by Heichashan Group. During the Luliang movement, the geosyncline closed and gave rise to NE-trending tight folds accompanied by intrusion of granite. They constitute the folded basement upon which the subsequent platform developed.

At the beginning of the Late Proterozoic (Sinian) the region stepped into a platform stage. The Sinian was its initial-stability period and the platform was the western part of the broad North China paleoplatform of post-Luliang age. During this period, though the region was in the process of flattening out, the land surface still preserved an undulation to some extent. Marine transgression took place in a limited area, and monomineralic sandstones (below) and siliceous carbonate rocks (above), which rest unconformably on older structural layers, were deposited only in the central part. At the beginning of the Cambrian, the platform entered into its maximum-stability period. At this time the land surface was flat and the earth's crust was subsiding slowly. The center of sedimentation lay in Tongchuan and sea water advanced in a planar pattern. This situation continued to the Middle Ordovician when a carbonate formation with uniform facies and persistent thickness was laid down. Later, with the withdrawal of the sea, the whole region experienced uplifting, and thereafter came a long phase of denudation to the Early Carboniferous. In the Middle Carboniferous, the region stepped into the residual-stability period, during which the amplitude, speed and frequency of oscillatory crustal movement increased. This found expression in frequent marine transgressions and regressions alternately occurring in the Middle and Late Carboniferous and producing an alternately marine and continental coal-bearing formation. Up to the Permian, along with the uplifting of the crust, the sea withdrew completely and produced a continental coal-bearing formation. During the Late Permian to the Early-Middle Triassic, topographic differentiation was apparent and basins took shape. The climate became semi-arid and a red salt-bearing formation appeared. In Late Triassic to Early Jurassic time, though the Chinese paleoplatform largely died out and was transformed successively into a series of diwa regions, the region still maintained its relative stability and appeared as a microplatform surrounded by diwa regions. In contrast with the differential uplifting of the adjacent diwa regions due to block faulting in the early period, the region was relatively subsiding to form a broad basin. But the sediments still lay parallel to the underlying Paleozoic rocks, indicating obviously the continuation of the platform stage. At this time, lakes were widespread in the region and the climate again turned to damp. Meanwhile, at the edge of the platform there were small intrusions of alkaline rocks. The above-mentioned alternation of an arid with a damp climate has continued up to now. Structural deformation took place weakly and strata in the center of the platform, including Mesozoic-Cenozoic rocks, lay flat, almost horizontally (Fig. 2.10). Only isolated non-linear anticlines or structural noses can be seen (Fig. 2.11). Neotectonic movement is also not severe; seismicity is low. Obviously, the region has not yet lost its relative stable platform character.

Generally speaking, the region showed a platform feature very markedly in Paleozoic time. After the middle Mesozoic, though affected by diwa activity in adjacent areas, it has still maintained principal features of a platform and is dominated by epeirogenetic movements. No striking unconformity is seen between various strata, including the Mesozoic and Cenozoic; rocks dip gently and remain

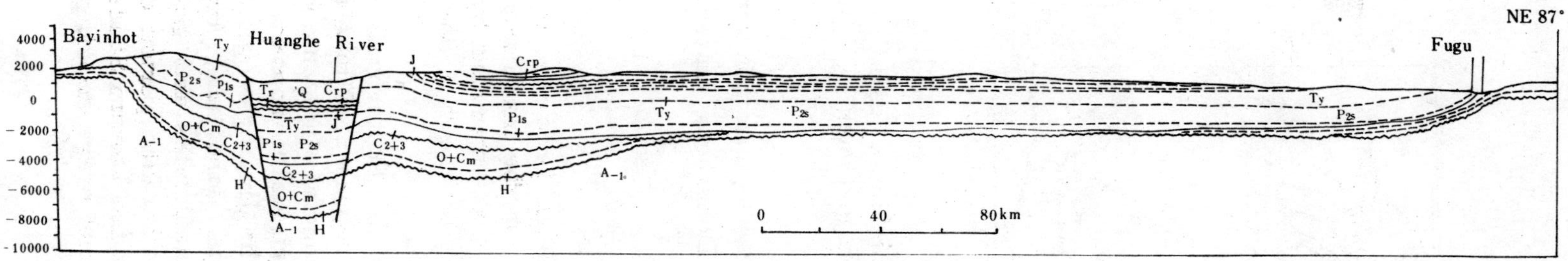

Fig. 2.10. Cross section from Bayinhot, Inner Mongolia, to Fugu, Shaanxi Province (after Xi'an Geological Section, Ministry of Petroleum Industry).

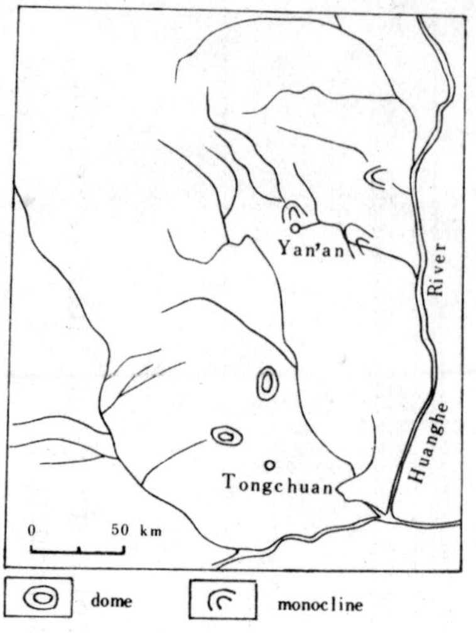

Fig. 2.11. Non-linear folds on the Yishan platform.

essentially horizontal. Magmatic activity was insignificant except on the border.

Third-order subregions are: the Dongsheng anteclise, the North Shaanxi syneclise, and the Luliang anteclise.

2. Platform regions of post-Jinning age

The Sichuan platform region[1]

Situated in the central part of Sichuan Province, it is linked to the Central China and Yunnan-Guizhou diwa regions on the north and southeast, respectively, without striking tectonic boundaries, but is separated by the Hanzhong-Guanxian fracture zone from the central segment of the N-S diwa region on the west and northwest[56, 99, 100].

The known geologic record began in the Proterozoic which was a geosynclinal stage. The early period of this stage is represented by a structural layer composed of schist, sandy slate and acid-to-intermediate volcanic rocks of Proterozoic age. These rocks, 9,000 m thick, were originally deposited in a deeply subsiding geosyncline. The geosyncline closed in the Jinning orogeny which resulted in folding and metamorphism of the geosynclinal structural layer, with accompanying intrusions of basic rocks, diorites, granodiorites and granites. At the beginning of the Sinian, this region stepped into the late period of the geosynclinal stage and a molasse formation represented by the Nantuo Sandstone, 100-1,600 m thick, was laid down. After that the geosynclinal stage ended, giving place to a platform

1. Wei Zhouling.

stage.

The initial-stability period of the platform stage is marked by the appearance of tillite which is followed by the deposition of clastic and carbonate rocks of the Doushantou and Dengying Formations. At the end of the Sinian, the region was in a uplifted state which was responsible for a stratigraphic break or disconformity. At the beginning of the Cambrian, the region entered into the maximum-stability period of the platform stage, during which slow crustal sinking affecting a large area took place and steady shallow-marine sandstone, shale and carbonate rocks were deposited. At the end of the Silurian, the sedimentary cycle was interrupted by a general uplift. In the Devonian and Carboniferous, crustal subsidence and uplift repeated several times, resulting in stratigraphic breaks, that is to say, the Lower Devonian, Lower and Upper Carboniferous were missing. During the Dongwu movement in the Early Permian the region underwent a general uplift again. The Early Permian saw the outpouring of extensive basaltic lava which marks the beginning of the residual-stability period of the platform stage. Afterwards, upward and downward movements affecting a large area occurred more frequently. In the Late Permian, an alternately marine and continental coal-bearing formation, one of the important coal-bearing formations in the region, and a marine carbonate formation were deposited. After a sedimentary hiatus, crustal subsidence affecting a large area occurred again in the Early Triassic, and a carbonate formation represented by the Daye Group was deposited. During the late Early and Middle Triassic, an alternately shallow marine and lagoon carbonate formation was deposited. Beginning in the Late Triassic, affected by diwa activity in adjacent areas, crustal movement in this region became more intense and the sea retreated completely. Moreover, in contrast with the uplift of surrounding diwa regions, this region was relatively subsiding to form a basin. The crustal movement, however, still maintained an epeirogenic character, and the whole region was dominated by vertical movement. No Indosinian and Yanshanian orogenic movements, so strongly displayed in the adjacent diwa regions, manifested themselves in this region. The relationships both between the Triassic and Jurassic and between the Jurassic and Cretaceous are disconformity. Only at the close of the Tertiary, during the Sichuan movement, could occur horizontal (orogenic) movement at the eastern edge resulting in a pre-Tertiary unconformity. In the basin, a continental coal-bearing sandstone-shale formation (Xiangxi Formation, an important coal-bearing formation in the region) and a red sandstone-shale formation were deposited. During the Cenozoic, crustal movement became more intense. Within the region, especially on the border, a sandstone-conglomerate formation and sand-gravel-mud accumulations were laid down and unconformities appeared several times. But structural features formed in most places remained simple, except for the eastern part which is characterized by the extensive development of comb-like folds forming a transitional zone to the Yunnan-Guizhou diwa region. In the western part discontinuous folds continue to predominate, domes and structural noses are frequently seen, and strata dip at low angles or even lie horizontally (Fig.2.12). Neotectonic movements are not intense, mainly differential upwarping and downwarping and tilting; earthquakes occur infrequently and weakly, and are concentrated on the border (e.g., in Dayi and Mabian). It can be, therefore, concluded that since the Triassic this region has experienced to some extent the influence of the adjacent diwa regions and that it no longer keeps the original stability of the Paleozoic platform and is only a residual fragment of it. As regards its principal features, however, this region can be regarded mainly as a small platform.

Third-order subregions are: the Chongqing anteclise in the east, the Chengdu syneclise in the west and the Micang anteclise in the north.

3. Platform regions of post-Hercynian age

The Songliao platform region[1]

It corresponds approximately to the Songliao Plain [110, 129, 158]. The whole region is largely covered by the Quaternary. The schist complex hidden in the central part of the plain is probably the pre-Sinian crystalline basement[56, 99, 100].

Slightly metamorphosed Silurian-Permian strata crop out in the southwest of the platform. Their characteristic features bear a resemblance to those of the synchronous strata exposed in the eastern part of Inner Mongolia, NE diwa region. Both are of geosynclinal type. The above-mentioned strata are also seen in wells drilled in the inner part of the platform. It can be, therefore, inferred that the region was at a geosynclinal stage during Middle and Late Paleozoic time. It was then a part of the broad Northeast China Hercynian paleogeosynclinal region, which was largely transformed, after Triassic time, into the NE diwa region. But the Songliao region still kept its relative stability, severed itself from the NE diwa region and became a post-Hercynian platform. It subsided due to faulting and, as a result, a thick platform structural layer formed.

No Triassic rocks have been found in the platform structural layer. The Jurassic coal-bearing formation is seen only on the border of the platform. The Cretaceous system is well-developed in the inner part of the platform and consists of a limnic detrital formation, chiefly argillaceous, with a thickness of 5,700 m. It outcrops in the eastern part of the platform, stretches in a NNE direction and gently dips to WNW. It includes the following divisions from bottom to top: the Denglonku, Quantou, Qingshankou and Yaojia Formations (Fig. 2.13) of Early Cretaceous age; the Fulongquan and Sifangtai Formations of Late Cretaceous age. Tertiary limnic deposits are distributed in the central and northwestern parts of the platform. The Lower Tertiary consists chiefly of argillite with a thickness of 700 m and rests conformably on the underlying Cretaceous system. The Upper Tertiary comprises an alternation of sandstone, conglomerate and mudstone, about 250 m thick, occurs horizontally and lies unconformably on the Lower Tertiary and Cretaceous strata slightly folded in a NNE direction to form discontinuous-type structures such as ramparts. Quaternary sediments vary in type, including early Pleistocene drift sheet, middle and late Pleistocene loess and Holocene black soil.

The borders of the region have experienced the influence of adjacent diwa regions and have a greater tectonic mobility. Magmatic activity and earthquakes are closely related to fractures developed at the edges. Along the E-W trending Nemor fracture at the northern edge, for instance, there is a volcano swarm named Wudalianchi which burst out in late Pleistocene time and was still active in 1720. Along the NNE-trending Siping-Changchun fracture zone at the eastern edge lie Mesozoic granite and early Pleistocene basaltic volcanic cones; here earthaquakes

---

1. Wen Shanji.

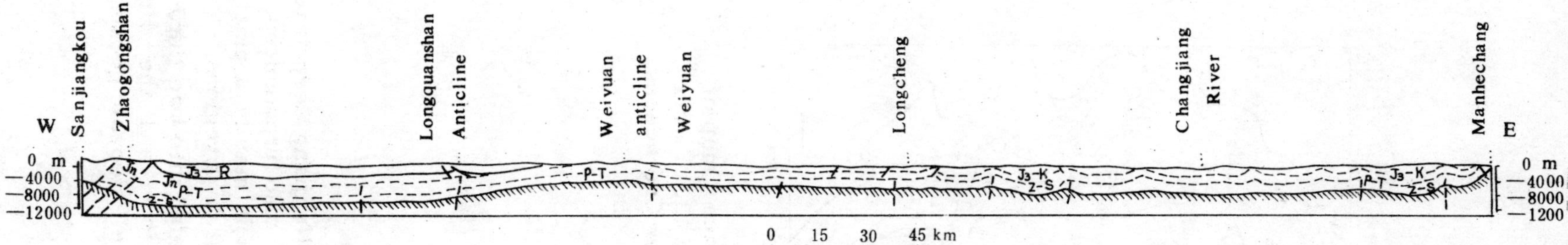

Fig. 2.12. Geologic-structural section from Sanjiangkou to Manhechang in the Sichuan platform region.

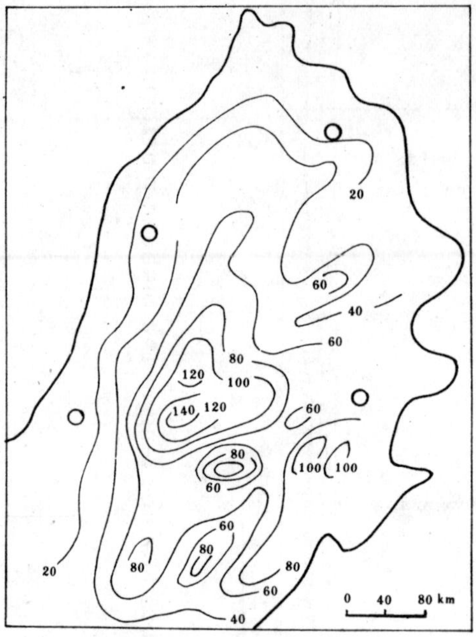

Fig. 2.13. Isopach map of members 2 and 3, Yaojia Formation, Lower Cretaceous, Songliao Basin.

often occur. Along the NNE-trending Nenjiang deep fracture at the western edge, seismicity and magmatic activity are also pronounced. From these features we may conclude that the Songliao region is mainly a platform in character, but has experienced the influence of surrounding diwa regions.

For the time being the region is divided into two third-order units: the Song-Nen syneclise occupying most of the platform and the Hure anteclise situated in the southwest and having outcrops of the Paleozoic geosynclinal structural layer.

III. Diwa regions

Diwa regions in China are well-developed, totalling 14 in number in the mainland and belonging to Cathysian and Central Asia ages, respectively. There is another diwa region at sea.

1. Diwa regions of Cathysian age

(1) The SE diwa region[1]

It is situated in southeast China and occupies southeast Yunnan, southeast Guizhou, Guangxi, Hunan, Jiangxi, Guangdong, Fujian, Zhejiang, south Jiangsu, south Anhui, and the western half of Taiwan Strait [9, 11, 16, 20, 22, 23, 24, 27, 33, 44, 49, 53, 56, 80, 81, 99, 100, 102, 112, 132].

Little is known about the pre-geosynclinal stage. Geophysical data indicate that under the Banxi Group, at a depth of 5,000 m in Xuefeng Mountain, exists an older basement. The Archean (?) deeply metamorphosed complex (gneiss, schist, hornblende schist, marble, etc.) exposed in some places (east Jiangxi and west Fujian) probably represents the pre-geosynclinal structural layer.

The geosynclinal stage began in the Proterozoic. At this time, a thick (up to more than 10,000 m) sequence of shallow-marine sandstone and shale accumulated and frequent submarine volcanic eruption of intermediate-to-basic composition occurred in north and southeast Guangxi, central Hunan and north Jiangxi, making up flysch and spilite-keratophyre formations. The Xuefeng (Jinning) movement has resulted in folding and regional metamorphism of rocks in this region (especially in its northern and northwestern parts). A geosyncline in north Jiangxi closed partly, with accompanying large intrusions of granodiorite. In early Sinian time, a molasse formation (Luokedong Formation) of clastics mixed with volcanic material was laid down. Afterwards, in late Sinian time, the north Jiangxi area stepped into a platform stage, but the other areas continued their geosynclinal development and received a thick (up to more than 10,000 m everywhere) flysch formation of early Paleozoic age composed of sandstone and shale intercalated with volcanics in some places. This formation is well-developed in west Guangdong, southeast Guangxi, Nanling Mountain, Jiangxi, west Zhejiang, and south Anhui. During the Caledonian orogeny at the end of the Silurian, the geosynclinal regime came to an end in the whole region, resulting in linear tight folds and regional metamorphism. A fold belt was thus formed. In addition, we observed Caledonian intrusions of granite or granodiorite in north Guangxi, Guangzhou-Zengcheng-Boluo, Xingning, Guanyang, Penggongmiao, Mount Wanyang, Guangning, Taibao and Huaqiao. With the erosion of the fold belt, the geosynclinal stage terminated in the region.

The platform stage began in the Middle Devonian (or Early in Guangxi), except for north Jiangxi where it commenced in late Sinian time. In the initial-stability period of the platform development, topographic contrast remained sharp, giving rise to the deposition of sandstone and conglomerate (e.g. Lianhuashan Group in Guangxi, Guitou Group in Guangdong, Tiaomajian Formation in Hunan). During the maximum-stability period (Middle and Late Devonian to Early Carboniferous),

---

1. Wei Zhouling.

the land surface was so plain that extensive sandstone-shale, coal-bearing sandstone-shale (e.g. Ceshui Formation of Early Carboniferous age), and carbonate formations were laid down. In the Early Permian, the Dongwu movement caused general uplift of the region, resulting in a regional disconformity between the Upper and Lower Permian and in the termination of the maximum-stability period. The residual-stability period (Middle Permian to Middle Triassic) is marked by coal-bearing sandstone-shale and carbonate formations which are less persistent than those formed in the maximum-stability period. Middle Triassic rocks, for example, vary markedly. They are limestone in north Jiangxi, absent in Xuefeng Mountain, central and south Hunan, sandstone and shale (Yangjia and Huangsha Groups) in east Hunan, and sandstone and mudstone with interbeds of tuff in southeast Guangxi. Generally speaking, crustal movements in this region during the platform stage were far weaker than those during the geosynclineal stage and were characterized primarily by upwarping and downwarping affecting a large area. Magmatic activity was also weak and simple, and basic eruptions and tuffs can only be seen in west and southeast Guangxi.

In the Late Triassic (or Middle in west Guangxi) the region entered into the diwa stage. At this time, structural features, landforms, sedimentary formations, magmatic activity and metamorphism showed characteristics peculiar to a new type mobile region. In the initial-mobility period of the diwa development, arcogenesis, folding and faulting began to liven up, tectono-topographic contrast became sharp, diwa basins began to appear, and a coal-bearing conglomerate-sandstone-shale formation, poorly sorted, less persistent and greatly variable in thickness, was deposited in them. Examples of this formation are the Xiaojiangkou, Anyuan (Fig. 2.14), Jinji, Lishan and Xiaoping Groups, with a thickness of over 1,000 m, but 5,000 m in south-central Jiangxi, Shaoguan and Lian County, or even more in the Shiwan Mountains and in south Guangxi, where, in addition, intermediate-composition volcanic rocks exist. In some places of Guangdong Province, the above-mentioned diwa formation is intercalated with liparite-porphyry and liparite-tuff. In some areas appeared a marine transgression which was responsible for the Lantang Group (Jurassic) in Zijin, Guangdong Province, and the Pingerguan Group in west Guangxi. Marine interbeds of Middle and Late Triassic or Jurassic age have been found in north Guangdong, east Hunan and in Pingxiang, Jiangxi. They are products of sea water invading the diwa basins along mountain valleys. During the initial-mobility period there were intrusions of granite in the Nanling Mountains and on Hainan Island. The granite petrochemically distinctly differs from the Caledonian granite at the geosynclinal stage, being of diwa type (Tables 4.1 and 4.2). The Late Jurassic-Cretaceous was the maximum-mobility period, during which topographic contrast became sharper and a red sandstone-conglomerate formation was deposited everywhere. This formation is also characterized by poor sorting and great variability. The lower conglomerate of Lower Cretaceous age in Dachong, Chen County, Chaling diwa, Hunan Province, has gravels varying greatly in size and in composition, mostly angular in shape and poorly sorted. As estimated, gravels of the Upper Cretaceous conglomerate in Longmenkou, Yongxing County, consist of: 57% metamorphic sandstone, 16% quartzite, 6% vein quartz, 5% limestone, 6% granite, 1% chert, 2% slate, 2% phyllite and argillite, and 5% calcite. The roundness of the gravels is as follows: 10% I class, 40% II class, 35% III class, and 15% IV class. The thickness, facies and lithology change rapidly both along strike and dip. Sandstone, for instance, generally has a polymineralic character and preserves fresh rock fragments (quartzitic sandstone). The Lower Cretaceous of the Hengyang Basin has a rapid facies variation (Fig. 2.15). These features are characteristic of a tectonically active and topographically contrasting environment. The thickness of the above-mentioned formation is usually 5,000-10,000 m. Unconformities frequently occurred between beds. During the initial- and maximum-mobility periods, folding, mainly of broad type (Fig. 2.16), took place strongly, and overthrusts also appeared (Fig. 2.17). Magmatic activity during the maximum-mobility period took place violently and produced various magmas of diwa type (Tables 4.1 and 4.2). Most of the granites in the region were emplaced at this time. In Guangdong, Zhejiang, and Fujian there is, in addition, a great amount of acid and intermediate volcanic rocks. The Tertiary (or the close of the Late Cretaceous) was the beginning of the residual-mobility period. At this time sandstone-conglomerate and sandstone-mudstone formations were laid down. Generally speaking, the sandstone-conglomerate formation remained poorly sorted. Tertiary sandstone in the Heyuan Basin, Guangdong Province, for example, appears to be polymineralic, contains more feldspar and detritus and has a sorting coefficient of 2.15-7.00, or even up to 10. Sandstone in the sandstone-mudstone formation is also polymineralic (Figs. 2.18 and 2.19). Both formations show features of a continental flysch formation (Fig. 2.20). Structural deformations in the residual-mobility period were dominated by block-faulting, which resulted in many zones of dislocation metamorphism (Fig. 2.21) and tension basins (Fig. 2.22). In Quaternary time a few basic rocks were intruded and a great amount of basalt extruded on Hainan Island (Fig. 2.23), Leizhou Peninsula (Fig. 2.24), Guangdong Province, and in Sheng County, Zhejiang Province. Hot springs are widespread (Fig. 2.25). Neotectonic movements are very intense: terraces with 3-4 steps exist everywhere, and those with 6-8 steps are present in the Yuan River valley and in the vicinity of Dongting Lake. In Quaternary sediments tilting and faulting can often be seen. In the coastal area and on Hainan Island uplifted wave-cut terraces (Fig. 2.26)[111] and large earthquakes occur frequently. According to the studies by Chen Enmin (1979, 1981)[112], during the 1605 Qiongzhou earthquake a land with an area of 100 km^2 sank to form a sea, with a subsidence amplitude of 3-5 or even 10 m (Fig. 2.27).

The region can be divided into six second-order units: the Fujian-Zhejiang diwa system, the Zhejiang-Guangdong geodome system, the Jiangxi-Guangxi diwa system, the Xuefeng geodome system, the Yunnan-Guangxi diwa system, and the Qiong-Lei diwa system.

(2) The Yunnan-Guizhou diwa region[1]

Situated in the western part of south China, it occupies southwest Hubei, northwest Hunan and part of east Yunnan as well as the greater part of Guizhou. It is separated from the SE diwa region by a set of major fractures on the east, and from the N-S diwa region by the Zhaojue-Yiliang deep fracture on the west[56, 98, 99, 100, 111].

The region entered into the geosynclinal stage in the Proterozoic which is represented by the Fanjingshan (Kunyang) and Banxi Groups. At this time a series of geanticlines and geosynclines, chiefly NE-trending, formed under the action

---

1. Wei Zhouling.

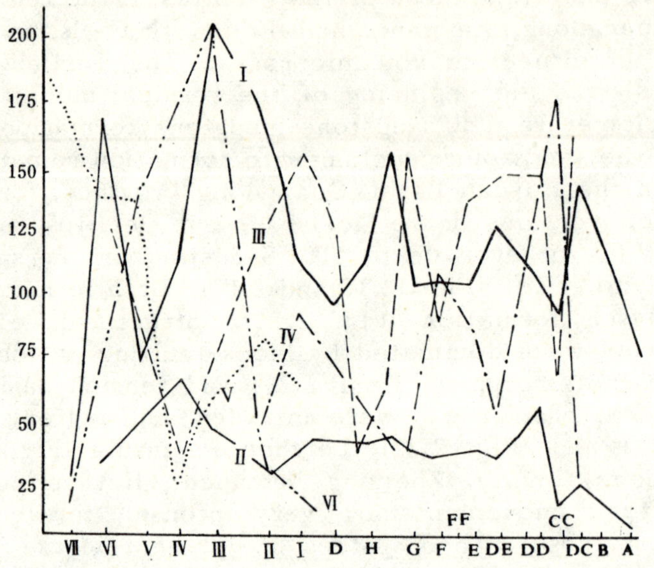

Fig. 2.14. Example of a great change in thickness of diwa sediments. Diagram showing the change in thickness of coal measures in a diwa-type coal-bearing formation, Pingxiang coal field, Jiangxi Province (After Pingxiang Coal Mine).

I, Zijiachong coal measures; II, Tianjishan coal measures; III, Laohuchong sandstone and shale series; IV, Sanjiachong shale series; V, Ailipo coal measures; VI, Sanqiutian coal measures. Vertical scale 1:2,500 (thickness), horizontal scale 1:100,000 (distance).

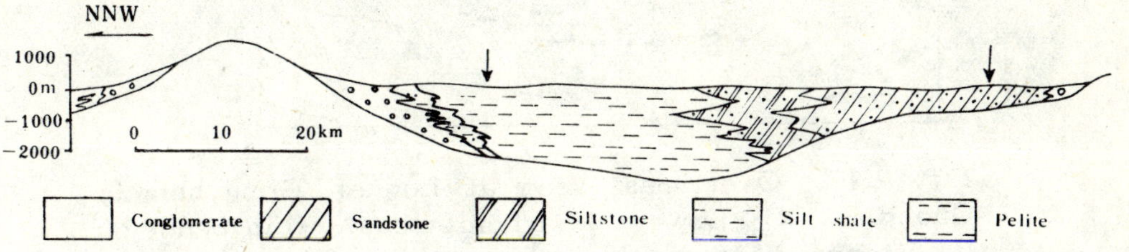

Fig. 2.15. Schematic cross section of early Late Cretaceous red beds in the Hengyang Basi.., Hunan Province, showing the rapid facies variation.

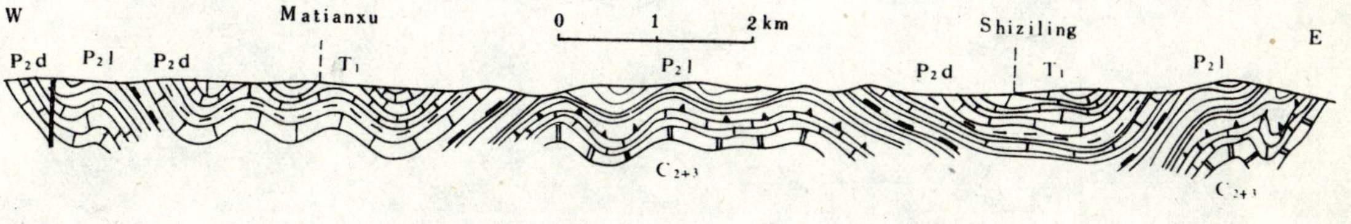

Fig. 2.16. Geologic section through Matianxu and Shiziling, Hunan Province.

Fig. 2.17. Overthrust seen at Luguei, Gongcheng, Guangxi. C_1-Lower Carboniferous thin-bedded limestone; γ-Mesozoic granite.

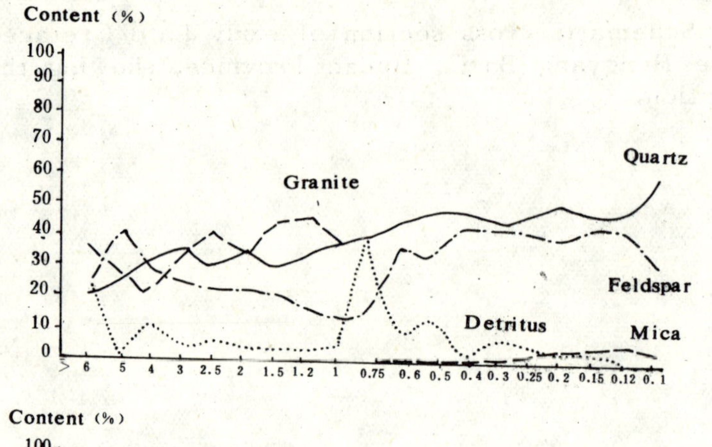

Fig. 2.18. Example of poor sorting of diwa sediments. Diagram showing the material composition and rock roundness of a Tertiary Danxia-type formation, north of the Heyuan diwa, Guangdong Province. The sample is a gravel-containing pebbly-gritty rock in the second major bed.

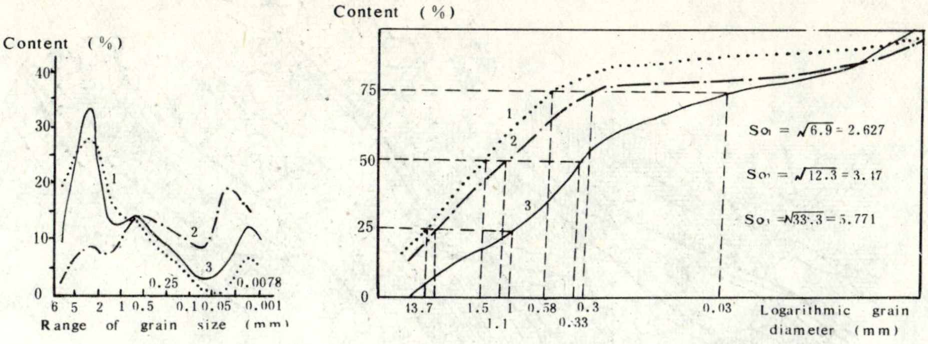

Fig. 2.19. Another example of poor sorting of diwa sediments. Grain size distribution curves and cumulative-frequency curves of three samples from a Tertiary Danxia-type formation, north of the Heyuan diwa, Guangdong Province. 1, Purplish red gravel-containing pebbly-gritty rock in the third major bed; 2, Gravel-containing pebbly-gritty rock in the second major bed; 3, Argillaceous pebbly-gritty rock in the fourth major bed.

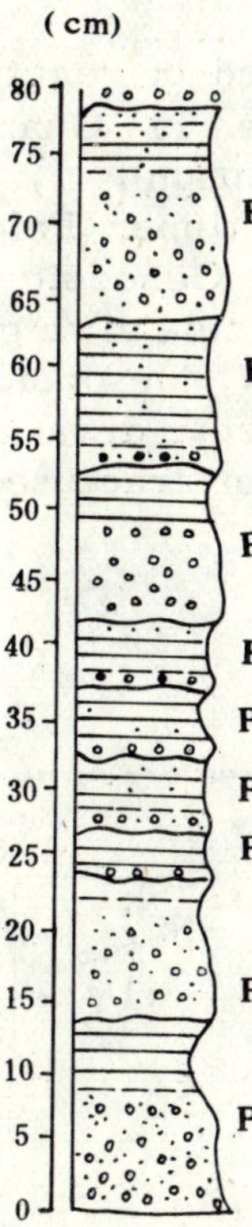

Fig. 2.20. Rhythmic features of a diwa sedimentary formation: example of a continental flysch formation. The Cretaceous at Bainiupang, Changhua County, Zhejiang Province (after Guan Yinwen).

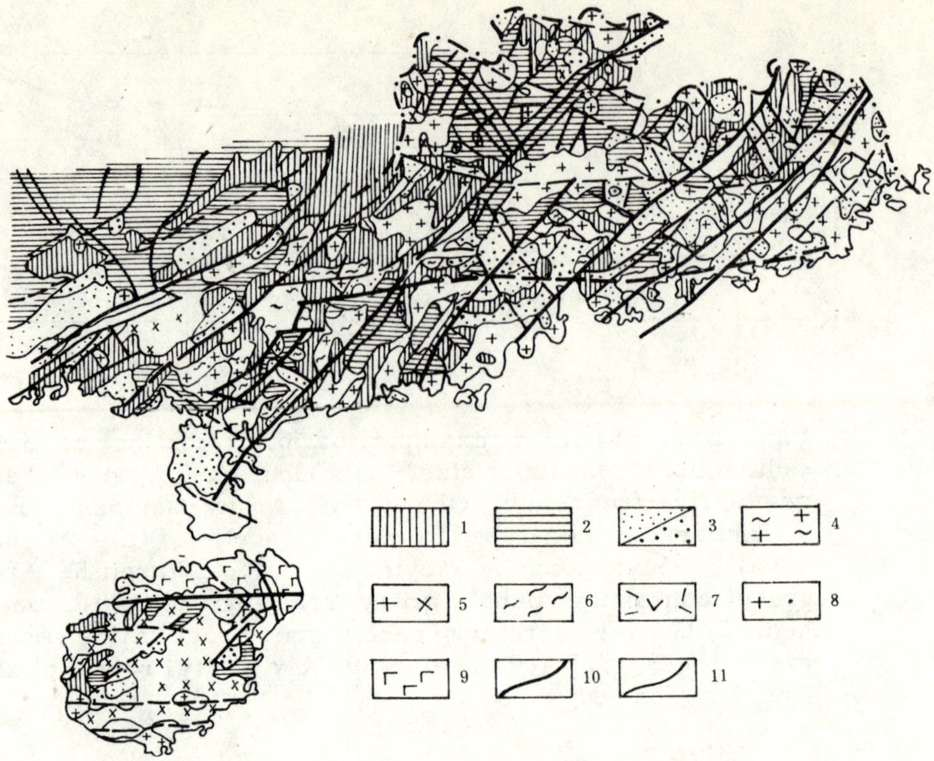

Fig. 2.21. Diagram showing fractures created or reactivated during the Mesozoic-Cenozoic diwa stage in Guangdong and Guangxi, SE diwa region (after Mo Zhusun). 1, Lower Paleozoic geosynclinal structural layer; 2, Upper Paleozoic platform structural layer; 3, Mesozoic-Cenozoic diwa structural layer; 4, Paleozoic gneiss; 5, Mesozoic diwa granite and granodiorite; 6, Mesozoic diwa gneiss; 7, Mesozoic diwa acid-to-intermediate effusive rocks; 8, Cenozoic diwa granite porphyry; 9, Cenozoic diwa basalt; 10, Major fracture zone; 11, Ordinary fracture.

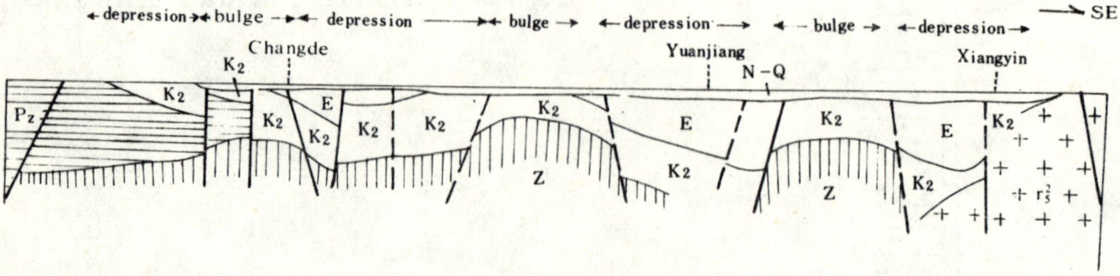

Fig. 2.22. Diagrammatic cross section through Dongting diwa basin.

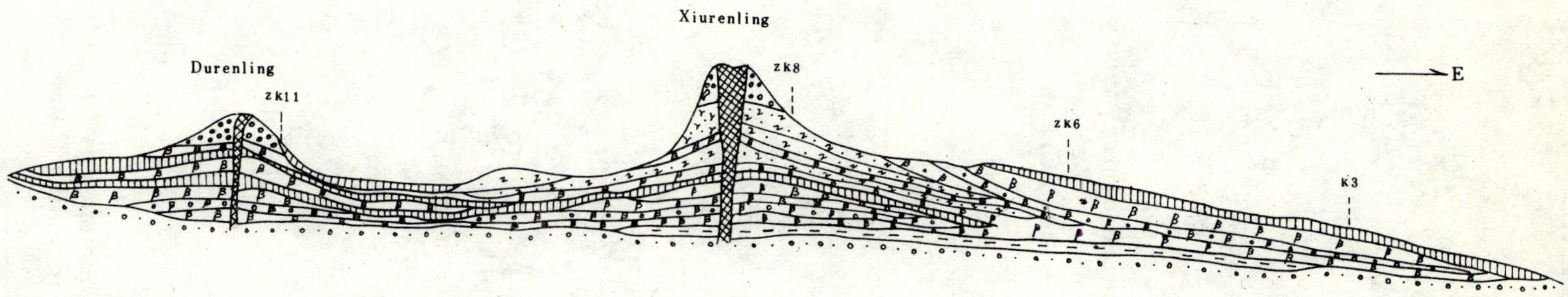

Fig. 2.23. Cross section through the Durenling and Xiurenling volcanic cones, Qiongshan County, Hainan Island, Guangdong Province.

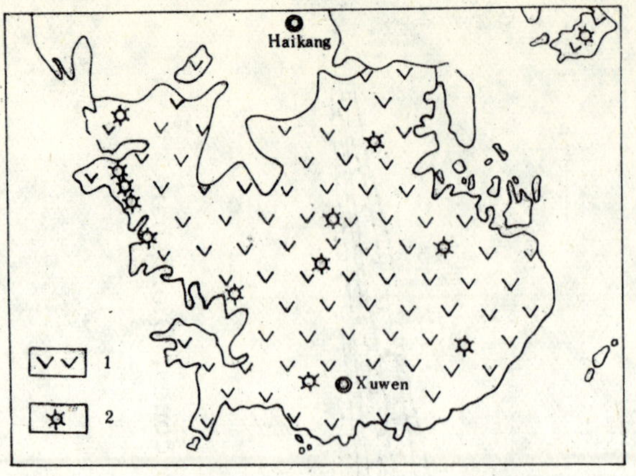

Fig. 2.24. Volcanos of fissure-central type controlled by intersecting fractures, Leizhou Peninsula, Guangdong Province. 1-Basalt; 2-Crater.

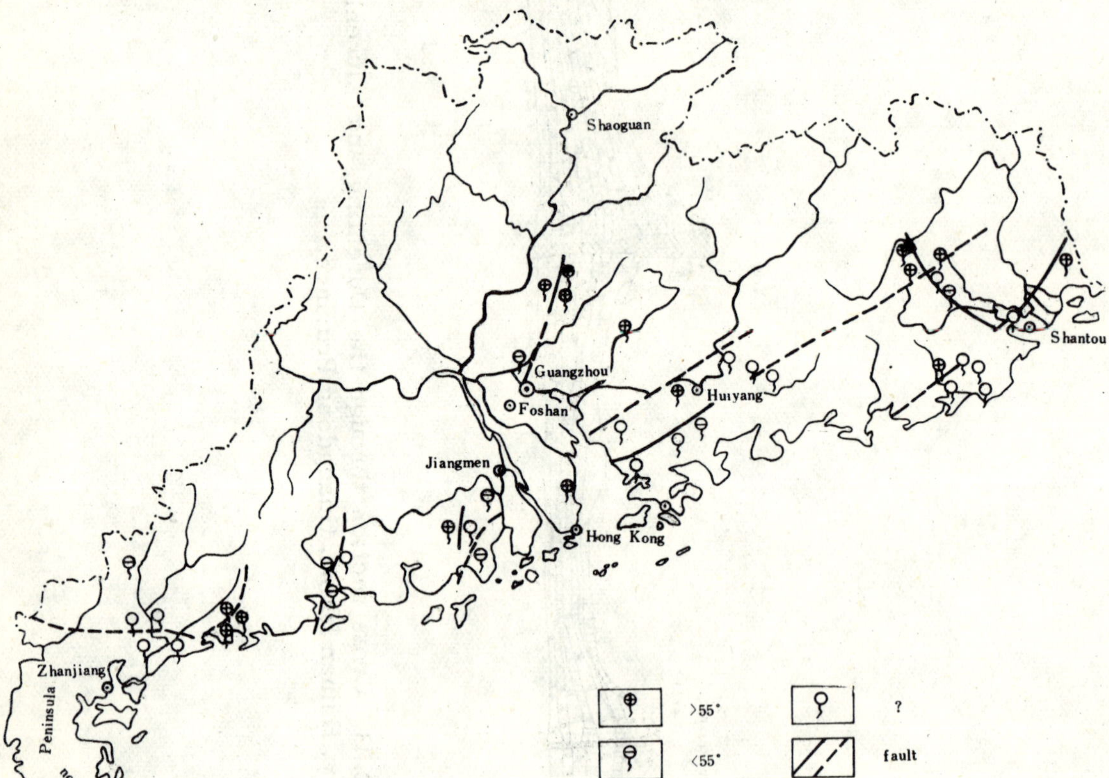

Fig. 2.25. Relationship of hot springs to faults along the coast of Guangdong Province (not including Hainan Island).

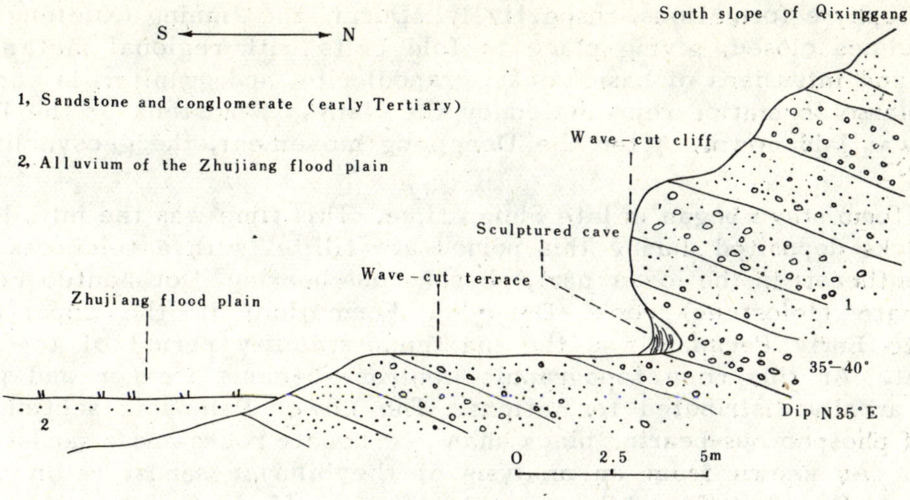

Fig. 2.26. Uplifted wave-cut terrace on the south slope of Qixinggang, southeast of Guangzhou, Guangdong Province.

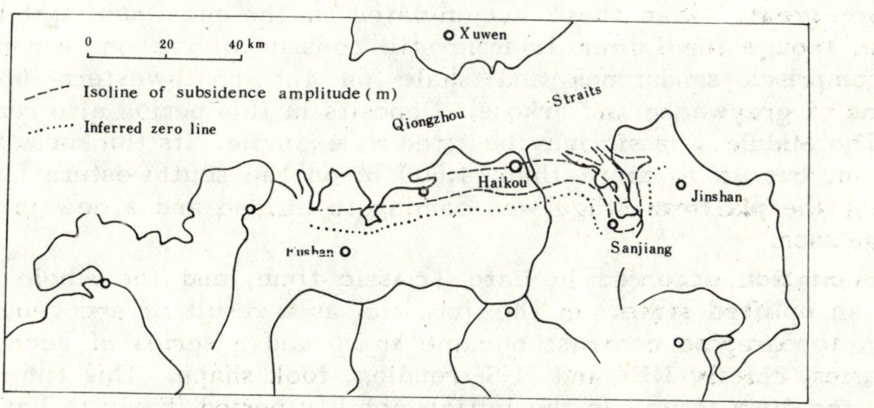

Fig. 2.27. Schematic map showing isolines of subsidence amplitude during the 1605 Qiongzhou earthquake (after Chen Enmin, 1981).

of horizontal compressive stress in a NW-SE direction. In deeply subsiding geosynclines, sandstone and shale with a thickness of up to 10,000 m were deposited and at the same time submarine basic volcanism took place producing flysch and spilite-keratophyre formations, respectively. During the Jinning (Xuefeng) orogeny the geosynclines closed, giving place to fold belts with regional metamorphism, tight folds and intrusions of basic rocks, granodiorite, and granite. In early Sinian time, a molasse formation represented by the Nantuo Sandstone or the Dengjiang Sandstone was laid down. After the Dengjiang movement, the geosynclinal stage terminated.

The platform stage began in late Sinian time. This time was the initial-stability period. Rocks deposited during this period are tillite (with a thickness of 4,300 m in the southeast) in the lower part, phosphorous-bearing (Doushantuo Formation) and carbonate (dolostone) rock (Dengying Formation) in the upper[98]. The Cambrian to Early Permian was the maximum-stability period of the platform development. At this time topographic contrast became weaker and gave rise to various areally distributed formations. The Lower Paleozoic section consists primarily of phosphorous-bearing black shale, carbonate rocks and a sandstone-shale assemblage. As known from an analysis of the Silurian sandstone in northwest Hunan, it is well sorted (Fig. 2.28) with a sorting coefficient of 1.5 on an average. Rocks deposited in Devonian to Early Permian time are a sandstone-shale assemblage, a coal-bearing formation and carbonate rocks. They lie conformably or disconformably one on another. This suggests that crustal movements occurring in the maximum-stability period were characterized chiefly by slow up and down movements affecting a large area. During the Dongwu movement at the end of the Early Permian, the whole region experienced uplifting and stepped into the residual-stability period. At this time basalt poured out extensively in east Yunnan and west Guizhou. Sedimentary rocks deposited in the residual-stability period changed more greatly than those accumulated in the maximum-stability period. For example, though the Lower Triassic still consists of carbonate rocks in most cases, it comprises sandstones and shale on the southwestern border, with intercalations of graywacke and arkose. Deposits in this period also reach a great thickness. The Middle Triassic may be cited as example. Its thickness is generally 1,700-2,310 m, but up to more than 4,000 m on the southwestern border. This indicates that the platform stage was coming to an end and a new mobile region would emerge soon.

No sedimentation occurred in Late Triassic time, and the whole region was probably in an uplifted state. In the Jurassic, as a result of arcogenesis, folding and faulting, topographic contrast became sharp and a series of geodome uplifts and diwa basins, chiefly NE- and N-S-trending, took shape. This time the region entered into the diwa stage. In the initial-mobility period (Early to Late Jurassic), conglomerate, a coal-bearing sandstone-and-shale assemblage, and a sandstone-and-shale assemblage, poorly sorted and greatly variable, were laid down in various diwa basins. Their thickness is generally more than 1,000 m, but varies from 370 m in northwest Hunan to 3,000 m in east Yunnan. The Late Jurassic to Cretaceous was the maximum-mobility period, during which crustal movement became more intense and topographic contrast sharper, giving rise to a sandstone-and-conglomerate assemblage and a salt-bearing formation with accompanying magmatic activity. Magmatic rocks include not only granodiorite, diabase, basalt and hornblendite, but also hidden basic-ultrabasic rocks in Dujiang

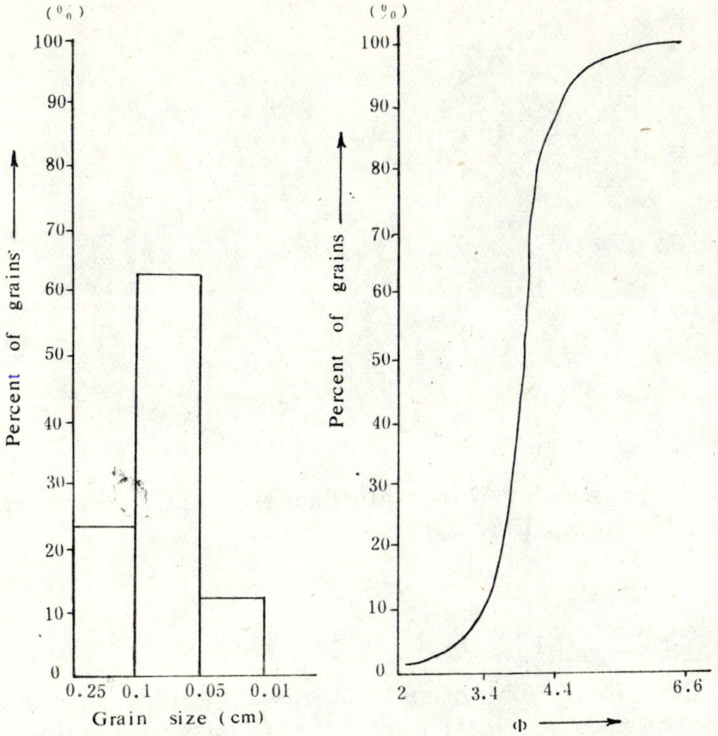

Fig. 2.28. Grain size histogram (left) and cumulative curve (right) of the Silurian Shamao Sandstone near Liushuta, Cili County (after Ding Chuanpu and Ou Yuanren, 1962).

and Leishan, east Guizhou Province. The residual-stability period began in the Tertiary. During this period a sandstone-and-conglomerate assemblage and a coal-bearing formation were laid down. Neotectonic movement is prominent. Block-faulting differential upward and downward movements take place strongly. River cut downwards deeply, forming V-shaped valleys and terraces with 3-4 steps. The famous Huangguoshu Waterfall is just located in the region (Fig. 2.29). Seismicity has a small intensity and a low frequency in the greater part of the region, but large earthquakes still occur on the western border.

The region may be divided into three third-order units: the East Yunnan geodome series, the Bijie diwa series, and the West Hunan-East Guizhou geodome series.

Fig. 2.29. Huangguoshu Waterfall and flat platform strata (after Zhou Haoyong, 1982).

(3) The Central China diwa region[1]

Situated between the North China and SE diwa regions in the form of an arc-shaped belt convex towards south, it occupies the eastern Qinling, Daba, Dabie Mountains and the middle and lower reaches of the Changjiang (Yangtze) River[56, 63, 82, 83, 99, 100, 134].

In the Archean, the region experienced a pre-geosynclinal stage. Its representative structural layer is the Dabie Mountain Group chiefly composed of a deeply metamorphosed sequence of gneiss, granulite and migmatite, more than 25,000 m thick, structurally complicated and intruded by granite.

In the Proterozoic, the region stepped into the geosynclinal stage. Under the action of horizontal bending force, arc-shaped geanticlines and geosynclines convex towards south originated. And a thick flysch formation with the predominance of sandstone and shale and sandy-argillaceous carbonate rocks accumulated in the geosynclines. There are also volcanics and spilite-kepatophyre formation in Dabie Mountain and in the northern flank of the eastern Qinling Mountains. During the Jinning orogeny the geosynclines closed in Dabie Mountain and in the middle and lower reaches of the Changjiang (Yangtze) River, where rocks underwent folding and regional metamorphism. In early Sinian time, a molasse formation composed of sandstone, conglomerate, and interbedded volcanic rocks (andesite, tuff and liparite) accumulated in the middle and lower reaches of the Changjiang (Yangtze) River and in Daba Mountain. But a geosyncline continued its development

1. Wei Bailin.

in the eastern Qinling Mountains, and in the Sinian to late Paleozoic a flysch formation composed chiefly of sandstone, shale and carbonate rocks (with intercalations of volcanic rocks in the Silurian) was deposited. Partial geosynclinal closure, folding, regional metamorphism and magmatic intrusion did not occur until the Caledonian orogeny at the end of the Silurian. At this time the geosyncline in the eastern Qinling Mountains migrated northwards and did not complete its development until the Middle Triassic. It is clear that the geosynclines in this region migrated from south to north.

The difference in the time of termination of the geosynclinal stage resulted in the difference in the time of beginning of the platform stage: Late Sinian in Huangling, in Dabie Mountain and in the middle and lower reaches of the Changjiang (Yangtze) River; Devonian in the greater part of the eastern Qinling Mountains. Nevertheless, the platform regime in various areas is all characterized chiefly by extensive upward and downward movements and, as a result, platform-type formations having a planar distribution and not experiencing regional metamorphism were widely deposited. For example, on the platform of post-Jinning age in the middle and lower reaches of the Changjiang (Yangtze) River and other areas, the lower Paleozoic section is composed, without exception, chiefly of carbonate rocks, and the sandstone-and-shale assemblage, coal-bearing formation and carbonate rocks of Upper Paleozoic age persist over a long distance. On the platform of post-Caledonian age in the eastern Qinling Mountains and other areas, the Gudaoling Limestone of Middle Devonian age already had a wide distribution. Magmatic activity during this stage was insignificant.

The Huaiyang orogeny at the close of the Middle Triassic was a turning point from the platform to the diwa stage. After that, horizontal crustal movement held sway again. Inherited from the geosynclinal arc-shaped structures of pre-Sinian age, diwa structures further developed on and reworked them. During the initial-mobility period (Late Triassic to Middle Jurassic) a coal-bearing conglomerate-sandstone-shale assemblage (Huangmaqing and Wuchang Formations) and clastic rocks composed of conglomerate, sandstone and mudstone was laid down. In Huanling they pass into a sandstone-and-mudstone assemblage intercalated locally with limestone. In the eastern Qinling Mountains they have been intruded by granite of Indosinian age. In southeast Hubei Province there are liparite and tuff of Middle Jurassic age. The Late Jurassic to Cretaceous was the maximum-mobility period, during which tectonic movement became more intense, with well-developed folding and block-faulting. Structural forms seen at present took shape just at this period. The red sandstone-and-conglomerate and sandstone-and-mudstone assemblages are the most prominent deposits. There are extensive outpourings of acid-to-intermediate and alkaline lavas and intrusions of various magmas in the middle and lower reaches of the Changjiang (Yangtze) River (Fig. 2.30). The residual-mobility period began in the Early Tertiary. It is marked by a sandstone-and-conglomerate assemblage, a sandstone-and-mudstone association and a gypsum- and salt-bearing formation as well as by basaltic effusives in some places. Neotectonic movements, chiefly differential upwarping and downwarping, are prominent. The Changjiang (Yangtze) River is cutting the rising West Hubei Plateau, forming the famous three gorges (the Yangtze Gorges)(Fig. 2.31). The Jianghan Plain has been sunk as deep as 1,000-2,000 m in Quaternary time. Seismicity is low.

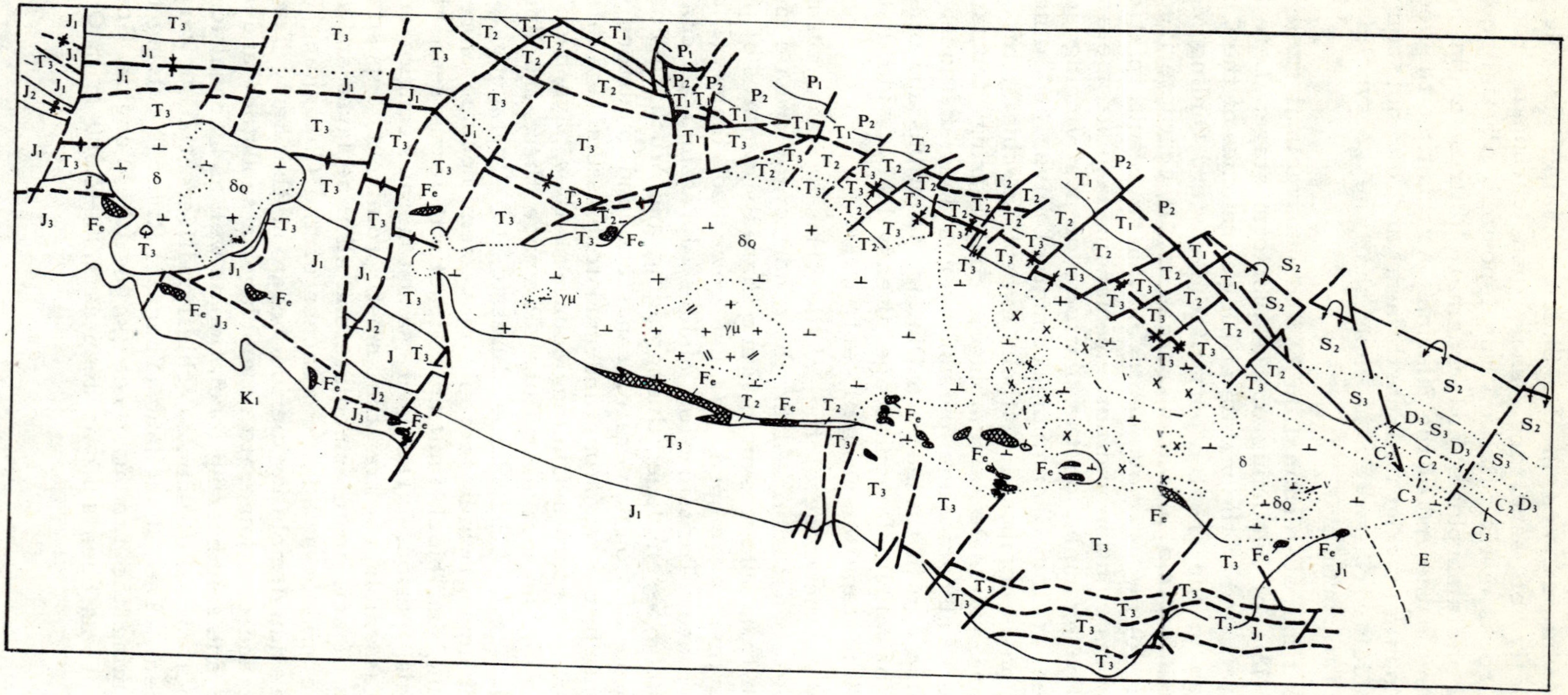

Fig. 2.30. Control of the Jinshandian Intrusion by folding in east Hubei Province. S_2, Luoreping Group; S_3, Shamao Group; D_3, Wutong Group; C_2, Huanglong Formation; C_3, Chuanshan Formation; P_1, Qixia Formation and Maokou Formation; P_2, Longtan Formation and Dalong Formation; T_1, Daye Formation; T_2, Jialingjiang Formation; T_3, Puyi Formation; J_1, Wuchang Formation; J_2, Ziliujing Formation and lower volcanics; J_3, Lingxiang Group; K_1, Upper volcanics; E, Red clastic rock; γ_u, monzonitic granite; δ_Q, quartz diorite; δ, diorite; v, amphibole gabbro; Fe, magnetite ore body.

Fig. 2.31. Sketch of the Wu Canyon in the Yangtze Gorges, looking east.

The region can be divided into three second-order units: the Jiangsu-Hubei diwa system, the Daiba geodome system, and the Qinhuai geodome system.

(4) The North China diwa region[1]

It is situated in a broad area of North China, east of the Fenhe-Weihe graben, south of the Bayanobo-Kaiyuan deep fracture, and north of the Xiangshuikou-Xuyi, Longmenchong-Nankang, and Luonan-Xinandian fractures[4, 5, 21, 25, 37, 42, 43, 48, 52, 55, 56, 59, 60, 61, 64, 79, 84, 99, 100, 106, 134, 142, 151].

The Archean was a pre-geosynclinal stage. At this time the earth's crust was already differentiated to some extent. In various areas gneiss and schist, such as the Fuping, Sanggan and Wutai Groups which underwent regional ultrametamorphism, granitization and migmatization in varying degree, were formed. Structural features are also different in different areas. In west Shandong Province and in the Wutai area tight folds can be seen, whereas in the Fuping area large domes have been developed. They constitute, in different structural forms, the crystalline basement of the region.

The geosynclinal stage was within the Proterozoic. It is characterized by the further differentiation of crustal structure and the origination of three geosynclines and two geanticlines of NNE trend. These are (from west to east): the Wutai-Luliang geosyncline, Fuping-Qinshui geanticline, Taihang-Yanshan geosynlcine, Hehuai geanticline, and Liaolu geosyncline. In the geosynclines,

1. Wei Bailin.

terrigenous rocks, flysch sediments and volcanics represented by the Hutuo and Zhongtiao Groups were deposited. During the Luliang orogeny the geosynclines closed and tight folding of NNE trend took place, accompanied by intrusion of basic-ultrabasic rocks in the early period and of granite in the late.

When the fold belt of Luliang age was subjected to erosion and reduced to a condition of peneplanation, the region stepped into the platform stage. In Sinian time when the initial-stability period just began, the influence of the geosynclinal development remained significant locally, giving rise to deep syneclises in the Yanshan-Liaoning and Henan-Huaihe areas, in which sandy-argillaceous and carbonate rocks in the lower part, approximately 10,000 m thick, and sandy-argillaceous sediments in the upper were laid down. In Bayanobo a syneclise was formed, in which deposition of iron ore took place. At the beginning of the Cambrian the region stepped into the maximum-stability period. During this period the terrain was flat and sea water invaded it in a planar pattern, forming flat nearshore and shallow-marine environments, in which carbonate rocks persistent in facies and thickness were deposited. This situation continued to the Middle Ordovician. At the close of the Middle Ordovician, along with a general uplift, the sea retreated completely and the region experienced a long-term erosion up to the Early Carboniferous. In the Middle Carboniferous, crustal mobility increased and the region entered into the residual-stability period. During this period oscillatory movements took place frequently, and the sea sometimes invaded the region and sometimes retreated, bringing about alternately marine and continental deposits and coal-bearing rocks. Up to the Early Permian the sea retreated and limnic deposits appeared. In Late Permian to Early Triassic time the climate became dry, with the deposition of gypsum and salt.

During the Yanshanian movement the diwa stage began. The Early Jurassic was the initial-mobility period. At this time old structural layers underwent arcogenesis and faulting, tectono-topographic contrast became sharper, and geodome uplifts and intervening diwa basins such as the Hunyuan diwa and the Mentougou diwa appeared. The principal deposits were coal-bearing rocks. Locally there were outpourings of intermediate-basic lavas. In the Middle Jurassic, various segments of the region successively stepped into the maximum-mobility period, earlier in the east but later in the west. Tectonic movement became more intense and tectono-topographic contrast sharper. The diwa basins uninterruptedly broadened, with the addition of the Zhangjiakou diwa, the Chicheng diwa and the Luanping diwa. Diwa-type deposits are mainly continental polymineralic sandstones and volcanic rocks. Magmatic activity was significant, with 2-3 phases of intrusion and 3-4 episodes of extrusion. Folds and faults had an extensive development. The former are of open type with a NE strike. The residual-mobility period began in the Tertiary. The most prominent feature is intense differential movement which occurred at the close of the Tertiary: Large-amplitude uplift in Shanxi, Hebei, east Liaoning and Shandong Provinces, but large-amplitude subsidence in the Hubei-Huaihe diwa system. In addition, there appear newly formed diwa basins such as the Fenhe-Weihe diwa, Datong diwa, and Yanqing diwa. They are all fault depressions filled chiefly with limnic detrital material and trend in a NE direction. Fractures extend deeply, and some of them may reach the Moho-discontinuity (Fig. 2.32). Volcanic activity occurred locally (Fig. 2. 33). Three episodes of extrusion have been recognized in Late Tertiary to Middle Pleistocene time. Heat flow value is high and the asthenosphere occupies a higher position in fracture zones (Figs. 2.34 and 2.35). Seismicity is high.

Second-order tectonic units are: the Liaoning-Shandong geodome system, the Hebei-Huaihe diwa system, and the Hebei-Shanxi geodome system.

(5) The NE diwa region[1]

It consists of all the three provinces of northeast China (except Songliao Plain) and north Inner Mongolia and lies north of the Dalad Banner, Guyuan-Zhangbei and Bayanobo-Kaiyuan deep fractures. It is a diwa region transformed, after erosion and peneplanation occurring in the Triassic, from a Hercynian geosynclinal fold region[37, 55, 56, 72, 99, 100, 110, 158, 159].

The crystalline basement of the pre-geosynclinal stage crops out sporadically in the region. North of Yinshan Mountain, Inner Mongolia[2], it comprises the Sanggan (early Archean), Erdaowa (middle Archean) and Majiadian Groups (Proterozoic), in varying metamorphic grades, and a magmatic complex dated at 16.67 m.y. Three similar series, namely the Mashan, Heilongjiang and Xingdong Groups, the age and stratigraphic sequence of which is still in debate, also exist in the Jilin area[3]. The crystalline basement is relatively young in the Hinggan area[4], e.g., the absolute age of the Xinghuadukou Group is as young as 950 m.y. The marked difference in the upper limit of the absolute age, structural style and metamorphic grade between the above-mentioned groups of different areas indicates that the crystalline basement within the region is by no means a single entity.

In Paleozoic time the region was in a geosynclinal condition. During the early Paleozoic, the NE-trending Ergun and Alshan geosynclines occurred in the Hinggan area, where Cambrian greenschist and carbonate rocks and Ordovician flysch deposits were laid down. The Bayanobo Group of the Bayanobo area, Inner Mongolia, extending in a E-W direction, is Cambrian-Ordovician in age. It is composed of slightly metamorphosed flysch deposits and submarine volcanic rocks accumulated in an early Paleozoic geosyncline. In the northern part of the Jilin area probably also existed a Sinian-early Paleozoic geosyncline.

As a result of the Caledonian orogeny, the above-mentioned geosynclines rose to form geanticlines. At the same time, an extensive area in Inner Mongolia north of Yinshan Mountain was invaded by a geosynclinal sea of middle to late Paleozoic age, which extended northeastward into the northwestern segment of the Xiao Hinggan Mountains. In the Jilin area the Zhangguangcai Mountain geosyncline, the maximum subsidence of which was in Jilin City and Panshi, and the Taiping Mountain geosyncline, the maximum subsidence of which is in Yanji and Hunchun, were taking shape; both strike in a NNE direction. Besides, Nadanhada Mountain also was a late Paleozoic geosyncline.

---

1. Wen Shanji.
2. It crops out widely south of the Bayanobo-Kaiyuan deep fracture, beyond the study region.
3. The Jilin area refers to the southeastern Xiao Hinggan Mountains, Zhangguangcai Mountain, Sanjiang Plain, Laoye Mountain, Taiping Mountain and Nadanhada Mountain.
4. The Hinggan area refers to the northern and central parts of the Da Hinggan Mountains and the northwestern segment of the Xiao Hinggan Mountains.

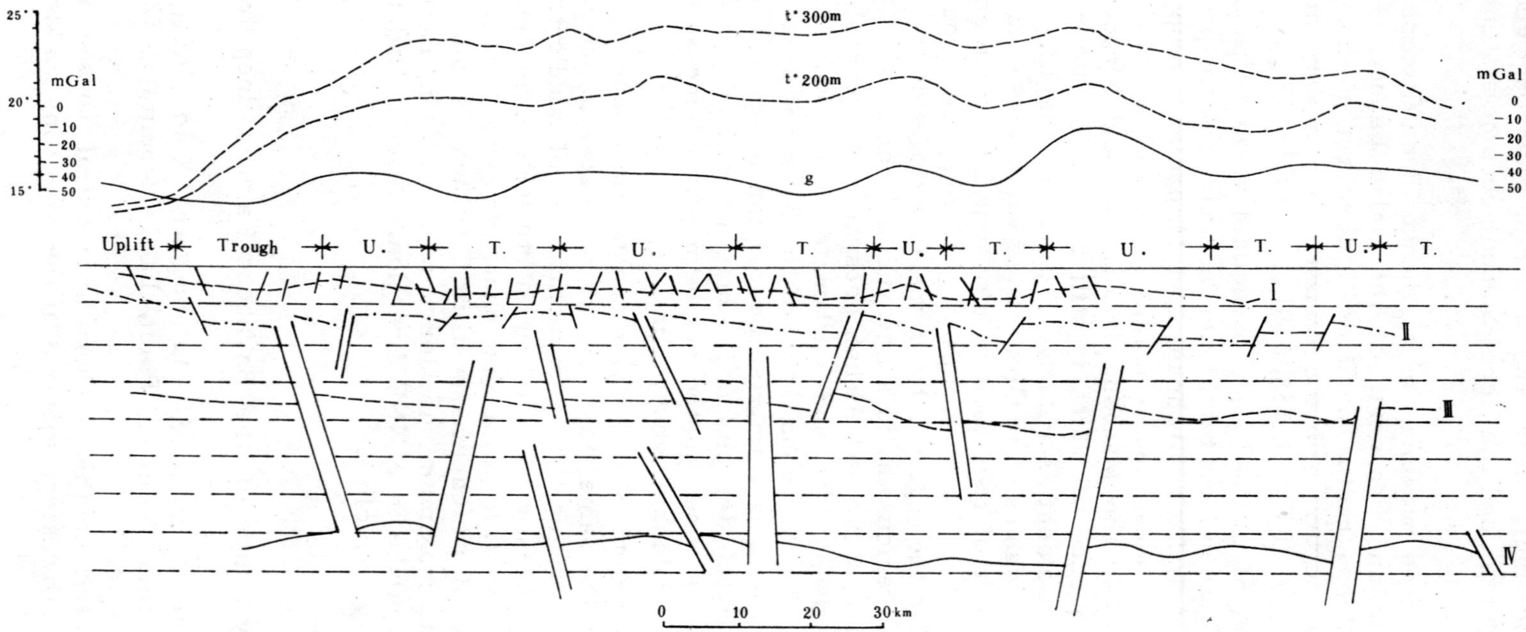

Fig. 2.32. Cross section through a deep structure from Yuanshi to Jiyang, North China diwa region. I, Top of the Upper Paleozoic; II, Top of the granitic layer; III, Top of the basaltic layer; IV, M-discontinuity (after the North China Oil Field).

Fig. 2.33. Langwoshan volcanic cone, Datong City, Shanxi Province.

Silurian and Devonian strata in this region have generally undergone medium- to low-grade metamorphism and contain abundant submarine volcanic rocks. Carboniferous and Permian deposits show a great facies change: In Inner Mongolia little volcanic material is known in the marine deposits, but in the Jilin area the role of acid and intermediate volcanic rocks and continental deposits has markedly increased.

Orogenic movements and magmatic intrusion occurred actively and widely at the geosynclinal stage and had a polyphasic character. At least four phases have been recognized, for instance, in Inner Mongolia. The Caledonian movement gave rise to an unconformity between the Middle and Upper Silurian and granite intrusions in the vicinity of Bayanobo. The unconformity under the Seribayanaobao Formation (Upper Devonian), the unconformity under the Zhesi Formation (Lower Permian), and the unconformity between the Zhesi (or Sanmianjing) Formation and the Jurassic represent three phases of the Hercynian orogenic movements. These orogenic movements were all accompanied by the intrusion of ultrabasic rocks and granite. The late-Hercynian orogenic movement, in particular, has made the Inner Mongolian Hercynian geosyncline to be folded and uplifted. Polyphasic orogenic movements and magmatic activity were also apparent in the Hinggan and Jilin areas. Among them the principal phase occurred at the close of the Early Carboniferous in the Hinggan area and at the close of the Early Permian in the Jilin area.

Fracture zones in the region lie mostly at the boundaries between paleogeosynclines and paleogeanticlines or adjacent paleoplatforms. The E-W-trending Bayanobo-Kaiyuan deep fracture is a good example. Yinshan Mountain, adjacent to it on the south, was a syneclise in the early Paleozoic, while the Inner Mongolian Plateau, adjacent to it on the north, was a geanticline. After the Caledonian orogeny, tectonic movements in the two areas changed their sign: the former became an anteclise, whereas the latter a geosyncline. That is also the case with the two sides of the Derbugan and Mishan-Dunhua deep fractures. This suggests that the fracture zones already existed in Paleozoic time and block-faulting movement already manifested itself at the geosynclinal stage.

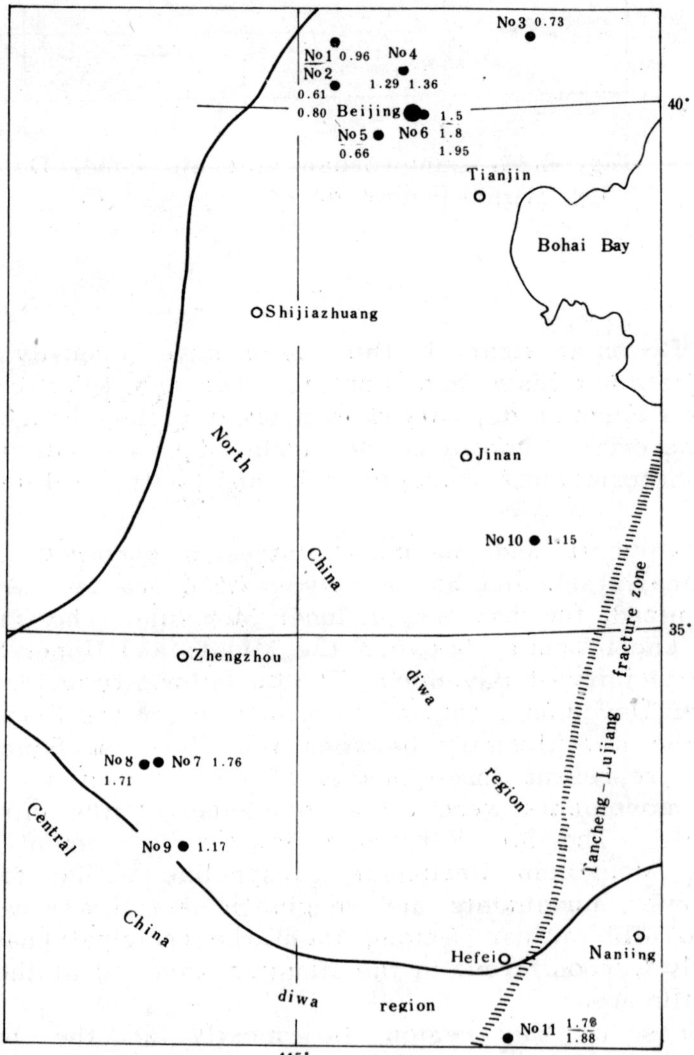

Fig. 2.34. Location of heat flow measurement points in the North and Central China diwa regions (after Institute of Geology, Academia Sinica).

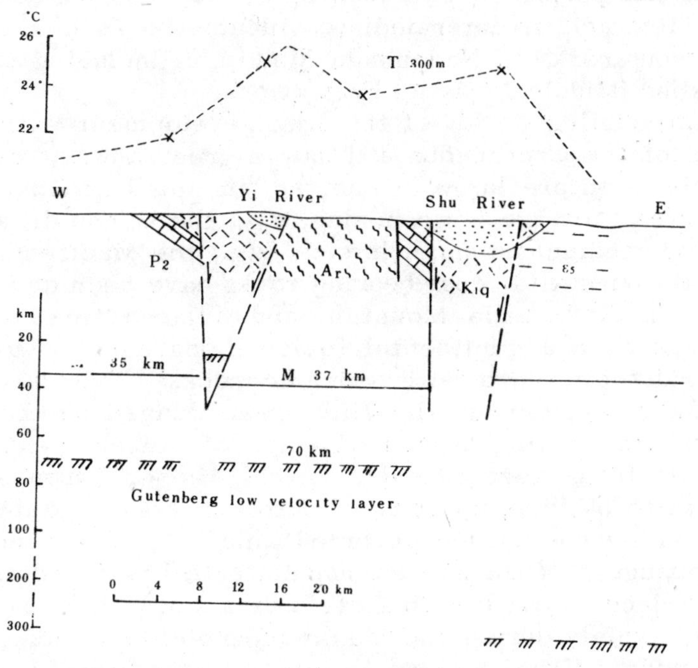

Fig. 2.35. Deep structure and earth temperature in the Yi-Shu fracture zone (after the North China Oil Field).

On these grounds it can be inferred that the NNE-trending Sunwu major fracture between the Hinggan and Jilin areas was seemingly present in the Paleozoic.

As a result of the Hercynian folding, the Paleozoic geosynclinal deposits were compressed into NE-trending anticlinoriums and synclinoriums complicated by faults. They united with the NW-trending structures taking shape simultaneously in the Tianshan and Altay Mountains to form a gigantic arc-shaped structure convex towards the south: the North Xinjiang-Hinggan arc.

The geosynclinal fold belt formed in the region during the Hercynian orogeny entered wholly into the diwa stage in Jurassic time after a short period of erosion in the Triassic. Many diwa basins have smashed the yoke of the Paleozoic geosynclinal structures, lying on several anticlinoriums and synclinoriums in the form of newly generated superimposed depressions. They trend obliquely to their structural line (e.g., the Oroqen diwa) or rest on Hercynian granite bodies exposed due to erosion (e.g., the Yichun diwa). This suggests that the Paleozoic geosynclinal fold belt was eroded to a condition of peneplanation and that the appearance of diwa basins heralded the dawn of a new stage, the diwa stage.

The initial-mobility period of the diwa development corresponds approximately to the Early Jurassic (in places it began in Late Triassic or mid-Jurassic time). Diwa deposits of this period are represented by coal-bearing rocks of Early Jurassic age, often with few acid-to-intermediate volcanic rocks in the lower part. There are marine siliceous rocks in Nadanhada Mountain, including the Zhenjiang (Upper Triassic) and Nalihe (Middle Jurassic) Formations.

The maximum-mobility period of the diwa development lasted from mid-Jurassic time to the end of the Cretaceous and saw a great amount of diwa basins. Diwa basins in the Jilin area are large in number but small in size, e.g., the Jixi diwa. In the Hinggan area they are large in size, e.g., the Genhe diwa. In Inner Mongolia they are rare and medium or small in size, e.g., the Manitemiao diwa. Extremely important is that continental coal-bearing rocks have been deposited in diwa basins of this period. In Nadanhada Mountain and adjacent areas, coal-bearing rocks of alternately marine and continental facies (Longzhaogou Group) have been laid down and generally alternate with volcanic rocks. This is quite typical in the Hinggan area and is common in the Jilin area. Judged by the region as a whole, during the maximum-mobility period, the age of volcanic extrusion migrated and its scale enlarged from west to east. In the Hinggan area acid-to-intermediate and intermediate-to-basic extrusions of Late Jurassic age predominated; in the north of the Jilin area extrusion occurred chiefly in Cretaceous time; west and east of Zhangguangcai Mountain intermediate-to-basic and acid-to-intermediate extrusions took place in Early and Late Cretaceous time, respectively. Igneous activity occurred widely during the maximum-mobility period, its age being early- and late-Yanshanian. Ultrabasic rocks are concentrated in the east Jilin area. Granites are seen everywhere as small bodies, chiefly of hypabyssal or ultrahypabyssal origin. They are controlled by fractures and occur generally like a string of beads. The granites are often of meta-alkaline type and contain more alkali metals in late-period bodies. They are petrochemically different from the Hercynian granites in the region (Tables 2.1 and 2.2). In some places potash feldspar is the dominant mineral, while dark-colored minerals are less. Such granites have been ascribed to alaskites. Assimilation and differentiation phenomena are unclear. The Yanshanian movement occurring in the maximum-mobility period of the diwa development has resulted in a major reconstruction of the tectonic framework of the region. The Mesozoic diwa structural layer has been crumpled into diwa fault-fold belts due to faulting and folding. Its principal structural lines trend in a NE direction and folds are of open type. In addition, older structural layers have also experienced the influence of the Yanshanian movement resulting in fractures and folds.

The residual-mobility period of the diwa development embraces the whole Cenozoic. During this period a series of newly formed diwa, intermontane depressions, appeared among the diwa fault-fold belts. These diwa are large basins in Inner Mongolia (e.g., the Qagan Lake diwa), in which lake deposits of post-Cretaceous age, chiefly gypsum- and salt-bearing red formations, have been developed. But they are mostly small basins in the Jilin area. In the residual-mobility period several episodes of basaltic extrusion occurred in the whole region (Fig. 2.36). Together with magmatic activities in the initial- and maximum-mobility periods, it forms a magmatic complex beginning with acid-to-intermediate (or intermediate-to-basic in the central and northern parts of the Da Hinggan Mountains) extrusion, climaxing with acid extrusion and granitic

Table 2.1. Comparison of average chemical composition of geosynclinal- and diwa-type granites in Inner Mongolia

Tectonic type	Age	Chief chemical composition (%)										
		SiO_2	TiO_2	Al_2O_3	Fe_2O_3	FeO	MnO	MgO	CaO	Na_2O	K_2O	P_2O_5
Diwa	early Yanshanian	74.94	0.10	13.59	0.59	--	0.10	0.16	0.51	3.95	5.70	0.03
Geosyncline	late Hercynian	59.48	0.75	14.38	6.54	1.59	0.16	3.68	6.73	3.22	2.13	0.60
China's average granite		71.27	0.25	14.25	1.24	1.62	0.08	0.80	1.62	3.79	4.03	0.16
World's average granite		70.18	0.39	14.47	1.57	1.78	0.12	0.88	1.99	3.48	4.11	0.19

Table 2.2. Comparison of petrochemical characteristics of geosynclinal- and diwa-type granites in Inner Mongolia

Tectonic type	Age	K_2O+Na_2O(%)	K_2O/Na_2O	Fe_2O+FeO+MgO (%)	Fe_2O_3+FeO (%)
Diwa	early Yanshanian	9.65	1.46	--	--
Geosyncline	late Hercynian	5.35	0.66	11.81	8.13
China's average granite		7.82	1.06	3.66	2.86
World's average granite		7.59	1.18	4.23	3.35

intrusion (as well as ultrabasic intrusion in Nadanhada Mountain) and terminating with basaltic extrusion.

Fig. 2.36. Tianchi Crater Lake, Changbai Mountains.

Tectonic movements occurring at the diwa stage have not only an inheritance character but also a distinct newborn element which has played a decisive role in the diwa development. The inheritance character consists in that the Mesozoic principal structural line has been inherited from the NE trend of the Paleozoic geosyncline, but folds are open, mostly arch-shaped. Some fractures originating at the geosynclinal stage were reactivated. These include: the Sunwu fracture which is active as a graben in Cenozoic time, the Derbugan-Jinhe fracture zone along which a series of granite stocks of Yanshanian age originated, and the Mishan-Dunhua fracture zone on which many diwa basins appeared in Mesozoic time like a string of beads and along which a great amount of basalt poured out in Cenozoic time and many deep-focus earthquakes occur at present. But a predominant place belongs to the newborn block-faulting movement. Some newborn fracture zones together with reactivated old fractures have carved up the whole region into many blocks of varying size, forming numerous diwa and geodomes. These fractures fall into three sets: NE-NNE, NW, and E-W. They are generally oblique to the direction of the older structural line at the place in question and have exerted control over the growth of diwa, over magmatic activity and metallogenesis at the diwa stage, and over the modern landform configuration. For example, an arc-shaped fracture zone, the Boli arc, which was developed in Laoye Mountain during the diwa stage in Mesozoic time, controlled the arrangement of diwa basins in the locality and the folds of strata. The landform configuration there is also related to the development of the Woken River arc-shaped structure. The NE-trending Yilan-Xitong fracture zone strictly controlled the zonal distribution of Cenozoic diwa basins. The Nenjiang deep fracture extends northwards into the Da Hinggan Mountains and cuts across the NE-striking old structure, forming the NNE-trending Mesozoic Oroqen diwa basin. This diwa basin and the Nenjiang deep fracture, in turn, have been intersected by the NW-striking fractures which occurred in Cenozoic time and along which valleys such as the Nuomin River Valley originated and basalt poured out. The E-W-trending Xar Moron River has been developed along a fracture zone occurring in Mesozoic time. This fracture has broken the framework of the NE-trending structure formed at the geosynclinal stage and serves as a natural demarcation line between different volcanic rock types on the two flanks. Finally, differential movement taking place along the E-W-striking fracture zone, which lies north of Daqing Mountain and was reactivated in Cenozoic time, has lifted the southern flank to become a rugged mountain, while the northern flank has been left to be the boundless Mongolian peneplain (Fig. 2.37). It can be concluded, therefore, that block-faulting movement reflected in fracturing is of utmost significance in the history of development of this region during the diwa stage. It should be mentioned in particular that, as a result of differential block-faulting movement, the Siping-Changchun and Nenjiang fracture zones between this region and the Songliao platform region began to separate the two regions since Triassic time. In the Cenozoic the differential block-faulting movement was again responsible for the creation of large uplifts and depressions, NNE-trending and manifesting themselves in mountains and plains, as can be seen today.

The region includes three second-order units: the Jilin geodome system, the Hinggan diwa system, and the Inner Mongolia diwa system.

(6) The N-S diwa region

It is an elongate tectonic region (Fig. 2.38) lying approximately longitudinally (the central segment deflecting to NNE) between longitudes 101° and 107°E and extending from Helan Mountain in the north through Longmen Mountain to Daxue Mountain in the south. Both its sides are bounded by faults or hidden fractures. This is where the longitudinal Yinchuan-Kunming deep fracture zone lies and is a geologic boundary or transition zone between the eastern and western halves of China. The transition character is evidently reflected not only in the tectonic evolution of the region and the characteristics of the present stage, but also in geophysics, tectonic systems, landforms, hydrogeology, crustal structure and thickness, depth to the Moho-discontinuity, crustal differential movements, metallogenesis and in seismicity. Having such distinctive features, this region occupies an important position in the tectonics of China (see Chapter 4) and has a wide areal extent. It is necessary to single out it as a first-order tectonic region[34, 37, 56, 62, 74, 89, 93, 99, 100].

The pre-geosynclinal stage began in the Archean and is represented by the crystalline basement composed of the Sangan (in Helan Mountain and Zhuozi Mountain of the northern segment) and Kangding Groups (in the southern segment). At that time folding took place intensely and magmatic activity occurred significantly.

The geosynclinal stage, unknown in the northern segment, began in the Proterozoic in the southern segment, lasted to the Early Paleozoic in the central segment and ended as late as Indosinian movement at the beginning of the Mesozoic in some areas (the northwestern zone of the central segment and its extension

Fig. 2.37. Mongolian Plateau (after Gao, 1980).

Fig. 2.38. Gongga Mountain, 7,556 m high, southern part of the N-S diwa region.

to the northern border of the Qinling Mountains). The lower part of the representative structural layer is made up of slightly metamorphosed rocks (phyllite and slate of flysch type) of the Kunyang (in the southern segment) or Bikou and Bailongjiang Groups (in the central segment). The Kunyang Group attains a maximum thickness of 20,000 m and has been intensely folded and intruded by acid-to-intermediate and basic rocks (in Luding and Yuanmou). It is overlain by conglomerate and gravel-bearing arkose of molasse type representing the deposits of late-geosynclinal development. The Bikou Group has intercalations of metavolcanics.

The platform stage, unknown in the northwest of the central segment, began in the northern segment in Sinian time (Fig. 2.10, west end), in the southern segment at the close of the Sinian and in the southeast of the central segment in Devonian time, and ended in Middle or Late Triassic time in all the segments. This stage is represented by carbonate and clastic rocks connected with platform-type sediments simultaneously deposited in adjoining regions. They have a planar distribution and their thickness changes less rapidly. Folding movement is not apparent. Magmatic activity is characterized chiefly by extrusion and intrusion of basic rocks occurring in the residual-stability period (the Emei Mountain Basalt of Permian age and the associated Panzhihua Complex).

The diwa stage began at the end of the Triassic or in the Jurassic and is represented by the diwa structural layer made up of Mesozoic-Cenozoic continental coal-bearing clastic rocks and red beds of basin type. Sediments, poorly sorted, generally assume a large and variable thickness (Upper Triassic to Lower Jurassic alone in the southern segment being 4,700 m thick) and evidently display the peculiarities of Cathysian-type formations. Magmatic activity occurred vigorously in the initial- and maximum-mobility periods. There are granites and granodiorites of Indosinian and Yanshanian ages in Luding and Xichang of the southern segment. Folding and faulting took place intensely. In Longmen Mountain overthrusts, klippes and imbricate structures occurred significantly. Neotectonic movement is apparent and seismicity is high.

According to the time of beginning of the maximum-mobility period, this region was originally a diwa region of Cathysian age, but the tectonic mobility has extended into the Cenozoic with active faulting and high seismicity. It is a transition zone between diwa regions of Cathysian age in east and those of Central Asia age in west China. Structural lines belong principally to the N-S tectonic system. Sometimes they are very striking (e.g., in the greater part of the northern segment and in the eastern zone of the southern segment). Sometimes they are not apparent on the land surface where only NW aligned structural lines outcrop. But judged by the distribution of earthquakes and the direction of the long axes of meizoseismal areas, it can be concluded that a set of N-S-trending basement fractures may lie hidden at depth, e.g., in the area between the northern and central segments, including the zone intersected by the Qinling Mountains. Meanwhile, the NE and NW aligned structural lines, which are widespread in east and west China, also exist in the whole region. In some cases, the two sets of structural lines look equal and are intertwined. In other cases, only one set is well-developed, leading to the deflection of the configuration of the tectonic region. For example, the central segment deflects to NE due to the dominance of the NE-trending tectonic system. There are some cases still, in which the two sets of structural lines gather together to form some minor southward-curving arcuate structures.

The Jintang arc at the southern edge of the central segment is a good example. The complex situation of these structural lines suggests that the structural lines in east and west China are intersected and interpenetrated just at this transition zone. These facts in combination with other peculiarities (see below) indicate the transition character of this region.

Three second-order tectonic units can be recognized from north to south: the Helan geodome system, the Longmen geodome system, and the Sichuan-Yunnan diwa system.

(7) The Altay diwa region[1]

It is situated in China, the Soviet Union and Mongolia, and stretches in a NW direction. In Chinese territory it lies in the northermost part of north Xinjiang and is separated from the Junggar diwa region by the Ertix deep fracture. It extends into the Soviet Union on the northwest and into Mongolia on the east and southeast[37, 45, 56, 99, 100, 133].

The Altay diwa region was assigned, together with the Junggar and Tianshan diwa regions, to diwa regions of Central Asia age. According to available data, it should be assigned to diwa regions of Cathysian age.

Nothing is known about the pre-Paleozoic situation of this region. It is inferred that the region entered into the geosynclinal stage at the beginning of the Paleozoic, being a constituent part of the ancient North China geosynclinal region. At that time, Middle-Upper Ordovician, Silurian, Devonian, and Carboniferous deposits were laid down. The Middle-Upper Ordovician and Silurian deposits are composed of clastic rocks of neritic and littoral facies, with a rapid variation in lithology, facies and thickness, the greatest thickness being in excess of 10,000 m. The Devonian and Carboniferous deposits consist of acid volcanic and pyroclastic rocks, also with a rapid lateral variation in lithology, facies and thickness, the greatest known thickness being over 6,000 m.

During the Caledonian movement, the region began to rise partly. This was probably due to the partial inversion of the geosyncline. In Devonian and Carboniferous times, deposits were absent in the greater part. The Hercynian movement was intense, giving rise to repeated folding and forming tight folds in a general NW direction as well as a series of fractures. These folds and fractures constituted part of the western wing of the Mongolian arc-shaped tectonic zone. The folding was accompanied by large-scale intrusions, which caused metamorphism of the enclosing rocks.

Little is known about the tectonic environment during the Permian and Triassic, because sedimentary formations of this age are not seen. Regional data indicate, however, that this period should be assigned to the geosynclinal fold belt period.

In the Late Triassic, the region probably passed through a short, relatively quiet period of denudation and planation, i.e., entered into the platform stage.

In the period from the end of the Triassic to the Jurassic-Cretaceous, the region entered into the diwa stage due to a great increase in terrestrial heat flow and rapid expansion of the crust. Folding was not prominent in the region, but block structure manifested itself evidently. There are many magmatic rocks distributed along fractures. The rocks have been dated at 0.86-2.30 million years, i.e., mostly Indosinian and Yanshanian, the latter predominating. Moreover, dislocation metamorphism is seen along fractures, indicating that Yanshanian

1. Huang Su et al.

age was the maximum-mobility period of the diwa development. Magmatic rocks of Yanshanian age are accompanied by ore deposits of rare and rare-earth elements.

In the Tertiary, the region entered into the residual-mobility period of the diwa development. This period was marked by a wave-like decrease in crustal mobility. Tectonic inheritance was apparent. NW-trending fractures were reactivated repeatedly. Block-faulting was prominent. Differential upward and downward movements were severe and occupied a dominant position, forming stair-like planation surfaces (Fig. 2.39).

Since the Quaternary, the region has been rising intermittently relative to the Junggar diwa region to the southwest. In Altay, multi-step terraces can be seen, the uppermost terrace being 150 m high. Earthquakes occur frequently. An earthquake of magnitude 8 struck this region in August, 1931. It was intimately related to the reactivation of the NNW-trending Haizikou-Kalaxianger fracture (Fig. 2.40) and occurred during the period of increased crustal mobility in the process of its wave-like decrease.

After becoming a geosynclinal fold belt, it was a positive unit of relative uplift and still is a positive unit at the present stage. Second-order subregions can not be singled out due to scarcity of geological data.

2. Diwa regions of Central Asia age

(1) The Junggar diwa region[1]

It is situated in the north of Xinjiang. The eastern and western ends extend into Mongolia and the Soviet Union, respectively. The southern limit is defined by the Manas River-Abinur Lake deep fracture, while the northern limit by the Fuyun deep fracture and adjoins the Altay diwa region. The region includes Jieshan Mountain, Junggar Basin, Karameili Mountain and Baytik Mountain[37, 45, 56, 76, 77, 99, 100].

The pre-geosynclinal structural layer does not crop out and, therefore, is unknown.

The Ordovician and Triassic sedimentary and volcanic rocks, observed on the eastern and western sides of the Junggar Basin, are the products of the geosynclinal stage. Lower Paleozoic strata, such as the Laba, Kekeshayi, Bulongguoer, Bulong and Kekexiongkudouke Formations, exposed in Mayili, Zayier, Sharburti, Sawuer and other mountains on the western side, and in Karameili, Baytik and other mountains on the eastern side of the basin, consist mainly of terrigenous and intermediate-basic volcanic rocks intercalated with carbonate and spilite-keratophyre formations. Lateral variation is rapid. The Upper Devonian Bulongguoer Formation, for example, is made up of conglomerate and conglomeratic sandstone with interbeds of carbonate rock in western Junggar, but the role of carbonate rock increases significantly in eastern Junggar, with interbeds of andesite porphyrite and tuff. The thickness varies from 0 to 295 and even more than 1,000 m. The Caledonian movement has clearly affected western Junggar, resulting in the local slight upheaval of the region and in the accumulation of clastics of alternately marine and continental facies. In the Carboniferous, marine geosynclinal deposits still predominated. A well-developed flysch formation of Carboniferous age exists in Zayier, Sawuer and other mountains. The general inversion of the

1. Huang Su et al.

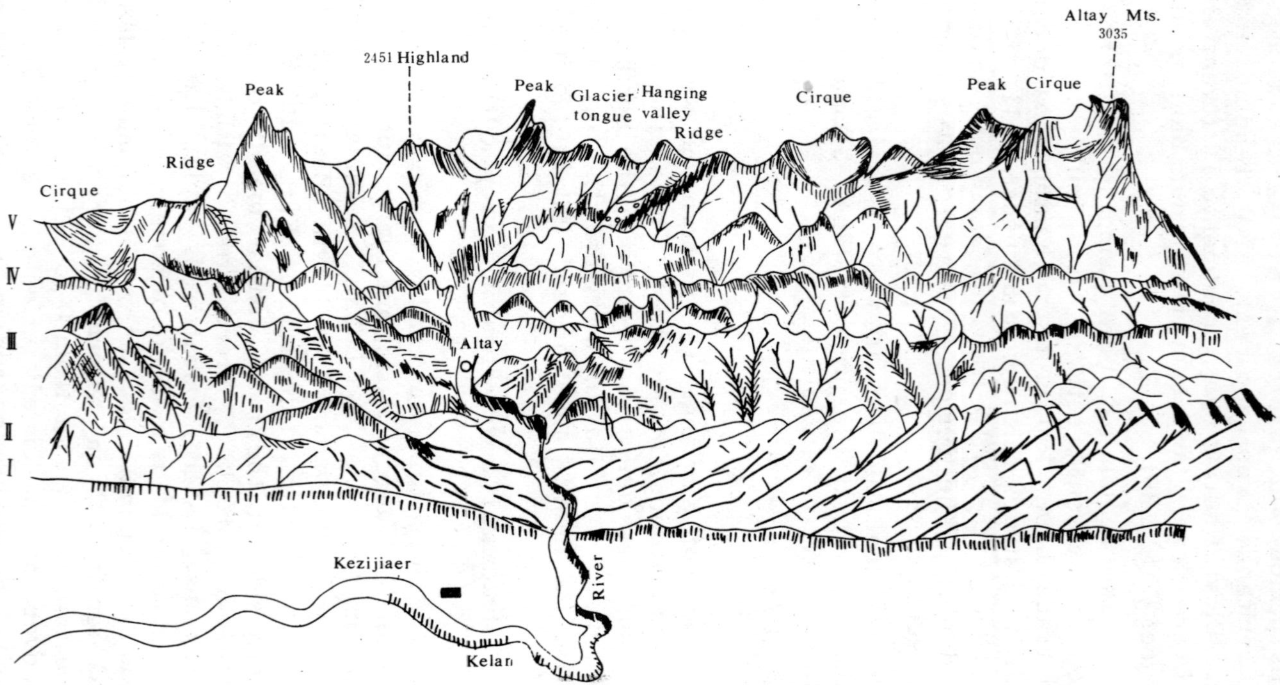

Fig. 2.39. Landform near Kelan River, looking from southwest to northeast (after the Regional Geological Survey Team, Xinjiang Geological Bureau).

Fig. 2.40. Haizikou-Kalaxianger fracture, a crack appearing during the 1931 earthquake (after Yin Hezhong, 1982).

geosyncline in this region occurred at the end of the Carboniferous, when numerous acid rocks were intruded, pre-Permian strata underwent slight regional metamorphism, and fractures and tight folds in a general NE-ENE direction took shape (Fig. 2.41)[103].

The Permian Wurhe Group and the Triassic Baikouquan and Karamay Formations are the products of the late period of the geosynclinal development. They are represented by a molasse formation distributed along the fold belt, with a transition from alternately marine and continental to continental facies and with a change from fine to coarse and again to fine grains. The middle part of the formation is rich in coarse-grained clastics, where sandstone and conglomerate are characterized by a mixture of gravels, a complex composition and an imperfect and variable roundness. The rocks grade upwards into sandstone and mudstone, which still have a complex composition and belong to feldspathic sandstone. The sorting coefficient of the sandstone is 2.2 or even greater.

From the end of the Late Triassic to the beginning of the Early Jurassic, the region underwent erosion, topographic contrast gradually decreased, and a peneplain took shape. Thus the region entered into the platform stage. The Early and Middle Jurassic should be assigned to the initial-stability period of the platform stage. At this stage, conglomerate, sandstone-conglomerate assemblage (Badaowan Formation), sandstone-mudstone assemblage (Toutunhe Formation) and coal-bearing deposits (Xishanyao Formation) were laid down. These rocks may be correlative at a long distance, but sometimes they have a high sorting coefficient (1.58-2.34)

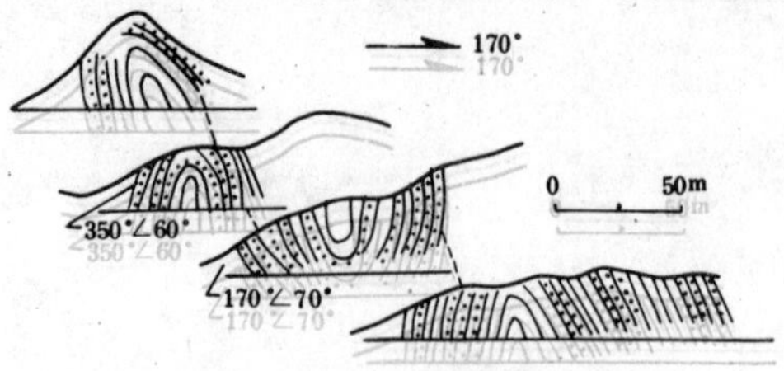

Fig. 2.41. Cross section through the west end of Alegelete Mountain, north of Wurhe, western Junggar. Tight folds composed of Carboniferous sandstone and mudstone are shown (after Huang Su, 1981).

and contain a great amount of feldspar. At the base of the Badaowan Formation in the Tailegula area southwest of Karamay, western Junggar, intermediate-basic volcanic rocks (e.g., andesite porphyrite, peridotite-basalt and pyroxene andesite porphyrite) developed. In the late initial-stability period of the platform stage, western Junggar was affected by horizontal movement, which gave rise to a local unconformity. This is an evidence for a considerable mobility in the initial-stability period of the platform stage. The Late Jurassic to the Cretaceous should be assigned to the maximum-stability period of the platform stage. At this stage sandstone and mudstone were deposited. Their material composition is relatively simple, with a sorting coefficient of 1.75-1.88. The Tertiary was the residual-stability period of the platform stage, when conglomerate often appeared in the formations (e.g., Honglishan, Suosuoquan, and Dushanzi Formations) and have an unequally grained structure. But the composition of sandstone and mudstone is simple. The sandstone of the Wulungu Formation, for example, can be used as glass sand.

At the beginning of the Quaternary, the region entered into the diwa stage. The Quaternary should be assigned to the initial-mobility period of the diwa stage. At this stage the earth's crust expanded, crustal movement became intense and topographic contrast increased, giving rise to the deposition of various rocks, mainly coarse-grained clastics. On the positive tectonic units of uplift, the Xiyu Formation, Wusu Group and Xinjiang Group of Pleistocene age are characterized by a wide development of conglomerate, sandy conglomerate and conglomeratic sandstone, which with a rapidly varying facies and thickness have been imperfectly sorted, but in the negative units (e.g., Junggar Basin) they are dominantly sandstone and mudstone. At the end of the Pleistocene, topographic contrast continued to increase, and some mountainous areas on the eastern and western sides of the basin rose above the limits of perpetual snow and received tillite. The intensification of crustal movement also resulted in the creation of various structures. Among them fractures predominate on the positive units, giving rise to block mountains. In the negative units, apart from fractures, broad folds can be formed (Fig. 2.42)[103].

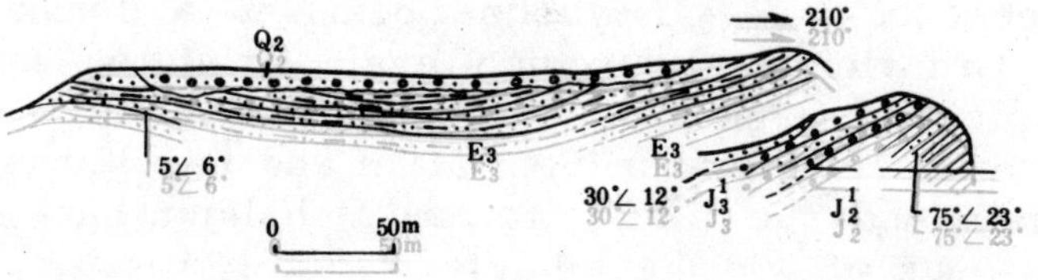

Fig. 2.42. Cross section across a quarry, southwest of the Heshen Tuoluogai, western Junggar (after Huang Su, 1981).

At present, neotectonic movement is quite intense. In eastern and western Junggar, erosion surfaces with 2–4 steps can be seen. Earthquakes occur frequently on the northern margin of Junggar. Some fractures of western Junggar (e.g., Darbute fracture) are still active today, and old strata (e.g., Carboniferous) may be thrust over Holocene sediments.

Second-order subregions are: the western Junggar geodome system, the central Junggar diwa system, and the eastern Junggar geodome system.

(2) The Tianshan diwa region

It includes the whole Tianshan Mountains both in China and in the Soviet Union[1]. In China, it extends from the Yining area, Xinjiang, in the west to north Ningxia in the east. It is separated from the South Xinjiang diwa region in the south by a fracture at the southern foot of the Tianshan Mountains, and from the Junggar diwa region in the north by the Manas River–Abinur Lake fracture[37, 45, 46, 87, 99, 100, 133].

The pre-geosynclinal structural layer is represented by the Archean of the Tianshan Mountains and the Wula Group north of Badain Jaran. They are made up of gneiss, schist, quartzite and marble, with a visible thickness of several thousand meters. Nothing is known about the tectonics at that time. The rocks have been generally reworked by metamorphism. Their lithology, facies and thickness greatly vary from place to place (e.g., the Archean on the northern and southern slopes of Keketongtao and Haliketao in the Tianshan Mountains). Magmatic activity was significant.

The Proterozoic to Triassic was the geosynclinal stage. During the Proterozoic, a series of WNW-trending geosynclines and geanticlines formed and a thick sequence of fine-grained clastics (siltstone and mudstone) was deposited, with minor acid

1. The western sector of the Tianshan Mountains in the USSR also should be assigned to diwa regions of Central Asia age. Here, the history of crustal development and the modern landforms are correlative with those of the eastern sector of the Tianshan Mountains in Chinese territory. In 1982 the author made an on-the-spot investigation of the western sector.

to intermediate effusive rocks and limestone; locally a flysch formation was laid down. The Sinian and Cambrian consist of tillite, carbonate, and siliceous rocks, which are correlative at a long distance. They may have been deposited in a quite tectonic environment or in a geosyncline being in a dormant state. From the beginning of the Ordovician, the region again exhibited manifestations of the great mobility typical of a geosyncline. The Ordovician to Carboniferous rocks, such as the Ordovician Henglanshan Formation and the Silurian Gongpoquan Group in Beishan Mountain and the Carboniferous Dahalajunshan group in Boluohuoluo Mountain, are made up of widespread intermediate to basic volcanic rocks, with a rapid lateral variation, and a flysch formation composed of sandstone and shale. The time spanning the Carboniferous to Triassic was the period of geosynclinal inversion and folding. At this time a poorly sorted and less persistent molasse formation was laid down, and rocks deposited in the early period of the geosynclinal development underwent regional metamorphism, which gave rise to phyllite, slate, metasandstone, marble, metavolcanics and schist. Simultaneously, linear tight folds and reverse faults, trending mainly in a WNW direction, took shape, and complex intraformational crumpling is frequently seen. Numerous acid rocks were intruded and small bodies of ultrabasic rocks emplaced along the NW-trending deep fractures. The Tianshan Mountains of the first generation were thus formed.

In the Jurassic, the region entered into the platform stage (Figs. 2.43 and 2.44). The Early Jurassic may be assigned to the initial-stability period of the platform stage. The Badaowan Formation at the base of the Jurassic consists of conglomerate and conglomeratic sandstone, which have a simple composition and a perfect roundness and are correlative at a long distance. The Middle Jurassic was the maximum-stability period of the platform stage. The coal measures of the Xishanyao Formation are correlative from the Honggon coal mine at the northern foot of the Northern to the Central and even Southern Tianshan Mountains. The Late Jurassic should be assigned to the residual-stability period of the platform stage. Rocks deposited at this time, though correlative, became poorly sorted. The Kalazha Formation composed of sandy conglomerate of proluvial origin, for example, has a complex composition. This heralded the transformation of the platform into the diwa stage.

The Cretaceous to Tertiary should be assigned to the initial-mobility period of the diwa stage [103, 104]. The Tugulu Group and Donggu Formation of Cretaceous age and the Ziniquanzi Formation, Anjihaihe Formation and Changjihe Group of Tertiary age consist poorly sorted and less persistent sandy conglomerate, conglomeratic sandstone, sandstone and silt mudstone. The Quaternary was the maximum-mobility period of the diwa stage. Sediments deposited at this time have a still poorer sorting and a less persistence. The Quaternary Xiyu Formation, for example, varies greatly in diameter of gravels, which usually are angular. Its thickness on the northern border of the Tianshan Mountains ranges from 0 to 2,066 m. Unconformities are frequently seen between various strata. As the Quaternary belongs to the maximum-mobility period of the diwa stage, crustal movement at this time is very severe, giving rise to the present-day Tianshan Mountains of the second generation.

Folds formed during the Yanshanian and Himalayan orogenies are common in Mesozoic-Cenozoic basins of the region, creating broad brachyanticlines. In the Central Tianshan Mountains, the Tertiary dips steeply and even vertically.

Fig. 2.43. 4000-m-high peneplain, Tiansha. the remnant land surface of the ancient post-Hercynian Northwest China platform (after China Pictorial, 1980).

Fig. 2.44. 2,000-m-high peneplain, Southern Tianshan, the remnant land surface of the ancient post-Hercynian Northwest China platform.

On the northern border of the Northern Tianshan Mountains, folds of Yanshanian and Himalayan ages are well-developed, with 3-4 rows of anticlinal structures. Locally the Tertiary dips steeply and even vertically.

At present neotectonic movement is very severe. For example, deep fractures at the northern and southern feet of the Tianshan Mountains have such a large amplitude of vertical displacement that the Tianshan Mountains have risen at an average speed of 20 mm per year since the middle Quaternary, while the Turpan depression within the Tianshan Mountains 4,000 m high above sea level has subsided to -154 m. The rapid rise of the Tianshan Mountains caused the accelerated cutting down of rivers and gave rise to multi-step terraces (Fig. 2.45) [103]. Moreover, the deep fracture at the southern foot of the Tianshan Mountains tended to cause the hanging wall, the Tianshan Mountains, to thrust towards the Tarim Basin on the southern side. Both the above-mentioned deep fractures are seismically active.

Second-order subregions are: the Northern Tianshan diwa system, the Central Tianshan-Southern Tianshan geodome system and the Badain Jaran diwa system.

(3) The South Xinjiang diwa region[1]

It lies in south Xinjiang and occupies the Tarim Basin and the Altun Mountains, separated from the Kunlun geosynclinal region and the Qinghai-Gansu diwa region on the south by the South-Tarim-Edge deep fracture and the Kegantag major fracture [37, 56, 87, 99, 100, 133].

Prior to the diwa stage, the tectonic history of the region varied from south to north. The Archean of the Tarim Basin and the Precambrian of the Altun Mountains constitue the pre-geosynclinal structural layer. Little is known about the tectonics at that time. Judged by the Archean rocks of Kuruktag and the deeply metamorphosed rocks of the Altun Mountains, crustal mobility was quite high and magmatic activity occurred frequently.

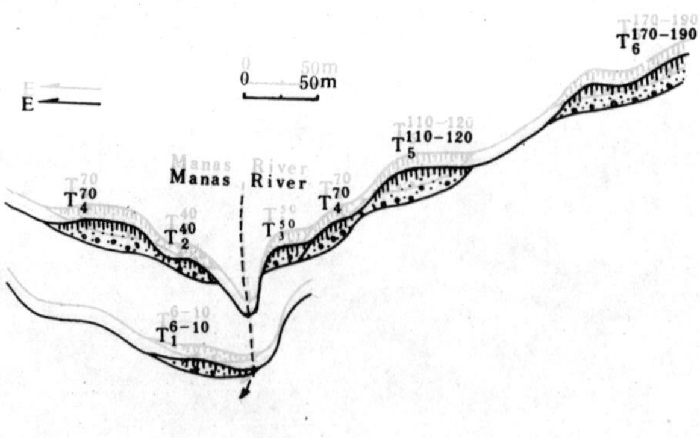

Fig. 2.45. Six-step terrace in the upper reaches of Manas River.

1. Chen Jiachao.

The Proterozoic of the Tarim Basin belongs to geosynclinal-type deposits. They are terrigenous calcareous sandstone and mudstone laid down in deeply subsiding geosynclines and displaying flysch features. At the close of the Proterozoic, the geosynclines closed and tectonism took place intensely, forming folds of tight type and igneous rocks exposed mainly in the Kuruktag area.

After Middle Sinian time, crustal mobility of the Tarim area progressively reduced and the platform stage started. The Beiyixi Formation of Sinian age and the Permian constitute the platform structural layer composed of,in ascending sequence: tillite (Middle Sinian), carbonate rocks (Cambrian to Ordovician), a sandstone-mudstone association (Silurian to Deveonian) and a coal-bearing formation. The tillite was formed in the initial-stability period of the platform development when the residual geosynclinal regime and sharp topographic contrast still had some influence on its deposition. The Cambrian to Permian was a period of crustal stability. The lithology, facies and thickness of the formed sedimentary rocks changed not abruptly in a broad area. Magmatic activity was insignificant. In the lower part of the carbonate rocks in Kuruktag and Kalpin there are a few intercalations of intermediate-basic volcanic rocks. Within the coal-bearing formation interbeds of basalt sometimes exist.

The Altun area was still at the geosynclinal stage in Cambrian to Carboniferous time. Strata deposited during this stage constitute the geosynclinal structural layer. The lower part of the strata is a flysch-like formation (Lower Proterozoic) grading in places into carbonate rock. Its lithology, facies, and thickness changed abruptly, with accompanying frequent extrusion of intermediate-basic magma, which tended to intensify eastward. Beginning in the Devonian, an upper flysch formation (Devonian) and intermediate-basic volcanics (Carboniferous) were laid down. But in the east, affected by the Caledonian geosynclinal folding in the adjacent Qinghai-Gansu region, the Upper Devonian and Carboniferous have changed into a molasse formation attended by volcanic eruption. During the late-Hercynian orogeny, the Altun geosyncline was folded and closed; tectonism and magmatism took place intensely, forming a linear fold belt. In the Permian, crustal mobility began to reduce and the region stepped into the platform stage, becoming one with the Tarim platform formed earlier. The eastern Altun Mountains received platform-type sandy-argillaceous deposits (Permian).

During the Triassic, no sedimentation took place and the region was probably still in the process of denudation. Afterwards, the whole region stepped into a diwa region. Beginning in the Jurassic, crustal mobility increased markedly. The original flat topography of the platform began to break due to intense block-faulting movement. The Tarim Basin was formed just in such an environment. In this large basin, coarse clastic rocks (Jurassic and Cretaceous), chiefly sandstone and conglomerate, were deposited. During the maximum-mobility period in Cenozoic time, the basin was further divided into many blocks receiving thick Tertiary red beds and Quaternary sediments. Covered by Quaternary and older diwa-type deposits, the Tarim Basin as seen on the present-day Earth's surface seems to lack tectono-topographic contrast. But available data show that since the middle Mesozoic, because of intense tectonic movements in this region, two sets of X-shaped fractures, NW- and NE-trending, have been formed. As a result, the basement of the basin was divided into a series of rhombic blocks of different size, and a strikingly undulating and sharply contrasting basin bottom topography appeared. As is known from geophysical data, on a NW-trending horst in the western part, the depth of the basin bottom is only two kilometers, whereas in the adjacent graben it attains 10 km, being 8 km in difference. Some downfaulted blocks (e.g., the Tuoyun Basin) were still invaded by sea water in the Paleogene. Near the western Kunlun Mountains and in the Tianshan, the Tertiary rocks alone attain a thickness of 5,000-6,000 m. On the northern margin of the basin, large lakes, such as Bosten Lake with an elevation of 1,048 m above sea level, can be seen (Fig. 2.46). Although the central part of the basin is a broad desert (Fig. 2.47), many fault depressions and uplifts also exist under the desert, the strongly uplifted areas being arid monadnocks. Modern neotectonic movements are strong, especially along fractures on the edges of the Tarim Basin, which appear to be important seismogenic structures. The two sides of the Kalpin fracture show a marked difference in elevation. A major fracture at the north edge of the Altun Mountains has cut through the Quaternary and has given rise, at its east end, to Quaternary basaltic lava. It is a belt of intense seismic activity.

Second-order tectonic units are: the Tarim diwa system and the Altun geodome system.

(4) The Qinghai-Gansu diwa region[1]

Situated in Qinghai and Gansu Provinces, it occupies the Qaidam Basin, Hexi Corridor, Qilian Mountain and part of Alxa. The region, the southern and southwestern limits of which are defined by the Tikanlik-Nomhon and Nanshan major fractures, adjoins the Kunlun and Bayanhar geosynclinal regions [6, 37, 50, 56, 68, 69, 87, 99, 100, 122, 146].

Pre-Sinian (partly pre-Cambrian) rocks, including the Dunhuang, Maxieshan, Longshoushan and Gaolan (or Niutouhe) Groups in Gansu Province and the Dakendaiban Group in Qinghai Province, represent the crystalline basement of the pre-geosynclinal stage. They have been deeply metamorphosed and have a complex lithology. Further studies are needed to clarify the tectonics during this stage.

In some areas of Qinghai Province, the Upper Sinian, represented by the Quanji Group, has been ascribed to platform-type deposits. It attains a thickness of 1,000-1,656 m and is characterized by slight metamorphism, insignificant magmatic activity and weak structural disturbances. But in other areas, Sinian to Devonian time was a geosynclinal stage. Its representative structural layer consists of the Xinglong (or Dunzigou) Group of Sinian age to the Laojunshan Group of Devonian age. These are, from bottom to top: lower terrigenous-carbonate (Sinian), spilite-kepatophyre (Cambrian to Ordovician), flysch (Silurian) and molasse formations (Devonian). Their facies and lithology change abruptly. The lower terrigenous-carbonate formation, for example, can pass into carbonate rocks, if calcium content increases, or may be replaced by pyroclastic rocks, if volcanogenic material increases. It can be divided into several subformations in a cross section. Owing to the difference in intensity and mode of tectonic activity, one subformation can grade in space into another. In Cambrian-Ordovician time, tectonic activity became more intense and magmatic extrusion occurred strongly, giving rise to a spilite-kepatophyre formation. The lithology, facies

1. Chen Jiachao.

Fig. 2.46. Bosten Lake, northeast of Karla, the largest fresh water lake in south Xinjiang (after China Pictorial, 1980).

Fig. 2.47. Taklamakan Desert, South Xinjiang diwa region.

and thickness of this formation change abruptly, and sometimes it wholly grades into a carbonate formation. During the Silurian, the earth's crust was still in the process of large-amplitude upwarping and downwarping. The process of downwarping was accompanied by frequent oscillatory movements, which caused a rhythmic succession of sandstones and mudstones constituting flysch deposits. The environments were NW-SE-trending geosynclines and geanticlines. The areas of subsidence and centers of sedimentation in the process of the geosynclinal development uninterruptedly moved outward. For example, at the beginning of the Sinian an early trough appeared in central Qilian Mountain and was uplifted at the close of the Sinian, when two new troughs, the northern and southern Qilian geosynclines, originated on its north and south sides. Later, in the Silurian, underwater geanticlines emerged again in the geosynclines and the centers of sedimentation moved outward once more. The geosynclines did not close until the end of the Silurian when foredeeps and intermontane depressions appeared. Geosynclinal sedimentation was accompanied by frequent and strong magmatic extrusion, especially in the northern Qilian geosyncline, which had an eugeosynclinal character. In the process of folding and rising of the geosynclines, both tectonism and magmatism were strong, forming numerous NW-trending folds of tight type and widespread magmatic rocks.

Carboniferous to Triassic time was the platform stage represented by the Chouniugou Formation of Lower Carboniferous age to the Triassic (Xidagou and Junzihe Formations). They are coal-bearing sediments characterized by the persistence of its stratigraphic position, lithology and facies, and can be easily correlated over a wide area. Among sandy sediments monomineralic sand predominates. Continental facies and shallow-marine carbonate facies are common, indicating that crustal movements were dominated chiefly by downwarping and upwarping which affected a large area. Magmatic activity was generally weak except on the borders, where, affected by the Hercynian magmatic intrusion in the adjacent Kunlun and Bayanhar geosynclinal regions, it was strong. Metamorphism was not apparent.

At the beginning of the Jurassic, the region stepped into the diwa stage which is marked by renewed crustal mobility. The representative structural layer consists of strata from the Lower and Middle Jurassic Longfengshan Formation to the Quaternary. At this time a sharply contrasting topographic relief began to appear on the original flat platform-type land surface. Mountains alternated with basins arranged along the NW-trending structural line, e.g., the Shule River basin, Datong River basin and basins on the north and south sides of Longshou Mountain. In the basins formations corresponding to the initial-mobility period of the diwa development and similar to the formations of Pingxiang type in east China bagan to accumulate, e.g., the Longfengshan Formation (Lower and Middle Jurassic) and Chijinbu Formation (Upper Jurassic). In addition, intermediate-basic magmatic extrusion occurred at Ela Mountain and Wahong Mountain in the north of the Chaoshui Basin. In Cretaceous time the diwa development entered into the maximum-mobility period. Some of the original lake basins broadened, others reduced and new downfaulted lake basins emerged one after another. Qinghai Lake is the largest surviving salt lake in the region. The lake received accumulation of red beds, among which Cretaceous deposits alone attain a thickness of several thousand meters (e.g., Huihuibao Group). The red beds are characterized by a great change in lithology, facies and thickness which cannot be correlative even in the same basin. Diwa basins in the region are dominantly grabens created by reverse faults and compressive in nature (Figs. 2.48 and 2.49). Magmatic activity was significant and produced a series of magmatic formations of Yanshanian age, chiefly granitic. Neotectonic movements are very apparent. The Qinghai Plateau now is still rising and the Huanghe (Yellow) River cuts down to form canyons (Fig. 2. 50). In Qilian Mountain, for example, narrow horst-type mountains alternate with broad graben-type valleys with major active fractures in between. These fractures constitute a seismic belt in the region. Quaternary sediments are also of piedmont facies, e.g., the Wuquanshan conglomerate (Fig. 2.51). Limnic sediments are rich in salt (Fig. 2.52).

According to the peculiarities in the distribution of diwa deposits three second-order units, the Corridor diwa system, the Qilian geodome system and the Qaidam diwa system, can be singled out.

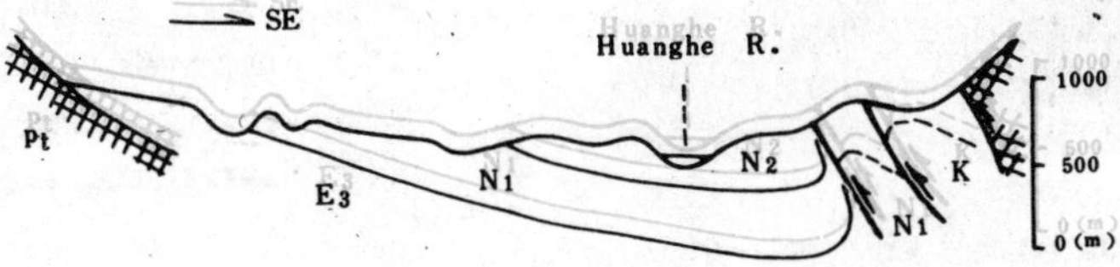

Fig. 2.48. Cross section through Xunhua Basin, Qinghai Province.

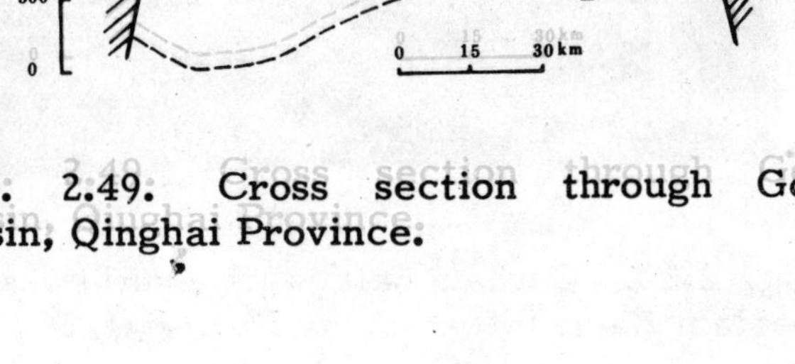

Fig. 2.49. Cross section through Gonghe Basin, Qinghai Province.

Fig. 2.50. Longyang Gorge, Huanghe River, Qinghai Plateau (after Zang, 1982).

Fig. 2.51. Example of poor sorting of roundness in diwa sediments. Wuquanshan Conglomerate. (Q), Lanzhou, Gansu Province.

Fig. 2.52. Qarhan salt with crushed surface, Qaidam Basin (after China Pictorial, 1981).

(5) The North Xizang (Tibet) diwa region[1]

It lies in the main part of the Qinghai-Xizang (Tibet) Plateau, adjacent to the West Yunnan diwa region on the southeast. It includes the Tanggula Range, Ali Mountain, and nearbv salt lakes. Its eastern segment includes also part of the Hengduan Mountains and its western segment probably extends into the Karakorum Mountains.[2] [7, 34, 37, 56, 65, 86, 99, 100].

Geologic records have their beginning in the Paleozoic. The Jinsha River Group (Lower Paleozoic) and the Shigu Schist (pre-Devonian) exposed along the Jinsha River, the gneiss[3] in the nuclear part of the Tanggula Range and the Qiling complex (Silurian to Devonian) of northwest Xizang (Tibet) constitute the lower part of the geosynclinal structural layer. They show that during the early Paleozoic, a geosynclinal trough took shape first in the east, along the Jinsha River and in the Tanggula Range, then the Xiyar trough appeared in the northwest. The former subsided deeply, and the Jinsha River Group alone is more than 10,000 m thick[4]; in the latter there probably occurred volcanic activity. At the end of the Caledonian orogeny, a phase of strong crustal movement began to take place and caused early-geosynclinal deposits to be metamorphosed for the first time. This phase of crustal movement lasted for a long time, reached its climax in Early Devonian time in the Jinsha River area, and continued probably to Late Devonian time in the Xiyar trough. Intercalations of liparite-porphyry, granodiorits, syenite-porphyry, liparite, and tuff in the Jinsha River Group of Jiaqia and Sanyan, south of Qamdo, are the products of magmatic activity accompanying this phase of crustal movement. Upper Paleozoic strata are a sequence of widespread and slightly metamorphosed rocks. Among them a carbonate formation occupies first place, a clastic formation comes second and in places of Wuli area a flysch formation can be seen. They include: the Xiagur Lake Series (Middle Devonian), Jian Lake Series (Upper Devonian), Tarma Limestone (Lower Carboniferous), Shalonglieding Series (Lower Carboniferous), Jiumixiong Limestone (Lower Carboniferous), Ganerbao Series (Lower Permian) in the central part, the Huoerbacuo Group (Carboniferous to Permian) and Tashilikekuer Group (Permian) in the western part, and Carboniferous limestones and Wuli coal measures (Permian) of Zadoi in the eastern part. This indicates that during the late Paleozoic, the geosynclinal trough was large in areal extent and its center was situated at Dageji Lake which has received 9,150 m of carbonate and clastic rocks. The Huoerbacuo Group in the western part consists chiefly of clastic rocks, but in the Zadoi and

---

1. Xue Jiamou.
2. The western limit of the region cannot be determined at present. The Karakorum and Ali Mountains in geologic histroy are linked with the region and, therefore, are tentatively included in it.
3. The gneiss near Amdo in the nuclear part of the Tanggula Range was considered pre-Sinian in the past. According to studies in recent years, it is tentatively ascribed to the Paleozoic.
4. It probably includes part of the pre-geosynclinal structural layer.

Wuli areas there is a flysch formation deposited probably at the margin of the geosynclinal trough. Volcanic activity occurring during this period is known as intermediate-basic magmatic extrusion (Permian) in the Jinsha River and in the upper course of the Nujiang River, forming spilite-keratophyre with pillow structure and belonging to submarine extrusion type. At the end of the Permian the geosyncline was wholly uplifted and folded, with accompanying acid-to-intermediate magmatic intrusion (Fig. 2.53). At the same time the geosynclinal structural layer underwent regional metamorphism.

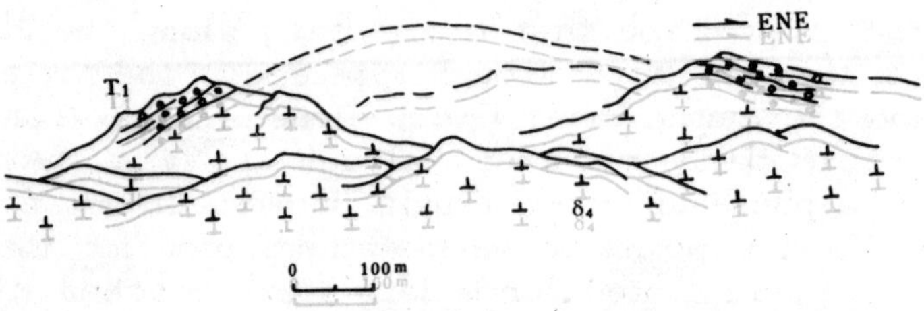

Fig. 2.53. Sedimentary contact between a quartz diorite intrusion (δ_4) of Hercynian age at the geosynclinal stage and the Pushuiqiao conglomerate of Permian age at the platform stage, Guse, Qamdo Prefecture, North Xizang diwa region (after Liu Guanjian, 1981).

Triassic to Jurassic time was the platform stage. At this time crustal movements were chiefly upwarping and downwarping affecting a large area; marine transgression covered almost all of the region and produced a platform sedimentary cover, which has a planar distribution and consists of carbonate rocks (Bangong Lake Series), epicontinental clastic rocks (in the Wenquan and Wuli areas), a coal-bearing sandstone-shale assemblage (in Togmegola) and a salt-bearing formation (in the eastern part). Breaks in deposition and alternately continental and marine strata are present, but sharp unconformity is absent. The rocks have not been subjected to metamorphism. No magmatic activity is observed in the main part (a few acid rocks were intruded in the eastern segment under the influence of magmatic activity in the adjacent geosynclinal region). Evidently these facts reflect a relatively stable platform environment.

Beginning in the Cretaceous, various parts of the region successively entered into the diwa stage marked by block-faulting movement and a great amount of fault depressions, and crustal mobility became intense again. A vast area north of the Tanggula Range was out of marine transgression. Strong block-faulting progressively disintegrated the paleoplatform formed in the Triassic and Jurassic and produced many fault depressions. The distribution and shape of these basins were controlled not only by the preceding structural trend (NW-SE), but also by newly-created fractures with a NE-SW trend. In the Cretaceous a trough remained on the southern border, but it disappeared at the end of the Cretaceous along with the uplift of the Gandise geosynclinal region adjacent to it on the south. The subsequent deposits are all continental red salt-bearing beds, clastic rocks and oil shale as well as recent marine, boulder and piedmont accumulations (Fig. 2.3, right segment). With the intensification of diwa tectonism there appeared intermediate magmatic extrusions of Cretaceous-Tertiary age in the central part, acid-to-intermediate magmatic intrusions in the eastern and western parts (Figs. 2.54 and 2.55), and basic and ultrabasic magmatic intrusions along some deep fractures (e.g., in Dengqen and Amdo). In the Cenozoic the whole region has been rapidly, differentially rising; lakes spread all over it; and modern seismic activity is intense. Great earthquakes occur frequently. Probably this region is now evolving into the maximum-mobility period of the diwa development.

According to the difference in geologic development, three second-order units can be singled out. They are the Bakehalu diwa system, the Ali-Tanggula geodome system, and the Siling Lake diwa system.

(6) The West Yunnan diwa region[1]

It lies in the southwest of Yunnan Province, west and south of the Red River deep fracture, till the boundary line, occupying an area in the shape of a triangle [37, 56, 74, 99, 100, 149].

The Ailao Mountain Group[2] and the Lancang Group outcropping in the Red River and west of the Lancang River respectively, and the Diancang Mountain Group west of Dali are a Precambrian metamorphic complex. Despite deep metamorphism of the rocks due to superimposition of later dislocation metamorphism and influence of magmatism, their original deposits may be correlative with the Kunyang Group in adjacent regions and are of geosynclinal type. They jointly constitute a geosynclinal structural layer. The Jinning orogeny caused most geosynclines in the region to be uplifted and Precambrian rocks regionally metamorphosed. South of Ailao Mountain, a local geosyncline remained till the Ordovician and Silurian, and deposits of geosynclinal type with 8,000 m thickness were laid down. This residual geosyncline did not come to an end until the close of the Caledonian orogeny.

Beginning in Cambrian (at least mid-Cambrian) time, various parts of this region progressively entered into the platform stage. The Middle-Upper Cambrian, Ordovician and Silurian in the Baoshan area are a complex of quite persistent and fossiliferous sandstone, shale, carbonate rocks and graptolithic shale showing evidence of platform-type deposits. In the late Paleozoic, the residual geosyncline lying south of Ailao Mountain was also converted into a platform linked to the original platform. But in other areas, some deep fractures were reactivated. In places (e.g., Nanding and Lancang Rivers) intermediate-to-basic volcanic extrusion occurred polyphasically. In the Lancang area, granite was intruded.[3] In the Late Triassic, crustal movement further intensified, the paleoplatform

1. Xue Jiamou.
2. The age of the Aliao Mountain Group is still a matter of controversy. Some geologists regarded it as Precambrian, but part of it may be formed by metamorphism of Cambrian to younger rocks.
3. Based on these facts, some authors considered that the geosynclinal stage in the Lancang and Nujiang Rivers lasted to the late-Hercynian orogeny. This problem needs to be further studied.

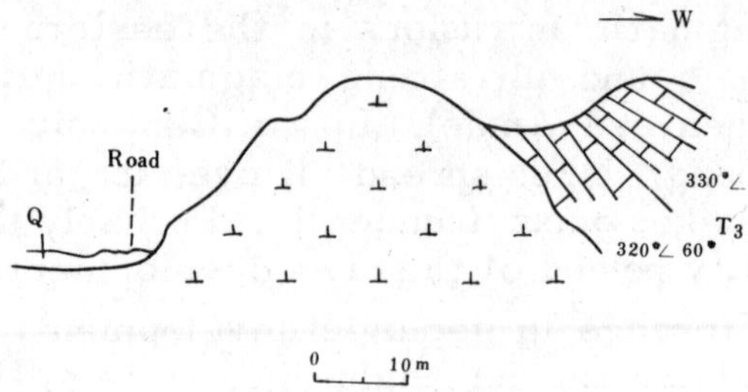

Fig. 2.54. A dioritic body of Indosinian-Yanshanian age at the diwa stage intruded into the Bolila Formation of Upper Triassic age at the platform stage, Jiangda, Qamdo Prefeture, North Xizang diwa region (after Liu Guanjiang, 1981).

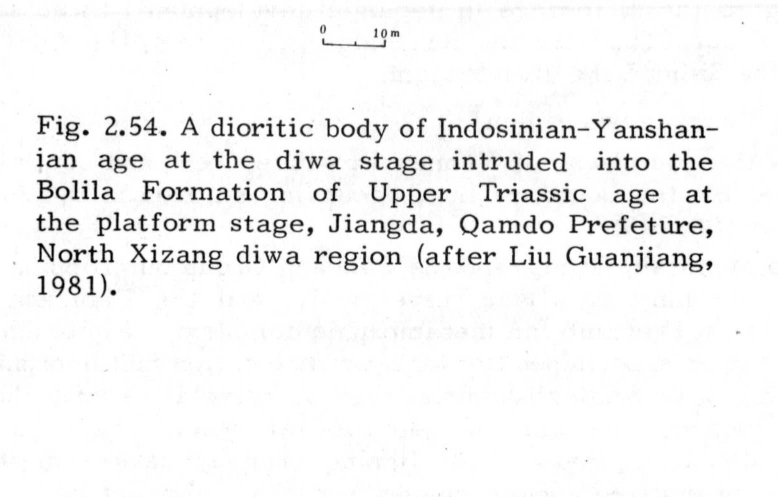

Fig. 2.55. Biotite microcline granite of Yanshanian age at the diwa stage intruded into limestone (C-P?) and phyllite (J-K?), Ralarasi, Qiangtang area, western North Xizang diwa region (after Liu Guanjiang, 1981).

was disintegrated at last, and the diwa stage characterized by block-faulting was introduced.

Diwa tectonism began with volcanic eruption of Late Triassic age, which in the north produced terrestrial acid-to-intermediate agglomerate followed alternately by continental and marine pyroclastic rocks (Waigucun Formation, approximately of Carnic age). With the intensification of block-faulting, large fault depressions took shape in Lanping and Pu'er. Many small fault depressions also appeared in Baoshan, Lancang and in Ailao Mountain. In late Triassic to Jurassic time, a marine transgression took place, which was responsible for the deposition of Late Triassic and Jurassic marine limestone, sandstone and shale in Lanping and Pu'er and the Liuwan Formation of Jurassic age on the east bank of the Nujiang River. Afterwards, the sea withdrew from the region and a lagoonal salt-bearing formation was laid down. Cretaceous and Lower Tertiary strata are continental red beds composed mainly of coarse-grained rocks. At the end of the Tertiary, folding and faulting occurred very strongly in the whole region and the structural lines formed belong chiefly to NW tectonic systems. At the same time, acid and alkaline magmatism and strong dislocation metamorphism took place and the maximum-mobility period of the diwa development began. A series of ultrabasic rocks emplaced along the Ailao Mountain deep fracture were also produced during this period. Late Tertiary deposits have a sporadical distribution. At present, the whole region is still in the process of rising with a sharp contrast, forming a plateau with an elevation of 3,000 m above sea level (Fig. 2.56). Large earthquakes frequently occur in Baoshan, Ximeng, Lancang, Gengma, Pu'er, Ruili and at the junction with the Bayanhar geosynclinal region. Therefore, the region is still in the maximum-mobility period of the diwa development.

Deep and major fractures played an important role in the tectonic evolution of the region. Repeatedly reactivated and jointly acting, deep fractures at the Red River, Lancang River and Ailao Mountain and major fractures at the Amo, Nujiang (Fig. 2.57) and Nanding Rivers not only controlled the boundaries of sedimentary basins, giving rise to a difference in sedimentary type in different parts, but also appeared as canals for the rising of magmas of various ages and ore-bearing hot solutions. More characteristic are large-scale zones of dislocation metamorphism along these deep and major fractures.

Second-order tectonic units are the Baolan geodome system and the Pu'er diwa system.

(7) The Southern Himalayan diwa region

It is situated on the southwestern border of China adjacent to neighbouring countries, and separated from the Yarlung Zangbo River-Himalayan geosynclinal region on the north by a line extending from west of Zanda through south of Zhongba and Kangmar to Moindawang. Both the southern part and the eastern and western ends stretch beyond the boundary line. The eastern end is separated from the North Xizang and West Yunnan diwa regions by the Nujiang fracture and includes an area extending from the Nujiang River, west Yunnan, to Gaoligong Mountain (Fig. 2.58) on the border of China and Burma. The region consists of the Higher Himalayas, the Lower Himalayas and the Siwalik fault depression (Fig. 2.59). So far as the part within Chinese territory is concerned, it includes the nappe tectonic zone of the northern slope of the Higher Himalayas, the imbricate crystalline

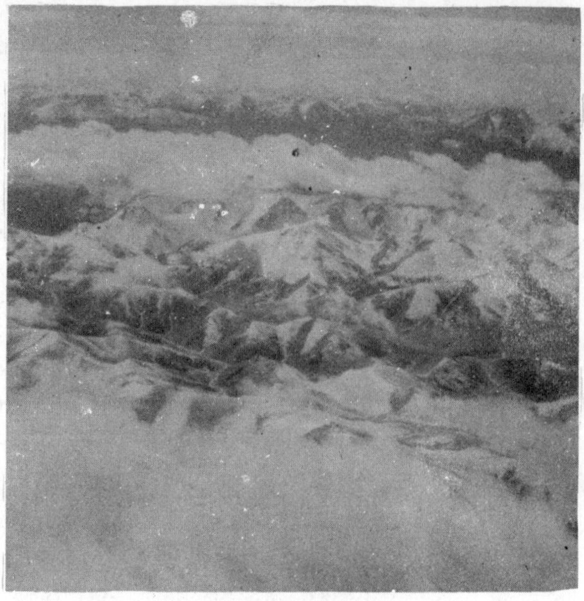

Fig. 2.56. 2,500-m-high peneplain, West Yunnan diwa region (after Zhang, 1982).

Fig. 2.57. Nujiang River valley, cut 2,000 m into the West Yunnan Plateau, the boundary between the West Yunnan and the Southern Himalayan diwa regions (after Zhang, 1982).

tectonic zone of the Higher Himalaya, and the nappe tectonic zone of the Lower Himalayas, arranged from north to south and distinguished by Zhou Yunsheng et al. (1981). The region corresponds to the Higher Himalayan, Lower Himalayan and Siwalik subprovinces in the stratigraphic division scheme proposed by Gu Qingge et al. (1979). In Chinese territory the latter two subprovinces are seen only east of Moindawang [2, 8, 37, 47, 56, 65, 85, 86, 99, 100, 109, 114, 116, 138, 141, 147, 152, 155].

The history of crustal evolution of the region has a complete record. The earliest record can be traced to the geosynclinal or even pre-geosynclinal stage, which is represented by a structural layer, the Qomolangma Group, composed of regionally metamorphosed crystalline rock series: schist, gneiss and minor marble. The group constitutes the lower part of the basement in the region, and the Purana Group of India and the Mogok Series of Burma are probably its equivalents. It is more than 20,000 m thick and, according to Gu Qingge et al., was originally a sequence of sandy-clayey rocks. As inferred from its rock composition, the Qomolangma Group is probably a rhythmically bedded clastic formation comprising graywacke and feldspathic sandstone and being contaminated by magma. The age of its metamorphism has been dated at 640-660 million years (according to the determinations by the K-Ar, Rb-Sr and U-Pb methods made by the Guiyang Institute of Geochemistry, Academia Sinica, 1979). Based on micropaleobotanic fossils, the Qomolangma group or its upper part has been assigned by Liang Dingyi et al. (1979) to the Sinian and is correlative with the metamorphic rock in central part of Nepal dated at 780 million years.

At Mount Qomolangma, there is another regionally metamorphosed sequence, the Rouqie Group, which rests on the Qomolangma Group and consists of slightly metamorphosed rocks: quartz schist and marble. It is probably Upper Sinian in age and constitutes the upper part of the basement in the region.

The Cambrian is probably absent. The basement is overlain directly by Ordovician and younger unmetamorphosed deposits forming a sedimentary cover. The lithological differences between this structural layer and the basement and the absence of the Cambrian indicate a phase of strong folding movement, dated at 640 million years and corresponding probably to the Xingkai movement. They marked the completion of the geosynclinal and the beginning of the platform stage. The absence of the Cambrian suggests an epoch of denudation and planation of a paleogeosynclinal fold belt. On entering into the platform stage, the region became the northern margin of the Indian Precambrian paleoplatform. Strata of Ordovician to Cretaceous age, representing the platform structural layer of the region, have a complete development. No unconformities have been observed between the strata, and only disconformities can occasionally be seen. Sedimentary formations are typical of the platform sedimentary cover. In most cases the thickness of an individual bed is small (ranging mostly from 20 to several hundred meters). Limestone is well-developed and interbedded with arkose, shale, sandstone, sandy limestone and occasionally conglomerate (seen in the Lower Carboniferous). Marine conditions predominated in the early period, and the Lower Permian (Selong Group) still retained marine facies. But in Dingri and Dinggye, the marine Lower Permian is intercalated with continental deposits containing glossopterid flora of Gondwanaland. This is also an evidence for that the region was at that time part of the Indian paleoplatform. The Selong Group is overlain, with a disconformity, by the Lower Triassic Tulong Group, indicating a long-term rising

Fig. 2.58. Plateau on Gaoligong Mountain (after Gao Mingyi, 1981).

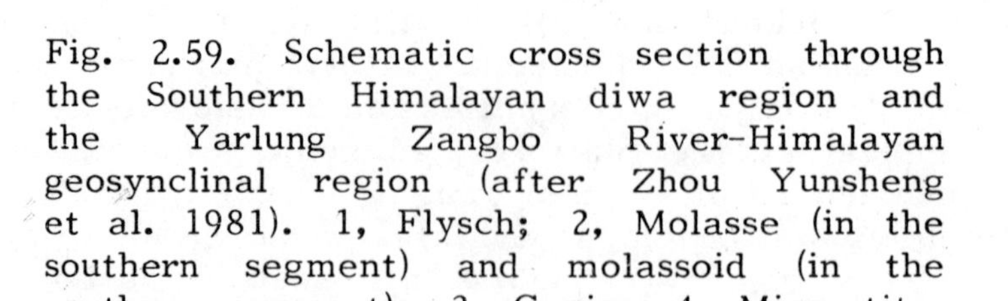

Fig. 2.59. Schematic cross section through the Southern Himalayan diwa region and the Yarlung Zangbo River-Himalayan geosynclinal region (after Zhou Yunsheng et al. 1981). 1, Flysch; 2, Molasse (in the southern segment) and molassoid (in the northern segment); 3, Gneiss; 4, Migmatite; 5, Schist; 6, Limestone and sandstone; 7,

Shale; 8, Granite; 9, Ophiolitic flysch; 10, Main thrust; 11, Basement of the Indian Shield. Main thrust: F_1, Boundary fault; F_2, Central thrust; F_3, North thrust; F_4, South Xizang hidden fracture; F_5, South Yarlung Zangbo River fracture. Tectonic division: I, Siwalik diwa system; II, III and IV, Southern Himalayan diwa region; V and VI, Yarlung Zangbo River-Himalayan geosynclinal region (including: V, Northern Himalayan miogeosynclinal fold system; VI, Yarlung Zangbo River eugeosynclinal fold belt-intermontane depression system).

movement affecting a large area. As a result, the Upper Permian is missing. Moreover, at the bottom of the Tulong Group dolostone was deposited. But marine regression on a large scale appeared as late as Cretaceous time. In the Paleogene, some residual sea basins still remained locally in Dingri and Gamba. This is how the platform stage ended.

In the Neogene, along with the activation of the Indian paleoplatform, the region entered into the diwa stage. At this time tectono-topographic contrast became sharp, and numerous strip-like geodome uplifts and diwa subsidences took shape. In addition, as the Indian and Eurasian crustal blocks moved towards each other, the region under consideration lying at the northern margin of the Indian crustal block was compressed and crushed most severely. As a consequence, multiple overthrusts and nappes were formed, resulting in an especially high position of small crustal blocks. Such a process was responsible for the rising of Mount Qomolangma, the highest mount in the world with an elevation of 8,848.14 meters above sea level, which was originally a small fragment on the northern margin of the Indian paleoplatform. Ordovician limestone at the top of Mount Qomolangma, lying flat, still retains the original structural feature of the platform sedimentary cover. At the same time, in diwa basins and valleys Tertiary and Quaternary continental sediments of intermontane depressions, mainly sandstone and conglomerate, were deposited. They vary greatly in lithology and thickness, belong to a molassoid formation, represent diwa sediments and constitute the newest structural layer in the region. The Pula Formation of Paleogene age in Pula region composed of diwa-type sandstone and conglomerate, with brown coal beds, attains a thickness of 800 m.

In the process of diwa tectonism, vigorous magmatic activity took place in the region (Fig. 2.60). Medium- and small-size granitic intrusions have a wide distribution, forming a granite belt which is the southern one of the two prominent granite belts in the Himalayas. Zhou Yunsheng et al. (1981) [80] called it "the Himalayan tourmaline granite belt". Granite and related rocks in the belt have been dated at 105-205 million years (Miocene to Pliocene) by the K-Ar method. Zhou Yunsheng et al. (1981) pointed out that "their intrusion was associated with nappe formation and overthrusting, rather than with strong folding movement. This fracturing has played an important role in the formation of granitic magma. At the depth of these fractures and in the underlying strata, probably partial melting of rocks occurred. The remelted magma produced by such a partial melting was rising along pre-existing faults to form granitic intrusions." These views are similar to those on the origin of S-type granite commonly observed in ordinary diwa regions. Moreover, petrochemically the granite is rich in silica and alkali, but poor in calcium, iron and magnesium, and abundantly contains boron, lithium, beryllium, niobium, tin and other trace elements. These petrochemical features are also in accord with those characteristic of diwa-type granite.

Neotectonic movement in the region is very intense, differentiation into uplifts and subsidences significant, and tectono-topographic contrast sharp. This consists in that geodome mountains are continuously rising, while diwa basins subsiding or relatively subsiding. The region shows high seismicity, with a magnitude of 6-6 3/4. It is an intense seismic region. Near Tengchong in southwest Yunnan intermediate to basic volcanic activity still took place as late as the Ming Dynasty (Figs. 2.61 and 2.62). Judged from the intensity of diwa activity, the region is probably entering into the maximum-mobility period and, therefore, has been assigned to diwa regions of Central Asia age. In terms of the history of its crustal development at the pre-diwa stage, the region is the diwa region in Chinese territory which has exclusively developed on the Indian Precambrian paleoplatform.

In regard to the history of crustal development and the cause of intense crustal movement in the region since the Tertiary, some geologists have proposed views similar to the author's idea. On the basis of an analysis of the stratigraphy of the region, Gu Qingge et al. (1979) [102] pointed out that the Xingkai movement in the region resulted in the Precambrian basement. All the overlying Paleozoic cover consists of platform-type formations. But since the Mesozoic, crustal mobility has increased, with a great change in the amplitude of subsidence under the condition of the unstable rigid basement". S. Sinha-Ray (1980) [88] of India revealed" the Himalayan core is a reactivated microcontinent, created by the fragmentation of North India. Mid-Tertiary collision between the Indian leading edge supporting the Himalayan microplate and Tibet formed the Trans-Himalayan and Himalayan foldbelts." Zhou Yunsheng et al. (1981) [80] also believed that "the Tethyan tectonic zone of the northern slope of the Higher Himalayas was continuously in a stable sedimentary environment from early Paleozoic to Eocene and had a platform-type block character. The Higher Himalayan imbricate tectonic zone is a Precambrian crystalline massif cut across by a series of thrusts and activated intensely during the Himalayan movement". Zhang Qi et al. (1981) [81] also held a similar view.

According to the peculiarities of geological development and the modern tectono-geomorphic differences, the Southern Himalayan diwa region within Chinese territory can be divided, from north to south, into the following six second-order subregions (some of which, for the most part, extending beyond the boundary):

1) The Xiangquan River diwa system. Situated at the western end of the Southern Himalayan diwa region, it is aligned in a WNW-ESE direction generally along the Xiangquan River. It is bounded by the Yarlung Zangbo River-Himalayan

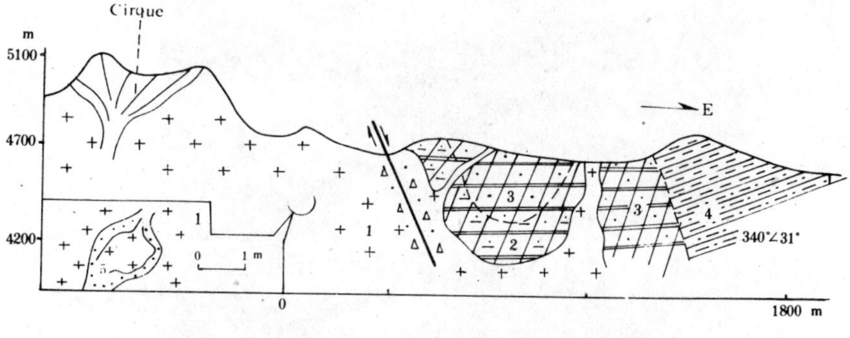

Fig. 2.60. Intrusive contact between the eastern part of the Yamarong, Cona County, tourmaline-bearing two-mica granitic body and the Paleozoic at the east end of the Higher Himalayan geodome system (after Zhou Yunsheng et al., 1981). 1 Tourmaline-bearing two-mica granite (K-Ar date 12.1 m.y.); 2, Amphibole hornstone; 3, Spotted hornstone; 4, Two-mica quartz schist; 5, Syngenetic pegmatite in the rock body (the partial illustration in the figure).

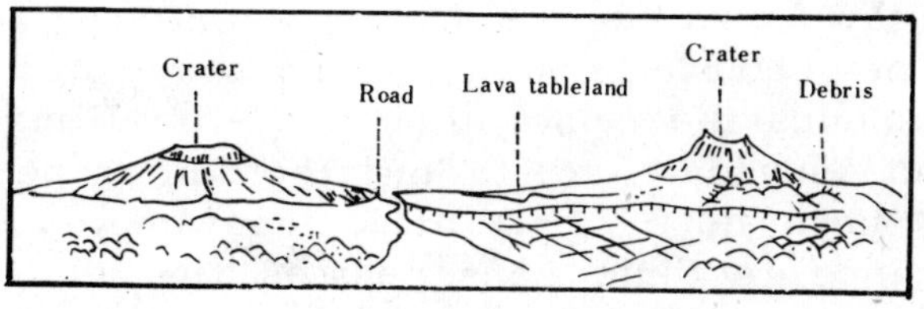

Fig. 2.61. Volcanic cones near Tengchong, Yunnan Province.

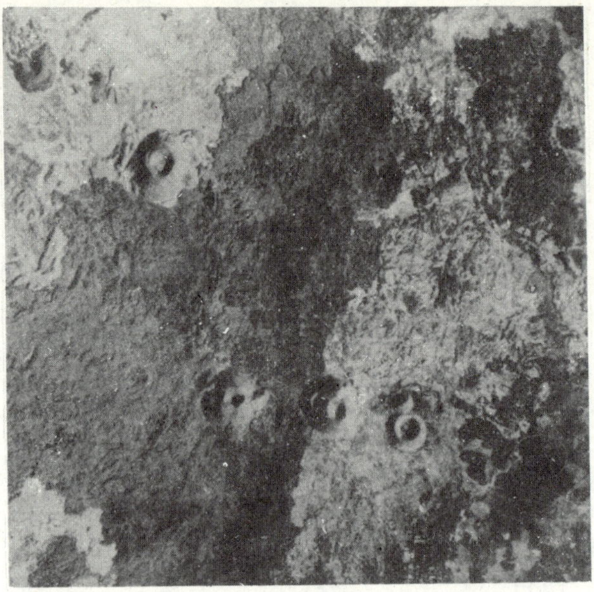

Fig. 2.62. Tengchong volcanic cluster, aerophotograph from 5,000 m height.

geosynclinal region on the north and by the Higher Himalayan geodome system on the south. Its western end extends beyond the boundary. The diwa system consists of a series of diwa basins (lakes) filled with Tertiary and Quaternary diwa deposits, mainly sandstone and conglomerate. In addition, there are many intrusions of ultrabasic rocks arranged like a string of pearls along a deep fracture zone.

2) The Pengqu River diwa system. It is situated east of the Xiangquan River diwa system and is also bounded by the Yarlung Zangbo River-Himalayan geosynclinal region on the north and the Higher Himalayan geodome system on the south. It lies just on the northern slope of the Qomolangma geodome, includes basins and valleys distributed west of Dingri to Dinggye and corresponds to the so-called zone of lake basins on the northern slope of the Himalayas. The basins are filled with Tertiary and Quaternary sandstone and conglomerate.

3) The Higher Himalayan geodome system. Situated west of Burang and east of Moindawang at the southwestern edge of China, it extends, for the most part, along the boundary with neighbouring countries, and some of its small segments pass through them. It includes the principal axis of the Himalayas and consists of "the Higher Himalayan imbricate crystalline tectonic zone" and "the nappe tectonic zone" on its northern slope. East of Moindawang, a small segment of "the Lower Himalayan nappe tectonic zone" passes through Chinese territory. The high peaks of the Himalayas, with Mount Qomolangma as the highest, stand within the geodome system, constituting the roof of the world. Here the Qomolangma Group of Precambrian age makes up the basement, belonging to the geosynclinal and pre-geosynclinal structural layers. It is overlain by unmetamorphosed strata, the platform structural layer, ranging from Ordovician to Mesozoic in age. Because of the tectonic process of imbricate nappes, the

geodome system has been rising, since Eocene time, at a speed of 10-50 mm (averaging 20 mm) per year, and at present this rising movement is still in progress. Gravimetric, seismological and magnetotelluric investigations show that the earth's crust at Mount Qomolangma is 48 km thick .

4) The Subanxi geodome system.

5) The Gaoligong Mountain geodome system.

6) The Siwalik diwa system. Situated at the southern foot of the eastern segment of the Higher Himalayan geodome system, it lies partly in Chinese territory, extends WSW-ENE along the border area of China and India, and corresponds to the Siwalik fault depression zone. It is filled with Upper Tertiary and Quaternary diwa deposits, mainly sandstone and conglomerate. The molassoid formation of the Siwalik Group, for example, is as thick as 5,000-7,000 m.

3. The South China Sea diwa region

Situated in the southermost part of China, it includes more than two hundred islands, reef banks and submerged sands, namely the Xisha, Dongsha, Zhongsha and Nansha Islands as well as adjacent sea areas [37, 56, 99, 100, 104].

The Precambrian belongs to the crystalline basement. On the Xisha Islands, the Precambrian underlying the Upper Tertiary is composed of granite-gneiss and migmatite (6.27 million years).

In the Paleozoic, as inferred from the geology of the continental shelf, the region probably was at the geosynclinal stage.

Little is known about the tectonic evolution in the early Mesozoic (Triassic). During this time, alternately marine and continental carbonate and clastic rocks accumulated locally. The problem whether they are products of the platform stage or not needs to be solved.

Beginning in the middle Mesozoic, crustal mobility of this region was very high, which marked the initiation of the diwa stage. The Jurassic and Cretaceous are dominantly deposits of strongly dissected fault depressions: continental clastic, acid to intermediate volcanic and pyroclastic rocks. The center of deposition in the depressions often migrated, with rapid thickness and facies variations. Being compressional in nature, the structures of middle and late Mesozoic age have NE, ENE and NNE strikes. In the Cenozoic, differential downward and upward movements caused by tension faults grew more intense. At the same time, there was also horizontal movement, giving rise to noticeable extension in some places. The heterogeneity of crustal movements caused unconformities between strata and apparent differentiation of sedimentary formations. According to the results of studies in recent years, the Upper Tertiary and Quaternary in the central part of the South China Sea, 500 m thick, are composed of globigerina ooze, volcanic ash and radiolarian ooze. The Upper Tertiary on the Xisha Islands is dominantly reef limestone: The Pliocene is coral-reef limestone, 140 m thick, while the Miocene is coral clastic limestone and coral-reef limestone, about 1,000 m thick. The Upper Tertiary and Quaternary on the Nansha Islands are also dominantly carbonate rocks with great and variable thickness. These rocks lie horizontally, correspond to the Pliocene and Quaternary, and thicken from northeast to southwest, with a maximum thickness of 2,000 m, while the Miocene has been gently folded and faulted, and thickens from southwest to northeast, with a maximum thickness of 1,500 m.

Magmatic activity of Yanshanian age was significant in the western part. It produced acid or acid-to-intermediate granite and granodiorite batholiths and stocks, with accompanying acid, intermediate and alkaline volcanic rocks. Magmatic rocks of Himalayan age are dominantly intermediate to basic volcanic rocks. In the central basin of the South China Sea there is an area of high magnetic anomalies, 100-200 km wide, with an amplitude of 250-400 gammas, a reflection of a submarine basaltic layer. Basalt in some places appears as a seamount, but on Gaojianshi it stands above sea level, being probably a volcanic neck.

The topography of the region is very complex. Various types of landforms such as seamount, valley, hill, tableland and plain attain an equal development, reflecting a sharp topographic contrast. At the base of the Dongsha, Xisha, Zhongsha and Nansha Islands exist steps arranged in a zonal or short-zonal pattern and oriented in the NE as well as E-W and NW directions. There are two steps on the Zhongsha and Dongsha Islands and three in the western part of the region. The width and depth of these steps vary from place to place. The upper step on the Dongsha Islands is 100 km wide and 300-350 m deep, while the lower step is 2,500-2,800 m deep. The step on the Xisha Islands is 102 km wide and 1,000 m deep. The upper step on the Zhongsha Islands is 37 km wide and 330-350 m deep, while the lower step is 2,000-2,400 m deep. The step on the Nansha Islands is about 320 km wide and 1,500-2,000 m deep. The three steps in the western part of the region are 440-450, 2,200-2,400 and 2,500-2,800 m deep respectively. Every step is cut across by NW-, NE- and E-W-trending deep tunnels and submarine canyons varying in depth with a maximum of 4,462 m. As a result of strong differential upward and downward movements, the central sea basin attains a maximum depth of 5,559 m. Thus the sea bottom topography with a great density and a great amplitude of relief came into being (Fig. 2.63).

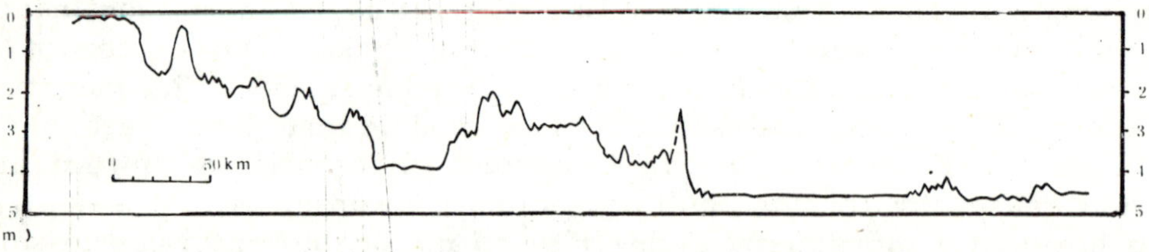

Fig. 2.63. Sea bottom topography from Dong Island, Xishan Islands, to the central sea basin (after Xie Yixuan. 1979).

The Bouguer gravity anomaly values on the Xisha and Zhongsha Islands are +40 - +179 mGal. As calculated from the gravimetric data, the thickness of the earth's crust on the Xisha Islands is 31 km, on the Zhongsha Islands 27 km. Southeast of the Zhongsha Islands, the thickness of the earth's crust decreases towards the central basin. The gravity anomaly values in the central basin increase up to +300 mGal, and the thickness of the earth's crust varies from 5.99 to 8.67 km. The thickness of the earth's crust in this region decreases from the margin to the central basin.

From the above-mentioned sedimentary and magmatic formations, modes of crustal movement and landforms, it can be concluded that the region is at the residual-mobility period of the diwa stage at present. In terms of the time and characteristic features of the diwa development, the region should be assigned to diwa regions of Cathysian type. However, as the extent of its stretching is greater and the thickness of its earth's crust thinner than the diwa regions of continental type, we can regard it as a diwa region of transitional crust.

The region can be divided into three second-order subregions, namely, from north to south: the Tongsha-Xisha geodome system, the Central Basin diwa system and the Nansha geodome system.

CHAPTER 3

HISTORY OF TRANSFORMATION BETWEEN MOBILE AND STABLE REGIONS IN CHINA'S CRUSTAL DEVELOPMENT

Based on the peculiarities in the geologic development of various tectonic regions, the history of China's crustal development and the process of transformation between mobile and stable regions can be summarized in the light of the following periods[36].

I. Tectonic outlines in the Archean

Little is known about the crustal development during this period in China. According to available data, it can be inferred that old sialic tectonic elements peculiar to the pre-geosynclinal stage existed at that time. The crystalline basement composed of the Anshan Group in northeast, the Taishan, Jiaodong, Sanggan, Fuping and Wutai Groups in north, the Maxie Mountain Group in northwest China and their equivalents in other areas is structural layers representative of these tectonic elements. At the end of the Archean the first geosynclines probably emerged in some areas.

II. Tectonic outlines in the Early Proterozoic (Fig.3.1)

In Chinese territory the Early Proterozoic was a period of geosynclines. South and north China and Tarim in northwest China already entered into the geosynclinal stage. Its representative structural layer consists mainly of the Hutuo Group or its equivalents such as the Luda and Penglai Groups in north China, and of the lower part of the Banxi Group or its equivalents in south China. In northwest China, the Proterozoic of Tarim are also the products of the geosynclinal stage. At the close of the Proterozoic, in some areas such as north China and Yishan (Ordos), geosynclines died out gradually, giving place to fold regions of Luliang age. But in other areas geosynclines continued their development.

In the Early Proterozoic some pre-geosynclinal tectonic elements remained locally in China. For example, the crystalline basement of Early Proterozoic (or pre-Sinian, partly pre-Cambrian) age exposed in Heilongjiang, Jilin, Nadanhada, southeast China (?), Kunlun, Qaidam, Gansu, Ningxia and in the Himalayas is their representative structural layer.

III. Tectonic outlines in the Late Proterozoic (Fig. 3.2)

In the Late Proterozoic, tectonic outlines in China changed greatly. During this period, some areas in north China, which were originally geosynclinal fold regions of Luliang age, were transformed into platforms. These are: all provinces in north China and the Yishan (Ordos) area. They constituted a primary platform,

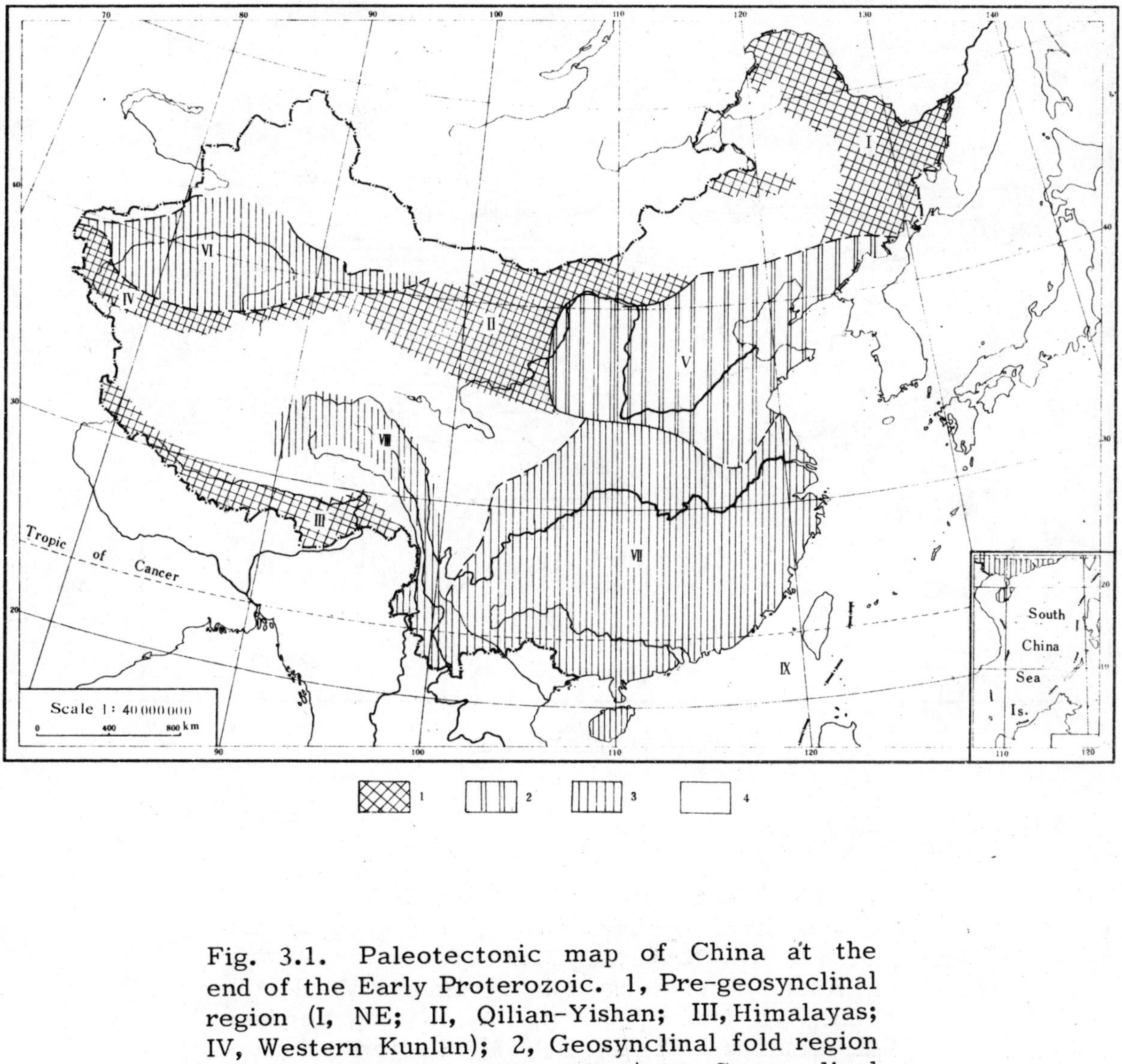

Fig. 3.1. Paleotectonic map of China at the end of the Early Proterozoic. 1, Pre-geosynclinal region (I, NE; II, Qilian-Yishan; III, Himalayas; IV, Western Kunlun); 2, Geosynclinal fold region of Luliang age (V, North China); 3, Geosynclinal region, unclosed (VI, Tarim; VII, south China; VIII, Xizang-Yunnan); 4, Region of unknown tectonic character (IX, Taiwan-South China Sea).

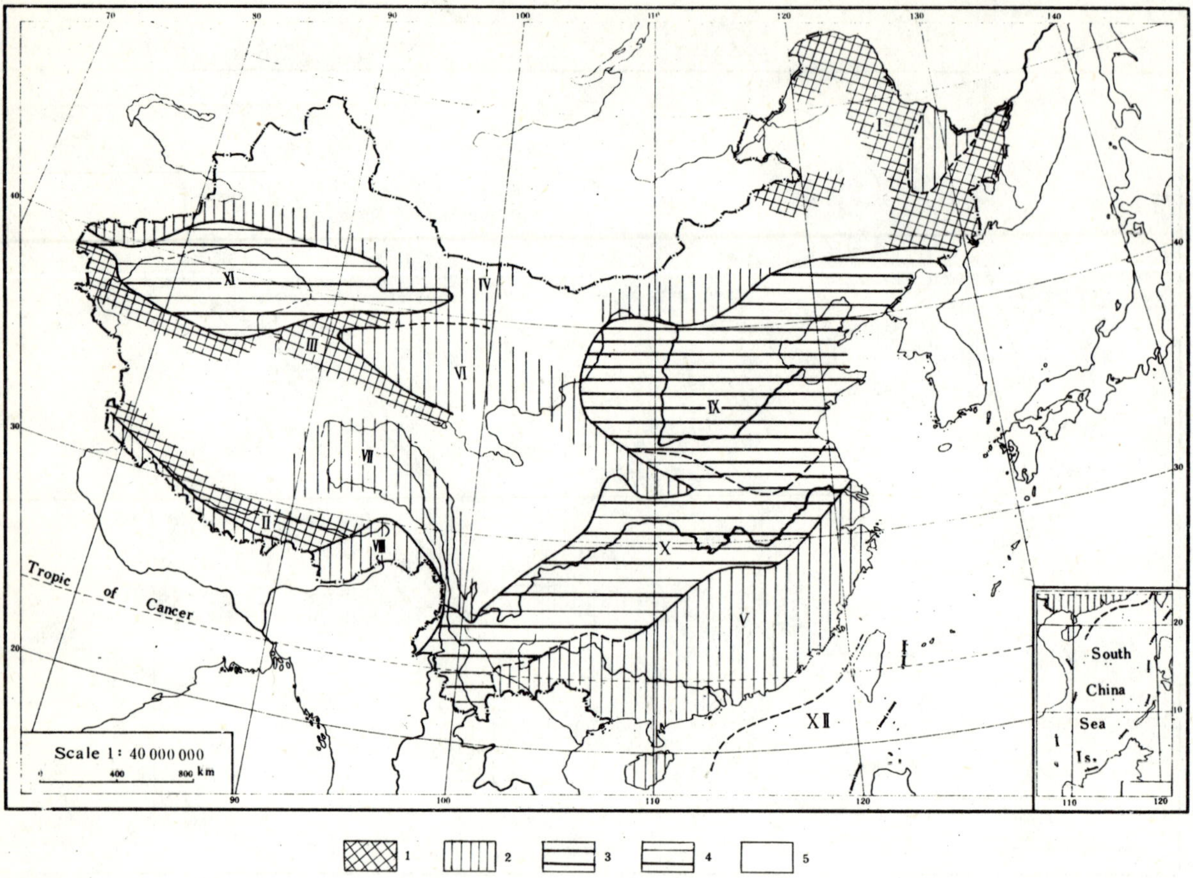

Fig. 3.2. Paleotectonic map of China at the end of the Late Proterozoic. 1, Pre-geosynclinal region (I, NE; II, Yarlung Zangbo River-Himalayas; III, Kunlun); 2, Geosynclinal region (IV, North; V, SE; VI, Qinling-Qilian; VII, Xizang-Yunnan; VIII, Southern Himalayas); 3, Platform region of post-Luliang age (IX, North China); 4, Platform region of post-Jinning age (X, Changjiang, XI, Tarim); 5, Region of unknown tectonic character (XII, Taiwan-South China Sea).

the North China paleoplatform, which later expanded into an extensive paleoplatform in the eastern part of China. During the Jinning orogeny, geosynclines in Tarim, Sichuan, Yunnan, Guizhou and in the middle and lower reaches of the Changjiang (Yangtze) River closed and became fold regions. As the Changjiang Sandstone of Sinian age in Yunnan Province possesses the features of a molasse formation, the late-geosynclinal stage (fold belt period) in its larger part lasted probably to the Early Sinian, i.e., the platform stage began in the Late Sinian. It is the same with the middle and lower reaches of the Changjiang (Yangtze) River, where the platform stage began later than in north China, but earlier than in the eastern part of south China.

In other areas of China, geosynclines in this period continued their development. At the end of the Sinian, geosynclines of southeast China were once in a dormant period (lasting to the Cambrian). During this period, carbonate rocks indicating a relative tectonic quiescence appeared.

In some areas such as the Altun, Kunlun and Yarlung Zangbo River-Himalayas, pre-geosynclinal tectonic elements represented by the Precambrian crystalline basement still remained in Late Proterozoic time.

IV. Tectonic outlines in the Early Paleozoic (Fig.3.3)

During the Caledonian orogeny, platforms in Chinese territory, on the one hand, were rapidly expanding, and extensive geosynclines, on the other hand, continued their development or began to originate. At that time the East Chinese paleoplatform already embraced all provinces of north China, Yishan (Ordos), east Ningxia, Sichuan, Yunnan, Guizhou and the middle and lower reaches of the Changjiang (Yangtze) River. In addition, Tarim and the Southern Himalayas were also occupied by platforms. Geosynclines were distributed in Inner Mongolia, Helongjiang, Jilin, the eastern part of south China, the larger part of Xinjiang (except Tarim), Kunlun, Qaidam, Qilian Mountain, Qinling, north Xizang, west Sichuan and west Yunnan. At the end of the Caledonian orogeny, some of the geosynclines lying in the eastern part of south China, in Qaidam, Qilian Mountain and in the southern part of the eastern Qinling Mountains closed one after another, giving rise to fold regions of Caledonian age. But the others continued their geosynclinal development.

V. Tectonic outlines in the Late Paleozoic (Fig.3.4)

During the Hercynian orogeny, platforms in north and central China and in the western part of south China continued their development; besides, the eastern part of south China also entered into the platform stage. As a result, the pre-existing East Chinese paleoplatform was further enlarged to embrace the larger part of east China. In northwest China, the Tarim paleoplatform still existed. In addition, Qinghai, Gansu, and west Ningxia, where geosynclines closed during the Caledonian orogeny, also entered into the platform stage successively at that time. These areas of platform character could not communicate with the Tarim paleoplatform due to the existence of the Altun geosyncline. On the contrary, they connected with the East Chinese platform, constituting jointly a vast paleoplatform across China.

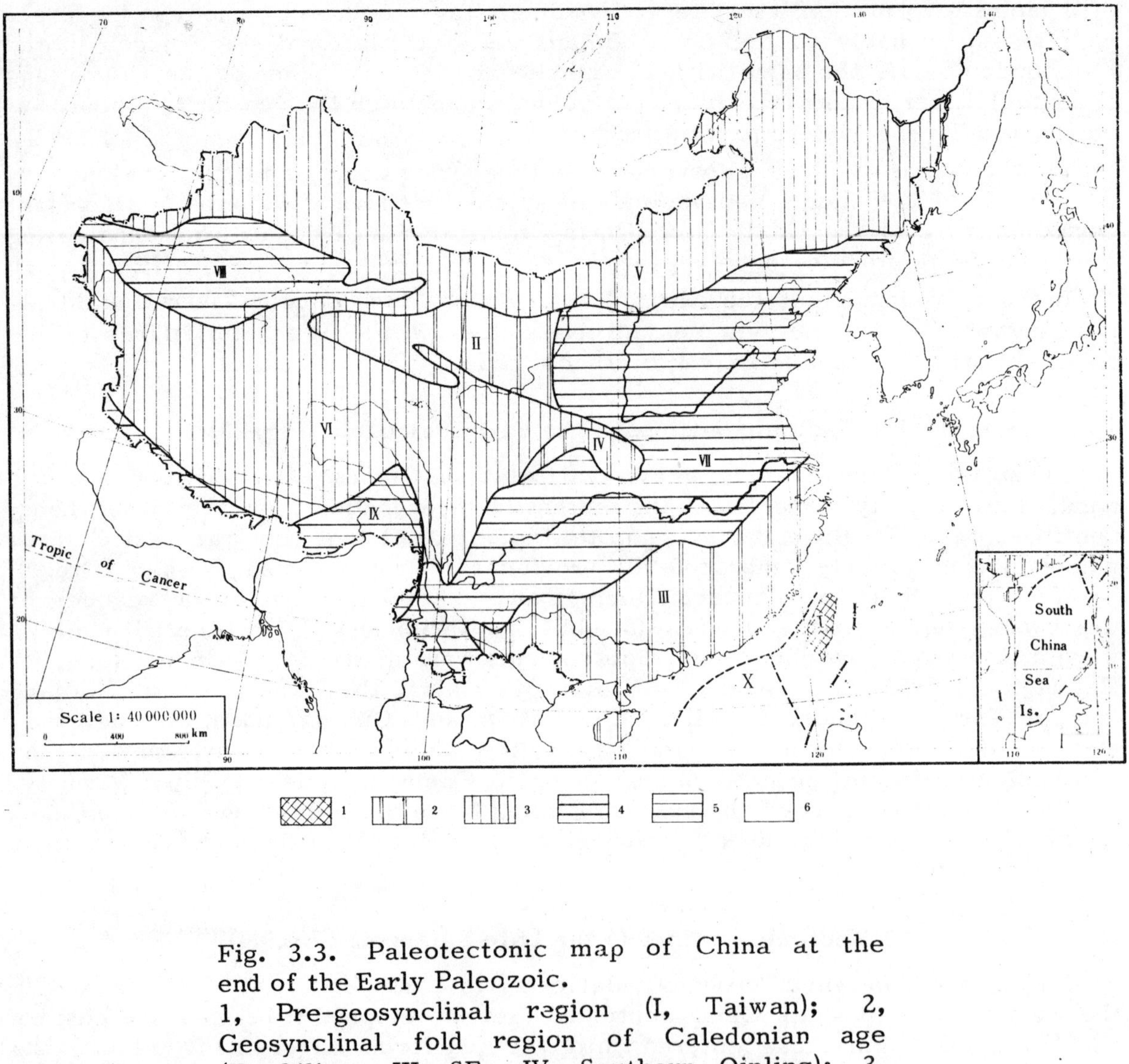

Fig. 3.3. Paleotectonic map of China at the end of the Early Paleozoic.
1, Pre-geosynclinal region (I, Taiwan); 2, Geosynclinal fold region of Caledonian age (II, Qilian; III, SE; IV, Southern Qinling); 3, Geosynclinal region, unclosed (V, North; VI, West); 4, Platform region of post-Luliang age (VII, East); 5, Platform region of post-Jinning age (VIII, Tarim); (IX, Southern Himalayas); 6, region of unknown tectonic character (X, South China Sea).

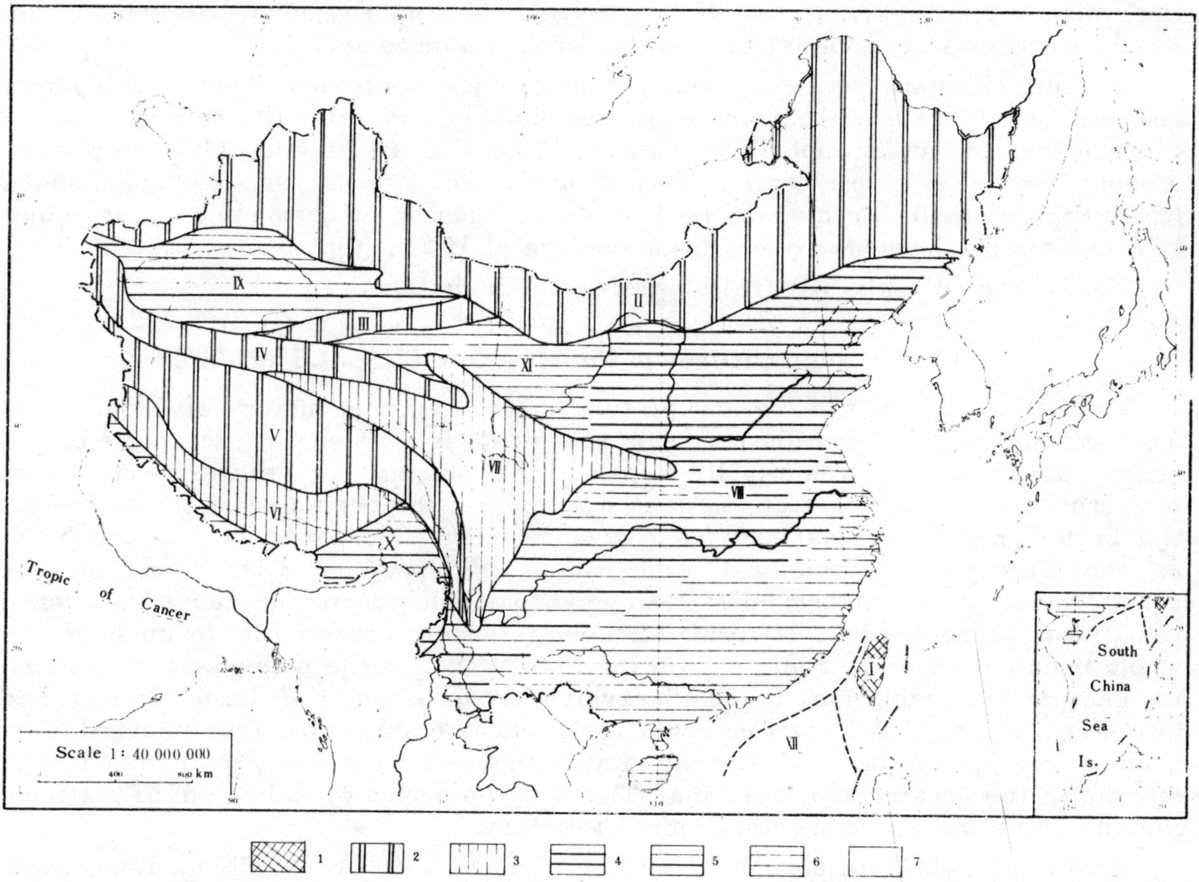

Fig. 3.4. Paleotectonic map of China at the end of the Late Paleozoic.
1, Pre-geosynclinal region (I, Taiwan); 2, Geosynclinal fold region of Hercynian age (II, North; III, Altun; IV, Kunlun; V, North Xizang); 3, Geosynclinal region, unclosed (VI, South Xizang; VII, Bayanhar); 4, Platform region of post-Luliang age (VIII, East); 5, Platform region of post-Jinning age (IX, Tarim; X, Southern Himalayas); 6, Platform region of post-Caledonian age (XI, Qilian); 7, region of unknown tectonic character (XII, South China Sea).

During this period, geosynclines continued their development in north Xinjiang, Inner Mongolia, Heilongjiang and Jilin, and did not close until the end of the Hercynian orogeny, giving rise to a geosynclinal fold region of Hercynian age, which stretches across north China in the form of a large arc.

In south Xinjiang, except Tarim, geosynclines continued their development and did not close successively until the middle and late Hercynian orogeny. Geosynclines in Kunlun and north Xizang closed also at the end of the Hercynian orogeny. In the western Qinling, Bayanhar, Garze, Gandise and Yarlung Zangbo River-Himalayas,the Hercynian epoch was just a period of continuing development of geosynclines. In the Southern Himalayas,the platform stage was continuing.

The Taiwan area was still at its pre-geosynclinal stage.

VI. Tectonic outlines in the Mesozoic (Figs. 3.5 and 3.6)

The Pacific epoch was an abrupt turn in the tectonic history of China. The most prominent event in this period is that diwa as a new type tectonic element began to appear and was rapidly expanding, replacing the greater part of the platforms and a minor part of the geosynclines. At the beginning of the Mesozoic, the East Chinese paleoplatform continued to exist. Moreover, on the one hand, as the Hercynian geosynclinal fold region of northeast China was already transformed into a platform (post-Hercynian paleoplatform), the above-mentioned vast East Chinese paleoplatform further extended northwards to include the whole mainland of east China; and on the other hand, as the Altun area terminated its geosynclinal evolution in the Hercynian orogeny and was later transformed into a platform, the Tarim paleoplatform became one with the enlarged East Chinese paleoplatform, constituting jointly the extensive Chinese paleoplatform embracing the greater part of China. This was the period of full bloom of platform regions with a maximum area in Chinese territory.

However, in the late period of the Late Triassic (southeast China, Kang-Dian, the southern part of central China, west Yunnan, and east Ningxia) or in the Early-Middle Jurassic (north China, the northern part of central China, Qinghai-Gansu, west Ningxia, Tarim and the larger part of northeast China), the long-lived and expanded extensive Chinese paleoplatform was broken up gradually from south to north and from east to west, and then transformed for the most part into diwa regions. The Indosinian (south China) or Yanshanian (north China) movement drew a line of demarcation between the dying of platform regions and the birth of diwa regions. Diwa regions coming into being as early as end Triassic are: SE (including the southern part of central China), Kang-Dian, Helan and west Yunnan. Diwa regions appearing successively in the Early and Middle Jurassic are: Yunnan-Guizhou, central China (the northern part), north China, N-S (its central segment, i.e., the eastern part of the Songpan area, Longmen Mountain, and the west end of the eastern Qinling Mountains), Qinghai-Gansu, south Xinjiang, Altay and NE. The Tianshan diwa region came into being in the Cretaceous. In north Xizang, the platform regime lasted to the Late Jurassic, but changed into a diwa one in the Cretaceous. At this time, of the original East Chinese paleoplatform only three small fragments, the Sichuan (its central part), Yishan (Ordos) and Songliao platforms, survived and continuously exited as residual platforms between diwa regions. In the Southern Himalayas, platform regime continued.

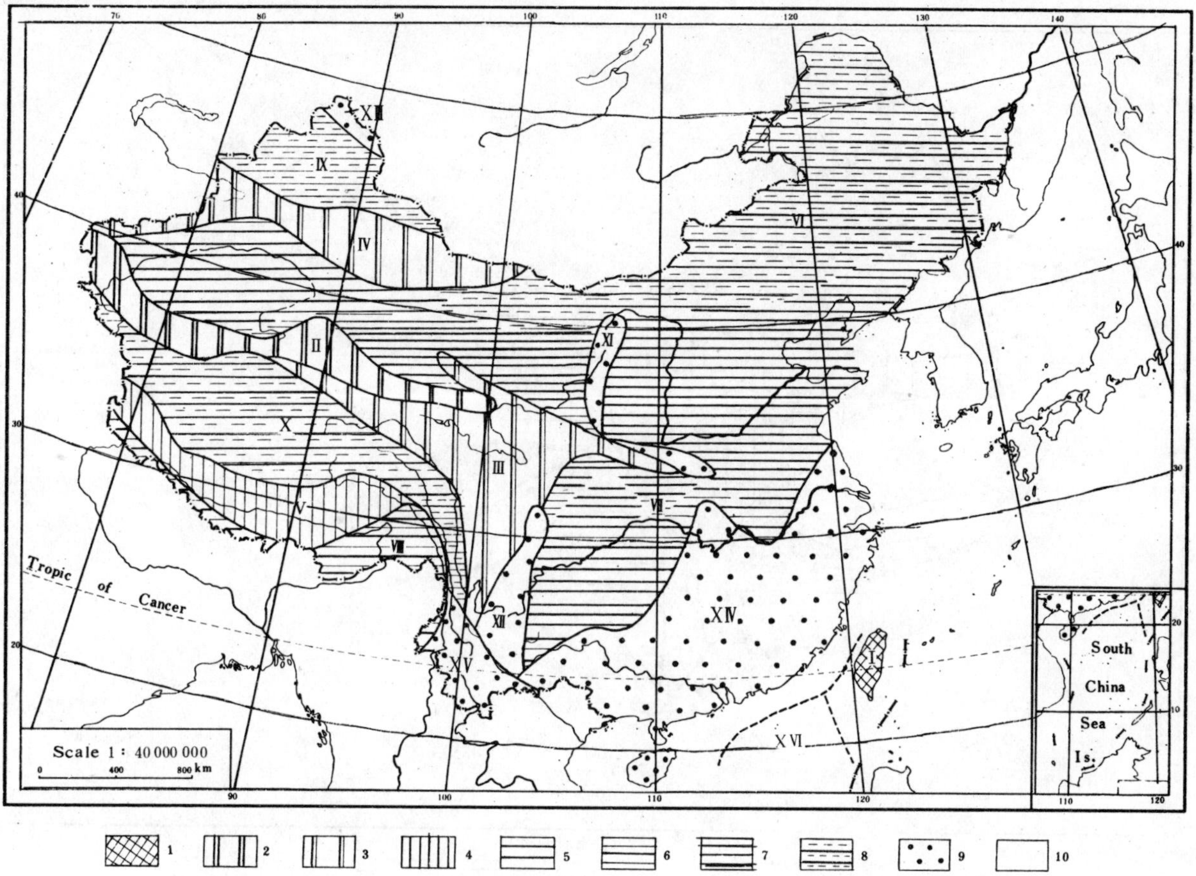

Fig. 3.5. Paleotectonic map of China at the end of the Triassic.

1, Pre-geosynclinal region (I, Taiwan); 2, Geosynclinal fold region of Hercynian age (II, Kunlun); 3, Geosynclinal fold region of Indosinian age (III, Bayanhar; IV, Tianshan); 4, Geosynclinal region, unclosed (V, South Xizang); 5, Platform region of post-Luliang age (VI, E. part of North); 6, Platform region of post-Jinning age (VII, Sichuan-Guizhou; VIII, Southern Himalayas; VI, NW. part of North); 7, Platform region of post-Caledonian age (VI, W. part of North); 8, Platform region of post-Hercynian age (X, North Xizang; IX, Junggar; VI, NE. part of North); 9, Diwa region, in diwa period (XI, Helan: XII, Kang-Dian; XIII, Altay; XIV, SE; XV, West Yunnan); 10, Region of unknown tectonic character (XVI, South China Sea).

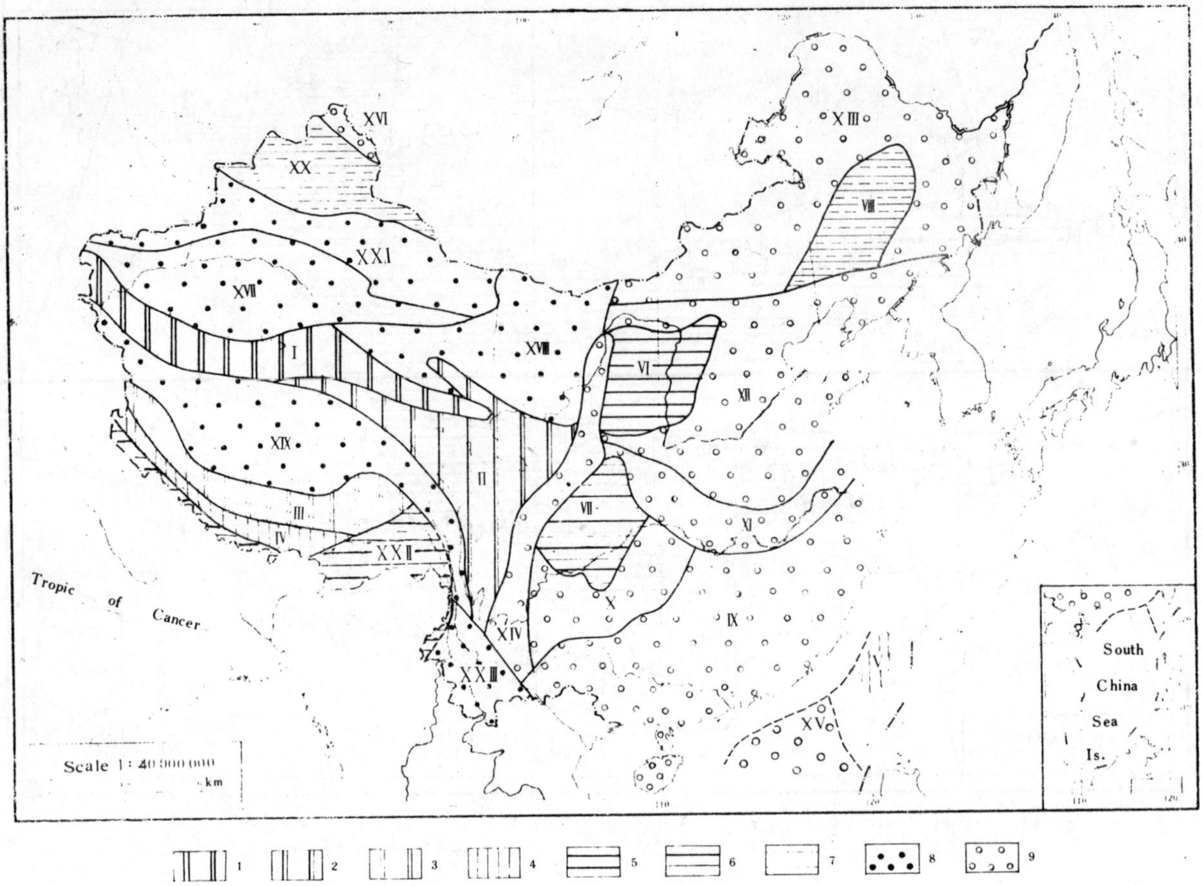

Fig. 3.6. Paleotectonic map of China at the end of the Cretaceous.
1, Geosynclinal fold region of Hercynian age (I, Kunlun); 2, Geosynclinal fold region of Indosinian age (II, Bayanhar); 3, Geosynclinal fold region of Yanshanian age (III, Gandise); 4, Geosynclinal region, unclosed (IV, Yarlung Zangbo River-Himalayas; V, Taiwan); 5, Platform region of post-Luliang age (VI, Yishan); 6, Platform region of post-Jinning age (VII, Sichuan; XXII, Southern Himalayas); 7, Platform region of post-Hercynian age (VIII, Songliao; XX, Junggar); 8, Diwa region, in diwa period (XVII, South Xinjiang; XVIII, Qinghai-Gansu; XIX, North Xizang; XXI, Tianshan); 9, Diwa fault-fold region of Cathysian age (IX, SE; X, Yunnan-Guizhou; XI, Gentral China; XII, North China; XIII, NE; XIV, N-S; XV, South China Sea; XVI, Altay).

Diwa regions in east China reached the maximum-mobility period one after another in the Middle and Late Mesozoic; they are typical of diwa regions of Cathysian age. The N-S diwa region, for the most part, also steped into the Maximum-mobility period at that time, but most of the diwa regions in west China had not reached the maximum-mobility period yet.

On the other hand, in Mesozoic time geosynclines still existed in some regions. In the western Qinling Mountains, Bayanhar and Garze, they closed in the Indosinian orogeny; in the Gandise Range, in the Yanshanian orogeny. The Yarlung Zangbo River-Himalayan geosynclinal region still had a development in Mesozoic time. Geosynclinal conditions in Taiwan Province began to appear in the late Mesozoic.

VII. Tectonic outlines in the Cenozoic (Fig.3.7)

During the Himalayan orogeny, the tectonic features of China, by and large, were more or less similar to the tectonic framework fixed during the Pacific orogeny. Diwa regions in east China had all experienced the peak of intense tectonic and magmatic processes taking place in the maximum-mobility period, and stepped into the residual-mobility period. But neotectonic movement in the whole was quite intense and the regions remained mobile. The three small platforms left over from the Chinese paleoplatform still remained at that time, continued their development and entered into the residual-stability period. In the neotectonic period, they are more active than before. In west China, the Altay, South Xinjiang, Qinghai-Gansu, North Xizang and West Yunnan diwa regions continued their development, and the Tianshan, Southern Himalayan and Junggar diwa regions came into being in the Tertiary and Quaternary respectively. In the neotectonic period, the above-mentioned diwa regions become more mobile, which is marked by rapidly occurring differential and large-amplitude subsidence and upheaval. As calculated from paleobotanic studies, the Xixiabangma peak in the Himalayas has risen from 2,500 m to 5,000 m above sea level during the period of more than two million years from Late Pliocene to present. The Jiabula and Nieniexiongla peaks at mount Qomolangma have risen from about 3,100-3,500 and 2,850-3,500 m to 5,000-5,200 and 4,900 m respectively, during the period of 200,000-300,000 years from Middle Pleistocene interglacial stage to present. The amplitude of the rising of this area is estimated to be 500 m during the period from Holocene Yali stage to present, averaging 20 mm per year. These diwa regions show volcanic activity (see Figs. 2.61 & 2.62) and high seismicity. At present they are still at the end of the initial-mobility period or at the beginning of the maximum-mobility period and are representative of diwa regions of Central Asia age. The N-S diwa region lying between the diwa regions of Cathysian age in east China and those of Central Asia age in west China is still tectonically active at this time, showing a transition zone character.

In the Cenozoic, a few geosynclinal regions remained in Chinese territory. At the end of the Early Tertiary, geosynclines in the Yarlung Zangbo River-Himalayas and in Taiwan Province closed one after another, giving rise to the newest geosynclinal fold regions in Chinese territory. At present these two Cenozoic geosynclinal regions are still tectonically active, which is expressed in extremely frequent and strong earthquakes as well as in rapid rising. The Gandise geosynclinal region gave place to a fold mountain belt at the end of

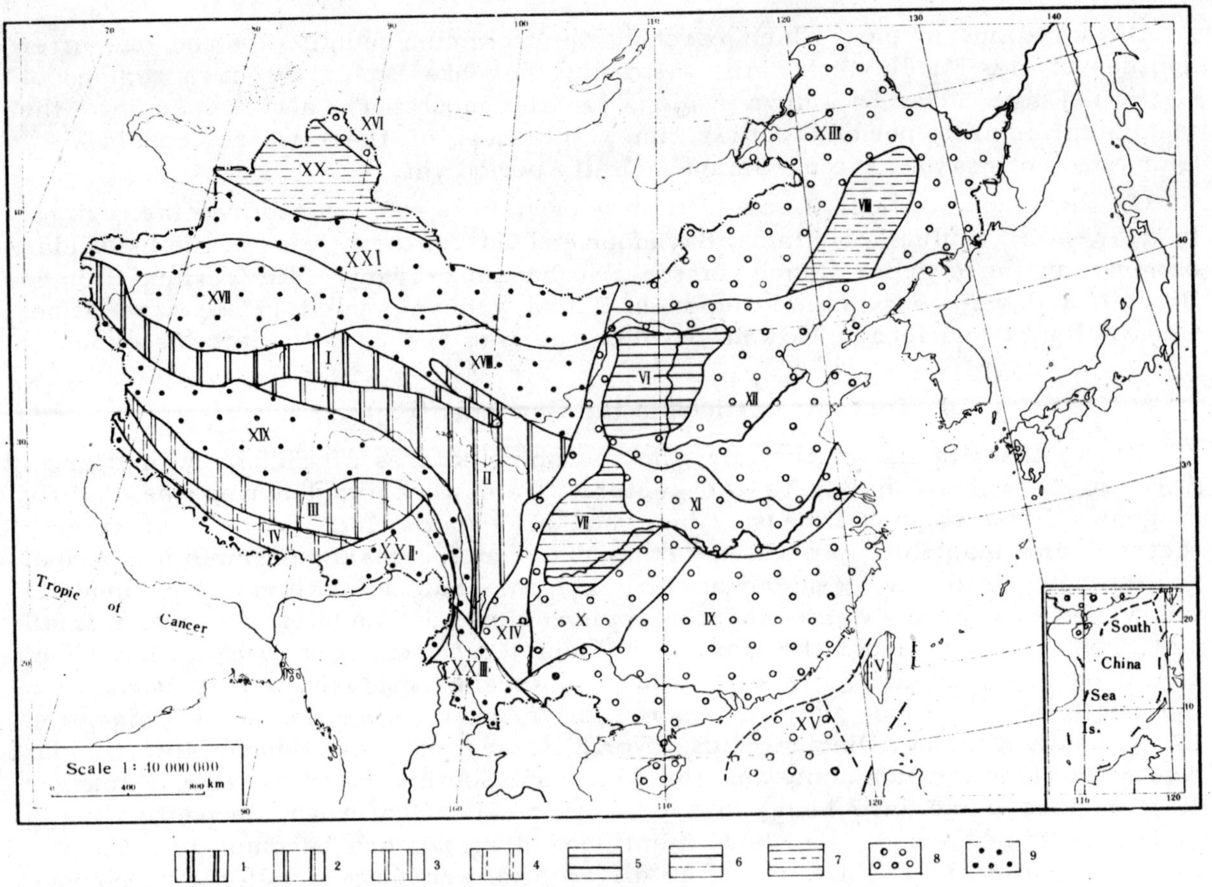

Fig. 3.7. Paleotectonic map of China at the end of the Tertiary.

1, Geosynclinal fold region of Hercynian age (I, Kunlun); 2, Geosynclinal fold region of Indosinian age (II, Bayanhar); 3, Geosynclinal fold region of Yanshanian age (III, Gandise); 4, Geosynclinal fold region of Himalayan age (IV, Yarlung Zangbo River-Himalayas; V, Taiwan); 5, Platform region of post-Luliang age (VI, Yishan); 6, Platform region of Post-Jinning age (VII, Sichuan); 7, Platform region of post-Hercynian age (VIII, Songliao; XX, Junggar); 8, Diwa fault-fold region of Cathysian age (IX, SE; X, Yunnan-Guizhou; XI, Central China; XII, North China; XIII, NE; XIV, N-S; XV, South China Sea; XVI, Altay); 9, Diwa fault-fold region of Central Asia age (XVII, South Xinjiang; XVIII, Qinghai-Gansu; XIX, North Xizang; XXI, Tianshan; XXII, Southern Himalayas; XXIII, West Yunnan).

the Mesozoic and now is in the residual-mobility period. The Kunlun and Bayanhar (including the western Qinling Mountains and the western part of the Songpan area) geosynclinal regions, whose geosynclines closed in the late Hercynian orogeny and at the beginning of the Mesozoic one after another, still have a high mobility in Cenozoic time. Further investigations are needed to clarify whether they still keep a geosynclinal character or have already been transformed into another type of tectonic regions.

The last phase of the Himalayan orogeny gave birth to the present tectonic outlines of China (Fig. 3.8).

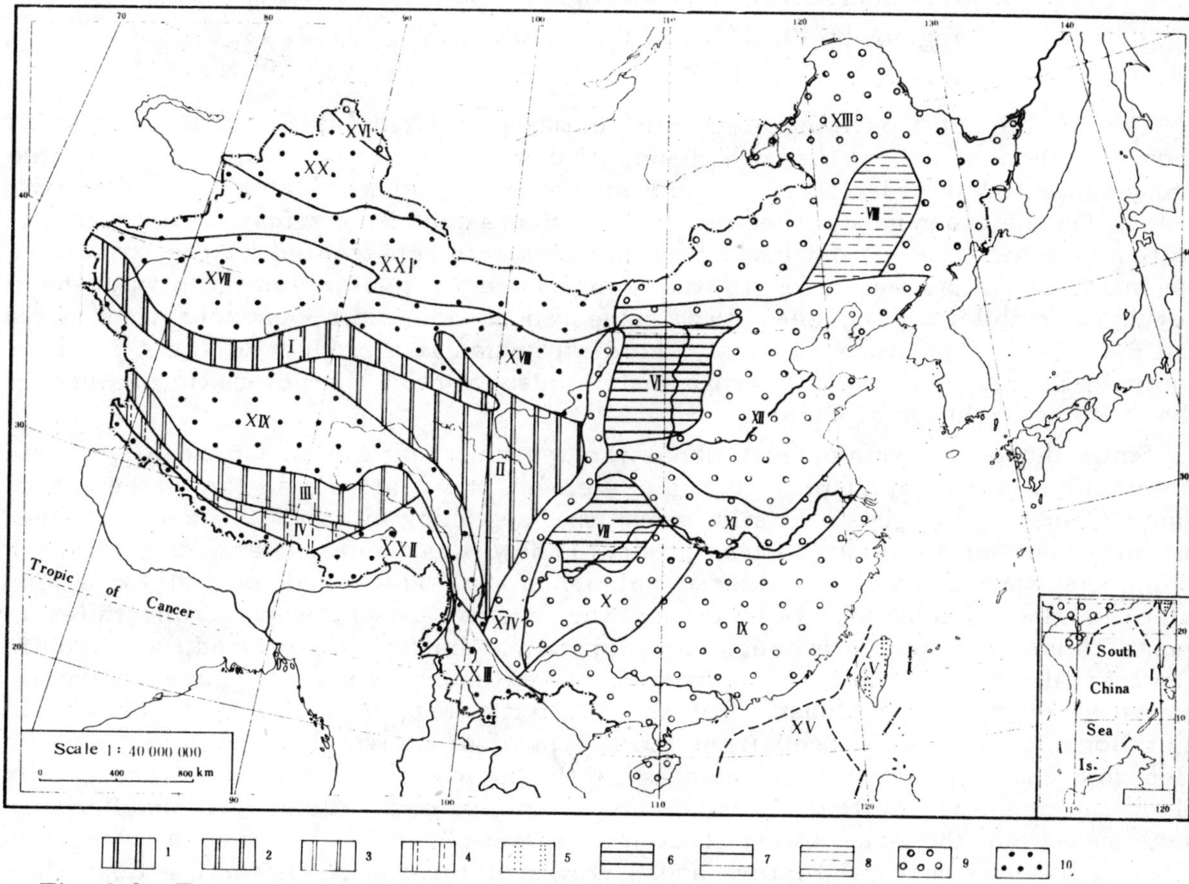

Fig. 3.8. Tectonic sketch map of China

1, Geosynclinal fold region of Hercynian age (I, Kunlun); 2, Geosynclinal fold region of Indosinian age (II, Bayanhar); 3, Geosynclinal fold region of Yanshanian age (III, Gandise); 4, Geosynclinal fold region of early Himalayan age (IV, Yarlung Zangbo River-Himalayas); 5, Geosynclinal fold region of late Himalayan age (V, Taiwan); 6, Platform region of post-Luliang age (VI, Yishan); 7, Platform region of post-Jinning age (VII, Sichuan); 8, Platform region of post-Hercynian age (VIII, Songliao); 9, Diwa region of Cathysian age (IX, SE; X, Yunnan-Guizhou; XI, Central China; XII, North China; XIII, NE; XIV, N-S; XV, South China Sea; XVI, Altay); 10, Diwa region of Central Asia age (XVII, South Xingiang; XVIII, Qinghai-Gansu; XIX, North Xizang; XX, Junggar; XXI, Tianshan; XXII, Southern Himalayas; XXIII, West Yunnan).

CHAPTER 4
SOME TECTONIC FEATURES OF CHINA

I. Wide distribution of diwa regions--new-type mobile regions [9, 10, 11, 12, 15, 19, 27, 30, 32, 33, 36, 37, 45, 46, 47, 50, 66, 67, 87, 90, 97, 104, 105, 158, 159, 161].

One of the most striking tectonic features of China is the multiformity of tectonic elements. On the one hand, two well-known tectonic elements, the geosyncline and the platform, not only appeared in geologic history, but also exist today. On the other hand, other tectonic elements can also originate. As previously stated, several types of old basic tectonic elements represented by the crystalline basement were present prior to the appearance of geosynclinal regions, and a new-type mobile region, the diwa, appeared in Mesozoic-Cenozoic time after the dying out of platform regions. The diwa region is a product of the continuous tectonic development of the crust and is also a new characteristic feature f the tectonic evolution of China.

Since the 1930s certain facts about the tectonic features of China in Mesozoic time and a new epiplatform tectonic element have attracted the attention of many Chinese geologists. Firstly, since the beginning of the Mesozoic, a series of intermontane tectonic basins, predominantly continental, have originated in China and spread over the Chinese paleoplatform, which had existed for a long time in the Paleozoic. Deposits in the basins are different from those in geosynclinal and platform regions. Secondly, during this period, a gigantic continental volcanic belt, accompanied by extensive zones of intrusive rocks, appeared on the east coast. Volcanic activity and its petro-geochemical and metallogenic aspects differed from those typical of geosynclinal regions. Finally, Mesozoic-Cenozoic crustal movements in the above-mentioned area were mainly block-faulting and faulting-folding, which, by means of their own peculiarities, have reworked the pre-existing tectonic framework and changed its character, forming a new structural feature and a new structural plan. After the liberation, with the rapid development of geological survey and exploration for mineral deposits, owing to the accumulation of many geologic data and achievements in scientific research, we could take further steps to study and clarify these peculiarities.

Mesozoic-Cenozoic diwa-type intermontane basins widely distributed in Chinese territory amount to about one thousand in number and cover a total area of more than one million square kilometers. Among them large diwa basins, e.g., the Tarim Basin and the North China Plain, can individually cover an area of several hundred thousand square kilometers. Medium-size diwa basins, e.g., the Qaidam and the Jianghan-Dongting, range from tens thousand to a hundred odd thousand square kilometers in area. Small diwa basins, e.g., the Turpan, Nanyang, Henyang and

Nanxiong, vary from several hundred to several thousand square kilometers in area. Major diwa basins generally contain minor depressions. In addition, there are numerous basins, ranging from tens to a hundred or less square kilometers in area and spreading over the greater part of China, except in the five geosynclinal regions and three small residual platforms [37].

As exemplified by east China, deposits in diwa basins include the Upper Triassic or its top (south China) or Jurassic (north and northeast China) to the Quaternary. Their sedimentary formations are chiefly coal-bearing deposits, red beds and volcanic rocks[11, 16, 95]. Most diwa basins are filled with two or more kinds of sedimentary formations, ranging from several hundred to several or even ten thousand meters in thickness. Diwa deposits have the following principal formational characteristics. On the one hand, they evidently differ from the underlying Paleozoic platform deposits at the same place in lithology, facies and thickness: They often contain piedmont sediments, poorly sorted and greatly variable. The deposits are formed in a mountainous environment with a sharp tectono-topographic contrast, laid down in intermontane basins and represent accumulations of a mobile region. On the other hand, although the deposits bear a strong resemblance to a molasse formation in the later period of geosynclinal development, they offer their own specific features: (1) Historically they are not at all deposited in the process of geosynclinal development, i.e., immediately after geosynclinal subsidence, but are laid down after dying out of a platform, in the conditions of gradual uplifting. (2) They have no direct relation to the formation of a geosynclinal fold region. The latter does not exert a structural control over their location and trend. They are scattered on a platform or a region of completed folding serving as a base for their development and controlled by the shape and trend of new-type basins. (3) They must have a basal conglomerate or other coarse-grained rocks at the bottom, resting with sharp unconformity (sometimes disconformity) on various older structural layers. In every stratigraphic unit constituent particles generally coarsen with increasing of depth, forming well-developed discontinuity-type rhythms of various orders. They are chiefly of continental origin (only in a few places marine beds are seen in the upper or lower parts) and never found in conformity with or transition to geosynclinal deposits. (4) They are often accompanied by strong magmatic activity; sometimes they have many intercalations of volcanic rocks. These characteristic features can distinguish them from the true molasse formation. Such deposits have been called formations of Cathysian type on account of their extensive development in China [19, 161].

Mesozoic-Cenozoic diwa basins in China and formations of Cathysian type in them are distributed systematically in a definite direction and grouped into a row [37]. This is common for all basins of mobile regions. In shape, arrangement and trend, they are somewhat different from and sometimes even have nothing in common with the structural lines formed at the geosynclinal stage of the same place. For example, in east China they form short belts, show an en echelon pattern and strike in a NE trend. But at the conjuncture between the North China and NE diwa regions, the structural lines formed at the Proterozoic or Paleozoic geosynclinal stage are aligned in long belts, in parallel with each other, and in a E-W trend. Thus it can be seen that in the period of development of Mesozoic-Cenozoic diwa basins and formations of Cathysian type, the stress field in the larger part of China underwent a great change in comparison with the previous one. This indicates that the region entered into a new stage.

Magmatism in diwa regions offers its own specific petrochemical features[82]. As exemplified by diwa granites in Hunan Province (Tables 4.1 and 4.2), most of them have a relatively high SiO_2 content which averages above 73% and sometimes reaches 77% at its maximum, thereby belonging to ultraacid rocks. The content of alkali constituents ($K_2O + Na_2O$) is also high, averages above 8% and sometimes even reaches 8.9%. As a result, ultraacid-alkaline granite peculiar to diwa regions has been formed. On the contrary, the oxides of Fe, Mg, Ca and Ti often have a low content. Among them $Fe_2O+FeO+MgO$ amount to a sum of < 4% and Fe_2O_3+FeO 3% on an average, but Fe_2O_3/FeO is relatively high and averages > 1%. TiO_2 content generally < 0.2% or even 0.02% in some cases. In regard to trace elements, diwa granite is characterized by high contents of non-ferrous metals (ore-forming elements) and rare, disseminated and radioactive elements (Tables 4.3, 4.4 and 4.5), directly indicative of an ore-forming mother rock. This phenomenon is most evident in the granite of Yanshanian age. As exemplified by the southeast side of the Qitianling plutonic body, Hunan Province, tin content ranges generally from 0.001 to 0.009% in granite and reaches 0.22% in alaskite. On the contrary, contents of V, Ni, Cr, Co and elements of the iron group are generally low (Tables 4.1, 4.2 and 4.3).

These specific petrochemical features are just contrary to those of geosynclinal granite, e.g., the Caledonian geosynclinal granite at the same place (see Tables 4.1, 4.2, 4.3, 4.4, and 4.5).

The high alkali content is also true for intermediate, basic and ultrabasic rocks of diwa regions. For example, diorite in the middle and lower reaches of the Changjiang (Yangtze) River contains a considerable amount of alkali metals. Quaternary basalt in east China, generally speaking, has a high alkali content. The diwa stage has yielded typical kimberlite belonging to alkaline ultrabasic rocks.

The above-mentioned tectonic features of China, beginning to appear in the Indosinian or Yanshanian orogeny, enable us to clearly distinguish the Mesozoic-Cenozoic diwa stage in China from the platform and the geosynclinal stage and to single out the diwa region as the third main tectonic element of the earth's crust.

Diwa regions have a wide distribution in China and cover three-fifths of the land area. In regard to their historical background, they have been developed mostly on the paleoplatforms of Eurasia and in little (in the Southern Himalayas) on the Paleoplatform of India. On the one hand, they are rich in exogenic deposits, such as oil, coal, oil shale, iron, copper and salt as related to diwa-type sedimentary formations. On the other hand, they yield important deposits of iron, which are of volcanic rock type or associated with igneous intrusions, as well as deposits of tungsten, tin, bismuth, molybdenum, copper, lead, zinc, beryllium, chromium, nickel, titanium, platinum and rare, rare-earth and radioactive elements as related to diwa-type magmatic formations. So far as both exogenic and endogenic deposits are concerned, diwa regions have their own characteristic features and are of great importance. Moreover, the complex construction of this tectonic element leads to the comprehensiveness, multiformity and complexity of diwa-type mineral deposits and to the superimposition of ores to form polygenetic compound ore deposits. As a consequence, the mineral deposits are rich, thick and varied. In

Table 4.1. Comparison of average petrochemical composition of diwa- and geosynclinal-type granitic rocks (Hunan Province)

Tectonic type	Tectono-magmatic cycle	Principal chemical composition (%)						
		SiO_2	TiO_2	Al_2O_3	Fe_2O_3	FeO	MnO	MgO
Diwa	Yanshanian-Indosinian	73.10	0.16	13.94	1.24	1.41	0.07	0.67
Geosynclinal	Caledonian	69.64	0.44	14.38	1.33	3.19	0.05	1.14
Average composition of China's granite		71.27	0.25	14.25	1.24	1.62	0.08	0.80
Average composition of world's granite		70.18	0.39	14.17	1.57	1.78	0.12	0.88

Tectonic type	Tectono-magmatic cycle	Principal chemical composition (%)					
		CaO	Na_2O	K_2O	P_2O_5	Loss	Total
Diwa	Yanshanian-Indosinian	0.98	3.36	4.69	0.10	0.71	100.43
Geosynclinal	Caledonian	2.41	2.70	4.08	0.12	0.83	100.31
Average composition of China's granite		1.62	3.79	4.03	0.16	0.89	100.00
Average composition of world's granite		1.99	3.48	4.11	0.19	0.84	100.00

Table 4.2 Comparison of petrochemical characteristics of diwa- and geosynclinal-type granitic rocks (Hunan Province)

Tectonic type	Tectonomagmatic cycle	K_2O+Na_2O	K_2O/Na_2O	$Fe_2O_3+FeO+MgO$	Fe_2O_3+FeO	Fe_2O_3/FeO
Diwa	Yanshanian-Indosinian	8.04	1.39	3.32	2.65	1.35
Geosynclinal	Caledonian	6.78	1.48	5.65	4.52	0.41
Average values of China's granite		7.82	1.06	3.66	2.86	0.76

Table 4.3. Abundances (p.p.m.) of some trace elements in diwa- and geosynclinal-type granitic rocks (Hunan Province)

Tectonic type	Tectono-magmatic cycle	Cu	Pb	Zn	W	Sn	Mo	V	Cr	C	Ni
Diwa	Yanshanian-Indosinian	10-55	30-210	55-395	10-90	40-395	4-25	10-32	10-20	10	10-20
Geosynclinal	Caledonian	10-21	20-70	26-50	50	5-8	2	22-40	20-30	36	10-26
Average values of granite		20	20	60	1.5	3	1	40	25	5	8

Table 4.4. Abundances (p.p.m.) of rare, disseminated and radioactive elements in diwa-type granite (Hunan Province)

Kinds of rocks	Ta	Nb	Li	Rb	Cs	Be	Th	U	TR	Rock bodies	Samples
Diwa granite	26.2	33.6	221.0	786	49.8	12.9	64.9	23.8	216.3	6	25
Average values of granite	3.5	20	40	200	5	5.5	18	3.5	-	-	-

Table 4.5. Comparison of the abundance (p.p.m.) of Be in diwa- and geosynclinal-type granites (Hunan Province)

Tectonic type	Be	Rock bodies	Samples
Diwa	7.8	25	217
Geosynclinal	2.8	8	64
Average values of granite	5.5	-	-

addition, because of the high mobility of diwa regions, earthquakes occurring in China show their own peculiarities and regularities. For this reason, further studies of diwa regions are of uttermost importance both in theory and practice.

II. Appearance of some evolution systems

In regard to the origination, development and dying out of old and young tectonic elements and their successive replacement, there exist some different gigantic evolution systems in different parts of China. They exhibit not only differences but also general regularities to some extent.

The direction and law of evolution vary from one tectonic stage to another and between different tectonic elements, but show an inheritance character to some extent. For the sake of convenience, we discuss the tectonic evolution systems of China in the following order: geosyncline→platform→diwa.

1. Geosynclinal-platform evolution systems

There are three geosynclinal-platform evolution systems in the history of the crustal development in China.

(1) The northern geosynclinal-platform evolution system

This system lay north of the Southern Tianshan foot-Bayanobo-Kaiyuan line. Prior to the Mesozoic, the broad area from north Xinjiang through Inner Mongolia to northeast China was in a geosynclinal condition for a long time. According to available data, geosynclines here came into being at the latest in the Sinian and closed in the Permian. They constituted a geosynclinal evolution system completing its development earlier than the other two in China. It formed a gigantic arc-shaped (convex towards south) Hercynian paleogeosynclinal fold region, which lay across north China and included the fold systems of Altay, Tianshan, Da and Xiao Hinggan Mountains, and Laoye Mountain, being the southernmost part of the colossal paleo-Asian Sea system of geosynclianl fold regions. This system began to evolve at its nucleus, the Siberian Precambrian geosynclinal fold region, from which geosynclines closed one after another in a southward order. Correspondingly, the beginning of the platform stage was delayed in the same direction and the platform progressively enlarged its extent from north to south. The Hercynian geosynclinal fold region formed in north China was part of the above-mentioned system of geosynclinal fold regions, where geosynclines closed at the latest. Here the platform stage began very late (at the beginning of the Mesozoic). It was in places not so apparent, lacking stratigraphic records and represented only by a regional erosion surface.

(2) The eastern geosynclinal-platform evolution system

South of the northern geosynclinal-platform evolution system lay the other two evolution systems, the eastern and the western, which were separated from each other by the Helan Mountain-Longmen Mountain-Daxue Mountain line.

The eastern geosynclinal-platform evolution system was part of the paleo-Pacific system of geosynclinal fold regions. Here the geosynclinal stage began long ago and some fold regions came into being as early as the Luliang orogeny. The system has lasted for a long time, and up to the Cenozoic a geosynclinal region still exists. It is a geosynclinal evolution system which has not completed its development yet. The system began to evolve at its nucleus, the North China Early Proterozoic geosynclinal fold region, from which, roughly in a N(W)-S(E) direction, geosynclines closed one after another in the following sequence: north China (Luliang age, the nucleus) → the main part of central China and Yunnan-Guizhou (Jinning age) → southeast China and the western part of central China (Caledonian age) → the southeast coast (Hercynian age) → Taiwan (Himalayan age). In the process of evolution a series of fold belts of different age, mainly NE-trending, were formed. At present Taiwan is still in the maximum-mobility period of the geosynclinal development and is part of the circum-Pacific seismic and volcanic belt. Correspondingly, the beginning of the platform stage was delayed in the same direction and the platform progressively enlarged its extent from north to south.

(3) The western geosynclinal-platform evolution system.

West of the Helan Mountain-Longmen Mountain-Daxue Mountain line lay the western geosynclinal-platform evolution system. It was part of the Tethys Sea system of geosynclinal fold regions. Here the geosynclinal stage began also long ago and some fold regions came into being as early as the Jinning orogeny. The system has lasted also for a long time and a geosynclinal fold region is still developing in the Cenozoic. The evolution line of this system was approximately parallel to that of the eastern system. Here the Tarim Precambrian geosynclinal fold region acted as a nucleus, from which, roughly in a N(E)-S(W) direction, successively younger geosynclines closed in a sequence little short of symmetry with one for the eastern system: Tarim (Jinning age, the nucleus) → Qinghai-Gansu (Caledonian age, but the Altun Mountains are of Hercynian age) → Kunlun and north Xizang (Hercynian age, but west Yunnan is of Jinning age) → Bayanhar (Indosinian age) → Gandise (Yanshanian age) → Yarlung Zangbo River-Himalayas (Himalayan age). They formed a series of fold belts of different ages, mainly NW-trending. The Yarlung Zangbo River-Himalayan geosynclinal region is still in the maximum-mobility period and is part of the Mediterranean seismic and volcanic belt. The beginning of the platform stage and the direction of the platform evolution were about the same.

To sum up, the three geosynclinal-platform evolution systems in China are all the component parts of the global geosynclinal-platform evolution systems. China lies just at the junction of these three colossal evolution systems.

2. Evolution systems of diwa regions

After entering into mid-Mesozoic time, because of the appearance of diwa regions, crustal development in China shifted into a new stage. As a result, the process of tectonic evolution changed remarkably. This was expressed in that the sphere influenced by the evolution systems of diwa regions and the direction of their development were other than previous: both inheritance and new were present. On the one hand, they followed the areal extent of the eastern and western geosynclinal-platform evolution systems but extend northwards, i.e., the above-mentioned three geosynclinal-platform evolution systems were replaced by two evolution systems, the Cathysian and Central Asia, which affected the eastern and western halves of China and were separated by the Helan Mountain-Longmen Mountain-Daxue Mountain line and its northward extension named the Yinchuan-Kunming axis. On the other hand, the beginning of the diwa stage and its maximum-mobility period were progressively delayed from south to north roughly, i.e., in a sequence contrary to one for the pre-existing geosynclinal-platform evolution systems. East of the Yinchuan-Kunming axis, the diwa stage began in the following sequence: southeast China and the southern border of central China (Late Triassic)→north China and the northern border of central China (Early Jurassic)→northeast China (Early to Middle Jurassic). The maximum-mobility period began in the following sequence: southeast China (Middle and Late Jurassic-Cretaceous)→north China (Middle and Late Jurassic-Late Cretaceous)→northeast China (Late Jurassic-Cretaceous). In this Period, a series of geodomes and diwa, chiefly NE- or approximately NE-trending, were superimposed discordantly on the eastern geosynclinal-platform evolution system existing in the Paleozoic. West of the Yinchuan-Kunming axis, the diwa stage began in the following sequence: west Yunnan (Late Jurassic)→north Xizang (Cretaceous)→Tianshan (Cretaceous)→Junggar (Quaternary), with the exception of south Xinjiang (Jurassic). The maximum-mobility period began mainly in the Paleogene or later. During this period, a series of geodomes and diwa, chiefly NW- or approximately NW-trending, were superimposed on the pre-existing western geosynclinal-platform evolution system.

The above-mentioned two diwa evolution systems appear not only in Chinese territory but also in adjacent areas. They are parts of two gigantic diwa evolution systems in Asia, and China lies just at their junction.

III. Formation and development of three crustal bodies

1. Outlines of crustal bodies formed by differential evolution

The existence and development of the above-mentioned tectonic evolution systems have led to a tectonic differentiation in China. As a result of the three geosynclinal-platform evolution systems, there existed prior to the middle Mesozoic three corresponding crustal bodies (northern, eastern and western). They were separated form each other by the latitudinal Southern Tianshan foot-Bayanobo-Kaiyuan and longitudinal Yinchuan-Kunming long-lived deep fracture zones, which were the sites of their junction (or transition). The former existed as the southern boundary of the northern crustal body and lay, from west to east, across whole China, with a length of over 4,000 km. The latter existed as a transition zone between the eastern and western crustal bodies, extended from east Ningxia to central Yunnan, lay across almost whole China, approximately at the Helan Mountain-Longmen Mountain-Daxue Mountain line, with a length of over 2,000 km, and consisted of several sets of parallel and criss-cross fractures (partly concealed). On account of its great width, this transition zone may be singled out as a first-order tectonic region.

2. Principal features of the three crustal bodies

As stated, each of the three geosynclinal-platform evolution systems shows a particularity as well as a generality, and was affected by the diwa evolution systems since the Mesozoic. As a result, the above-mentioned three crustal bodies also display specific historic-geologic and modern tectonic features each. The geosynclinal stage of the northern crustal body lasted long and did not come to a close until the Hercynian orogeny. The platform stage began after the Hercynian orogeny, lasted relatively short in the main part and was not typical. After mid-Mesozoic time, the eastern and western sectors became differentiated markedly. By this time the eastern sector had largely been converted into diwa regions, and only the Songliao region remained as a residual platform up to date. The diwa stage began mainly in the Early Jurassic (earlier or later in a few areas).

These diwa regions are of Cathysian age according to the time of appearance of the maximum-mobility period, or of Cathysian type according to strong magmatism and specific structural features (with the predominance of arcogenesis-faulting type). As regards the western sector, the Mesozoic was probably a platform stage. In the Cenozoic, the whole sector was converted into diwa regions, which are still in the maximum-mobility period at present and thus of Central Asia age. According to weak magmatism and specific structural features (with the predominance of block-faulting type), they are of Central Asia type. Nevertheless, the eastern and western sectors of the northern crustal body showed a generality at that time, i.e., the diwa stage and its maximum-mobility period, generally speaking, began later than those in other two crustal bodies.

The eastern crustal body is characterized by a short duration of the geosynclinal stage, which terminated, for the most part, at the end of the Early Proterozoic (north China, Yishan) or in the Late Sinian (Sichuan, Yunnan-Guizhou, the larger part of central China). In other places (southeast China, the western part of central China) the geosynclinal stage also ended after the Caledonian orogeny. At present, a geosynclinal region remains only in Taiwan. Where the geosynclinal stage ended relatively early, correspondingly the platform stage began also early and lasted long: from Sinian or its end to Middle and Late Triassic in most cases, and at the present time the Yishan (Ordos) and Sichuan regions are still in the platform stage. The diwa stage began relatively early everywhere. During the Late Triassic (southeast China and the southern part of central China) or the Early Jurassic (Yunnan-Guizhou, the northern part of central China, north China), the diwa regime was successively set up from south to north. The diwa regions began their maximum-mobility period also relatively early, being of Cathysian age. According to strong magmatism and peculiar structural features, they are of Cathysian type.

So far as the area proportion of tectonic regions of different character and types is concerned, the most widely distributed in the eastern crustal body are diwa regions, which account for the overwhelming majority of the area. Except for the Taiwan geosynclinal region and two small residual platform regions, the Yishan (Ordos) and the Sichuan, the eastern crustal body now is occupied all by diwa regions.

The situation is somewhat different in the western crustal body. Here the geosynclinal stage ended relatively late for the most part. It lasted to the Early (e.g., Qinghai-Gansu) or Late Paleozoic (Altun Mountains and north Xizang), except in Tarim and west Yunnan, where it ended at the close of the Sinian. In some places the geosynclinal stage has not yet ended in the Cenozoic, though the geosynclines closed in the Hercynian (Kunlun), Indosinian (Bayanhar), Yanshanian (Gandise) or Himalayan (Yarlung Zangbo River-Himalayas)orogenies. The platform stage began after the Caledonian (Qinghai-Gansu) or Hercynian (north Xizang) orogeny, except in Tarim and the Southern Himalayas, where it began relatively early, after Jinning and Xingkai age, respectively. Moreover, the platforms lived shorter than those in the eastern crustal body, except in Tarim where the platform regime existed from the end of Sinian to the beginning of Mesozoic. The diwa stage, for the most part, began also later than in the eastern crustal body. It began in the Late Triassic (west Yunnan), Jurassic (Qinghai-Gansu, South Xinjiang, and Altay), Cretaceous (north Xizang and Tianshan), Tertiary (Southern Himalayas) or Quaternary (Junggar). The maximum-mobility period of the diwa development began later than that in the eastern crustal body, i.e., as late as Cenozoic, and thus the diwa regions are of Central Asia age. According to weak magmatism and peculiar structural features, they are of Central Asia type.

In regard to the area proportion of various tectonic elements in the western crustal body, diwa regions have a wide distribution, geosynclinal regions are still active in a number of places, but present-day platform regions have not yet been recognized.

The narrow belt lying between the eastern and western crustal bodies, i.e., the N-S diwa region, as previously stated, is their tectonic boundary[101]. Furthermore, the history of its geologic development also has a transition character. The central segment of this tectonic region may serve as example. The Geosynclines in its southeastern part closed in the Caledonian orogeny. The platform stage lasted from Devonian to the beginning or end of Late Triassic. The diwa stage began in the Late Triassic or in the Early Jurassic, and its maximum-mobility period started early. This part, therefore, belongs to the diwa regions of Cathysian age and bears a strong resemblance in development history to the western extremity of the Central China diwa region in the eastern crustal body. Geosynclines in the northwestern part, however, closed in the Indosinian orogeny, as did the Bayanhar geosynclinal region in the western crustal body. The platform stage is not distinct. The diwa stage began in the Jurassic and its maximum-mobility period appeared relatively late. Therefore, this part belongs to the diwa regions of Central Asia age and is similar to the western crustal body.

Next, within the above-mentioned crustal bodies, the development of various parts shows a difference as well as a generality. In the process of their long evolution they were differentiated into minor crustal bodies. Owing to the difference in geologic development, within the northern crustal body two minor crustal bodies, the eastern and the western, gradually came into being in the Mesozoic and were separated by the north end of the Yinchuan-Kunming axis. Similarly, on account of the difference in geologic development, the eastern and western crustal bodies were also divided into two minor ones, the northern and the southern, each separated by a transition zone trending latitudinally. Thus, there are three minor crustal bodies on each side of the Yinchuan-Kunming axis. They show a close symmetry in their arrangement and development. East of the axis, the northernmost part is the NE crustal body (NE diwa region and Songliao platform region). South of it lies the North China crustal body (North China diwa region and Yishan platform region). These two crustal bodies are separated by the eastern segment of the southern border of the northern crustal body, the Bayanobo-Kaiyuan deep fracture. Still south lies the South China crustal body (Yunnan-Guizhou, SE and South China Sea diwa regions, Sichuan platform region and Taiwan geosynclinal region). The transition zone between it and the North China crustal body is situated along the middle and lower reaches of the Changjiang (Yangtze) River and assumes an arc convex towards the south. This zone is a sharp paleogeographical, stratigraphical, paleontological, magmatic and metallogenic boundary between the northern and southern parts of the eastern crustal body. West of the Yinchuan-Kunming axis, the northernmost part is the North Xinjiang crustal body (Altay, Junggar and Tianshan diwa regions). South of it lies the Qinghai-Xinjiang crustal body (South Xinjiang and Qinghai-Gansu diwa regions). They are separated by the western segment of the southern boundary of the northern crustal body, i.e., by the Southern Tianshan foot deep fracture. Still south lies the Xizang crustal body (North Xizang and West Yunnan diwa regions,

Bayanhar, Gandise and Yarlung Zangbo River-Himalayan geosynclinal regions, and Southern Himalayan diwa region). The transition zone between it and the Qinghai-Xinjiang crustal body is situated at the Kunlun geosynclinal region and its WNW-ESE extension. This zone is a sharp paleogeographical, stratigraphical, paleontological, magmatic and metallogenic boundary of the northern and southern parts of the western crustal body in China.

3. Evolution since Mesozoic time

Because of the establishment of diwa evolution systems, the history of crustal development after the middle Mesozoic in China entered into a new stage. As a result, the tectonic outlines underwent a great change. At that time two diwa evolution systems, represented by a newly formed structural layer and other characteristic features, were superimposed on the three Paleozoic geosynclinal-platform evolution systems and reworked them. Accordingly, the relationship between the above-mentioned three crustal bodies has changed significantly[97]. Firstly, the eastern and western sectors of the northern crustal body showed a marked tectonic difference in Mesozoic-Cenozoic time, whereas the tectonic differences between them on one hand, and the eastern and western crustal bodies on the other hand diminished respectively. Secondly, the eastern and western halves of China, demarcated by the Yinchuan-Kunming axis, displayed a sharp contrast in geologic, geomorphologic and geophysical features, in the thickness of the crust, in the depth to the M-discontinuity, in the development history of geosynclinal and platform regions, in the beginning of diwa tectonism and its maximum-mobility period as well as in metallogeny. Thus the three crustal bodies, formed due to the differences between the three geosynclinal-platform evolution systems in the Paleozoic, were turned into two greater crustal bodies embracing the eastern and western halves of China. Their geological differences have been described above. Here we only point out other differences.

(1) Differences in landforms

Firstly, being at different stages of tectonic development and having a different tectonic contrast, the eastern and western halves of China show a different topographic contrast. The eastern half is composed mainly of medium to low mountains and seldom plains, attaining an altitude of less than 2,000-3,000 m above sea level. On the contrary, the western half consists of high mountains. There exist the Qinghai-Xizang Plateau, 4,000-5,000 m high above sea level, and the world's highest mountain ---the Higher Himalayan geodome system with 13 peaks being more than 8,000 m high and including 8,848.13 m-high Mount Qomolangma. The difference in elevation is conspicuous. Within the lofty Tianshan geodome system, more than 4,000 m above sea level (highest peak: 7,439 m), lies the deep Turpan diwa which is 154 m below sea level. Secondly, controlled by tectonic features, the shape of mountains is also different for the eastern and western halves of China. Mountains in the eastern half have a short line shape and an en echelon arrangement, which is associated with the extensive development of diwa-type structures in this region. Mountains in the western half usually have a long line shape and a parallel arrangement, which is thought to have resulted from the wide distribution of geosynclinal and plaeo-geosynclinal fold belts and their strong influence upon landforms.

Finally, affected by the principal direction of structural lines, the trend of mountains in the eastern is also different from that in the western half of China.

Mountains in the former generally trend NE, whereas in the latter NW. Mountains in both the halves extend towards their demarcation line, the Yinchuan-Kunming axis, and meet at this line, forming a feather-like symmetric pattern.

The N-S diwa region lying between the different landscapes of the eastern and western halves of China geomorphologically has an expression of division and transition[34].

1) It is situated in elevation amidst the two halves, and is generally 3,000-4,000 m above sea level, with over 7,000 m high Gongga Mountain in the southern segment.

2) It lies just at the meeting point of the mountains in the above-mentioned two halves. In its northern segment and in the eastern part of the southern segment exist N-S-trending mountains, whereas in its central segment and at the north end of the northern segment exist two groups of mountains, either intertwining with each other or occupying different positions in different segments: Longmen Mountain in the southeastern part of the central segment runs in a NE direction, whereas the east end of the western Qinling Mountains in the northwestern part of the central segment adopts a WNW trend.

(2) Differences in geophysical features

As a result of tectonic differentiation, different parts of the crust in China also vary in geophysical features. For example, gravity anomalies generally range from 0-100 in the eastern to -200-450 mGal in the western half. The N-S diwa region lying between the two halves is situated just at the gravity gradient zone from a gravity high in the eastern to a gravity low in the western half.

(3) Differences in the depth of the crust and in the position of the M-discontinuity

The distribution pattern of the above-mentioned gravity anomalies indicates that the eastern and western halves of China show a great difference in the depth and structure of the crust. As inferred from surface wave frequency divergence data, the depth of the crust in the western half is approximately 50-70 km (more than 60-70 km in the Xizang Plateau), whereas in the eastern it is thin, generally 30-35 km (thick in part of diwa regions, e.g., 45 km in Shanxi Province; thin on platforms, e.g., less than 30 km on both the Sichuan and Yishan platforms). The crust in the N-S diwa region lying between the two halves evidently has an intermediate depth (40-50 km in the known part) (Fig.4.1).

As calculated from the difference in the depth of the crust, in combination with the variation in the height of the earth's surface, the position of the M-discontinuity also greatly varies for the eastern and the western half. Taken sea level as a datum, the depth to the M-discontinuity is small in the eastern, but large in the western half. The N-S diwa region is just situated at the junction of the two halves, being a transform zone of this discontinuity from shallow in the east to deep in the west.

As stated above, the earth's surface of the western half of China is higher than that of the eastern, but the M-discontinuity under the former is lower than under the latter; besides, negative gravity anomalies in the former are higher than in the latter. These data in combination with the differences in diwa development and in intensity of upward movement in the eastern and western halves have led us to the following preliminary conclusion: One of the geophysical features of diwa regions is the predominance of negative gravity anomalies in most cases, which indicates that the crust there expands and, therefore, decreases

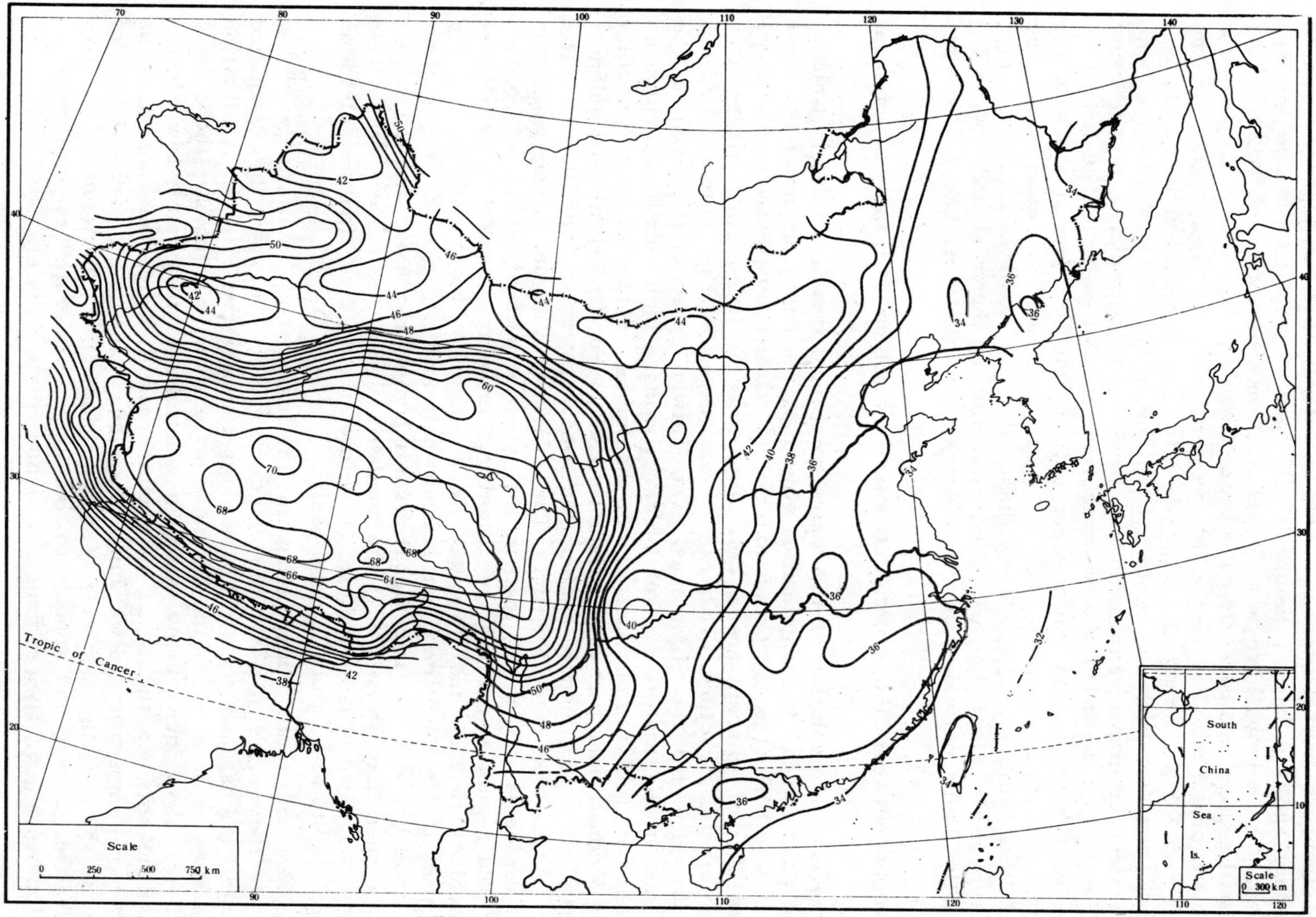

Fig. 4.1. Isopach map of the crust in China (after Huang Peihua, 1981).

in density. As a result, the earth's surface has risen and diwa deposits are all or chiefly of continental origin. Diwa regions in their maximum-mobility period undergo the strongest crustal expansion and, therefore, have the highest negative gravity anomaly and the largest height of the earth's surface. In the western half of China, diwa regions have a wide distribution, but some geosynclinal fold regions are still present and recently have been strongly rising. These diwa regions are in their maximum-mobility period at present, whereas diwa regions in the eastern half of China have all stepped into the residual-mobility period. Consequently, compared with the eastern half, the western half must have been subjected to stronger crustal expansion and have a larger thickness of the crust resulting in a greater depth to the M-discontinuity. However, as the height of rising of the earth's surface in the western half is small in comparison with the values of expansion and thickening of the crust, the M-discontinuity under the western half lies, below sea level, lower than under the eastern. In other words, though the top of the crust appears to be ascending in the west and descending in the east, the M-discontinuity shows a "descent" in the west and an "ascent" in the east.

4. Relationships between the crustal structure of China and that of adjacent areas

The formation and evolution of the above-mentioned crustal bodies in Chinese territory not only embody the regularities governing the development of the crust in China, but also reflect to a certain extent the relationships between the crust of China and that of adjacent areas, i.e., the former occupies a specific position in a broader crustal structure. This can be illustrated by the tectonic positions of Taiwan Province and the Himalayas. The Taiwan geosynclinal region is the only active one in the eastern half of China. Having a high mobility peculiar to a young Cenozoic fold region (just at the beginning of the maximum-mobility period), it is a constituent part of the gigantic circum-Pacific seismic and volcanic belt and lies close to the modern conjunction zone between the two global crustal bodies, the Asian and the Pacific. The Yarlung Zangbo River-Himalayan geosynclinal region is the newest one in the western half of China. It rose to be a young fold region as late as the Himalayan orogeny (at present still in the maximum-mobility period) and now has a particularly high mobility. It is a constituent part of the Mediterranean seismic belt and lies close to the modern conjunction zone between the two global crustal bodies, the Asian and the Indian. Thus it can be seen that China lies at the northern edge of the modern conjunction zone between the three crustal bodies. Therefore, the tectonic features of China are closely related to these global crustal bodies.

Diwa regions of south China lying in the east show increasingly strong magmatism as they move towards the conjunction zone between the Asian and Pacific crustal bodies, and appear to be high-temperature diwa mineralization zones (e.g., diwa regions of Fujian-Zhejiang style along the coast); those lying in the west also show increasingly strong magmatism as they move towards the conjunction zone between the Asian and Indian crustal bodies, and appear to be moderate-to-high-temperature diwa mineralization zones (e.g., diwa regions of Hubei style in Kang-Dian and west Yunnan); those lying between them show relatively weak magmatism and appear to be low-to-moderate-temperature diwa mineralization zones (e.g., diwa regions of Guizhou-Guangxi style).

IV. Five tectonic systems and their stress analysis

1. Principal types of tectonic systems

In Chinese territory structural lines expressed by geological structures of different dimensions vary from place to place and intersect each other, but at the same time they have their internal relations and regularities in their distribution. Being subject to regional and local stress fields, they constitute several tectonic systems each appearing in definite areas. The position of these tectonic systems, their shape, attitude and change at various times have controlled the localization, configuration and trend of geosynclinal and diwa regions (or geosynclines, geanticlines, diwa and geodomes in them) as well as the distribution of sedimentary formations and their subsequent deformations.

Tectonic systems are defined here as those composed of various structural components [1] or construction elements produced by the same force in a certain tectonic region during a certain stage of tectonic development. They are regularly arranged and grouped, constituting the tectonic framework of China[41]. The main tectonic systems in China are as follows

(1) Arc-shaped tectonic systems

These are represented by the North Xinjiang-Hinggan arc, convex towards south, situated in the northern crustal body and lying across whole north China. Its eastern wing is in the NE diwa region, including the Da Hinggan Mountains, and its western wing is in north Xinjiang, including the Altay, Northern Tianshan and other mountains. The arc top is at Yinshan Mountain and its westward extension. Formed as the most external link of the gigantic paleo-Asian tectonic system at the end of the Paleozoic, this arc consists of folds and faults in the Hercynian geosynclinal structural layer. During the geosynclinal stage, the form of the arc was complete. After stepping into the diwa stage, however, the arc was deformed a little by the Yinchuan-Kunming deep fracture, which extended northwards into the arc and shifted the arc top. At the same time, fault-fold structures formed by diwa tectonism (especially during the Yanshanian orogeny) were superimposed on the geosynclinal structural layer.

There are also minor arc-shaped tectonic systems in other parts of China, e.g., the Qinling-Huaiyang arc, the North Guangdong arc[132] and the Guangxi arc.

(2) E-W tectonic systems

Of the E-W tectonic systems the Southern Tianshan-Yinshan tectonic system is the most outstanding. It lies at the boundary between the Tianshan and NE diwa regions on the one hand, and the South Xinjiang and North China diwa regions on the other hand, i.e., at the Southern Tianshan-Yinshan line. It stretches along latitudes 41° -- 42° N and forms a long belt. Produced by a set of N-S-oriented horizontal compressive forces, it consists chiefly of a series of E-W-trending parallel folds and various associated fractures, including the Southern Tianshan

---

1. Structural components denote geological structures of different dimensions (major, middle and minor), and of different forms. Middle and minor structures include, for example, folds, faults, joints, cleavages, flow lines and flow layers; major structures, geosynclines,geanticlines, diwa and geodomes.

foot-Bayanobo-Kaiyuan deep fracture zone. In both its eastern and western segments, i.e., in the Yinshan and Tianshan mountains, there is a crystalline basement composed of Archean rocks. It formed several geanticlines, E-W-trending, during the Proterozoic geosynclinal stage and constituted the oldest known structural layer of the tectonic system. Afterwards, the western segment of the tectonic system and the northern part of its eastern segment continued their geosynclinal development, which did not come to a close until the Hercynian orogeny when tight folds took shape. But the southern part of the eastern segment was converted, together with the north China region, into a platform during the Luliang orogeny. Since the Yanshanian orogeny, the whole tectonic system has been affected by diwa tectonism represented by block-faulting, open folding and thrusting in Yinshan Mountain. This E-W tectonic system is situated just at the conjunction zone between the southern boundary of the northern crustal body and the northern boundaries of the eastern and western crustal bodies, and has been active for a long time accompanied by extensive magmatism. Therefore, the structure is very complex.

There are also many other E-W tectonic systems scattered here and there in China.

(3) N-S tectonic systems

Of the N-S tectonic systems in China the most important is the Helan Mountain-Daxue Mountain tectonic system[34, 41, 62, 89, 92]. Situated in the N-S diwa region, it manifests itself mainly in the northern and southern segments of the Helan Mountain-Longmen Mountain-Daxue Mountain line. It trends N-S, with the exception of the central segment Longmen Mountain where it deflects to NE due to interference by a NE tectonic system. It stretches along longitudes 102°-106°E and forms a zigzag belt. Formed chiefly by a set of E-W-oriented horizontal compressive forces, it comprises principally linear folds and associated fractures, including the Yinchuan-Kunming deep fracture zone. In the Helan Mountain segment there exposed an Archean crystalline basement, which extends in a N-S direction, forms a long belt and appears to be the oldest constituent part of this tectonic system. It is directly overlain by the Sinian-Paleozoic platform structural layer, on which a diwa structural layer was superimposed in Mesozoic time. In the Daxue Mountain segment the Proterozoic geosynclinal structural layer has a wide distribution and is seen chiefly at its southern extremity. After the Jinning orogeny this segment entered into a platform stage. Diwa tectonism began at the end of the Triassic and affected the pre-existing N-S structural lines. The Longmen Mountain segment is composed of three structural layers: Caledonian geosynclinal, Late Paleozoic to Early Mesozoic platform and Late Mesozoic diwa. Folds and associated fractures of a NE strike are in predominance. The diwa tectonism has produced many thrusts which overthrust from northwest to southeast and afterwards formed klippes. This N-S tectonic system is situated at the conjunction and transition zone between the eastern and western crustal bodies and has been active for a long time. Therefore, it is also one of the very complex tectonic systems in Chinese territory. The accompanying magmatic activity was significant and is marked by the extensive development of basic and ultrabasic rocks and associated minerals.

N-S tectonic systems also exist sporadically in other parts of China.

(4) NE tectonic systems

These denote tectonic systems with a trend ranging from NNE to ENE. They are concentrated mainly in east China, i.e., east of the Yinchuan-Kunming axis, and have a wide distribution. Their structural lines began to develop early in the eastern crustal body, but revealed themselves late, chiefly at the Mesozoic diwa stage, in the eastern segment of the northern crustal body, where they rest on the eastern wing of the North Xinjiang-Xingan arc formed in the Paleozoic. These tectonic systems generally have an en echelon arrangement and have been created by NW-SE directed compressive forces derived from a set of sinistral and N-S-directed regional horizontal shear forces. They comprise long-line-shaped tight folds and associated faults occurring in the geosynclinal and pre-geosynclinal structural layers during the Luliang, Jinning, Caledonian and Himalayan orogenies respectively, and gradually developing from NW to SE with accompanying granitoids and other igneous rocks of corresponding ages. These tectonic systems are an important constituent part of the paleo-Pacific tectonic system. After a long period of denudation and peneplanation and a relatively quiet platform stage varying in duration from place to place, i.e., in Mesozoic-Cenozoic times, they became very active again in the form of diwa structures with an age delayed progressively from south to north. Structural components during these times are chiefly block-faulting and thrusting and short-line-shaped open (and tight) folds. They have been formed mostly in the Yanshanian and partly in the Indosinian and Himalayan orogenies, manifest themselves chiefly in the platform and geosynclinal structural layers, and rest on the geosynclinal and pre-geosynclinal structural layers, which have been rebuilt to be converted into diwa fault-fold belts. At that time, along with the intrusion of numerous acid-to-intermediate rocks, in the coastal zone occurs extensive volcanic extrusion. Within the NE tectonic systems only diwa structures are exposed on the surface due to a widespread distribution of the diwa structural layer.

In some areas structural lines belonging to the NE tectonic systems appear to be convex towards northwest or west to form minor arc-shaped structures, which can be observed in Xuefeng and Qiyang of the SE diwa region. Sometimes they have developed into counter-S-shaped structures.

NE-trending structural lines also exist in west China, e.g., in the Altun Mountains of the South Xinjiang diwa region. Whether they belong to the NE tectonic systems remains a question.

(5) NW tectonic systems

They consist of structural lines, chiefly NW- or WNW-, sometimes approximately NNW-trending, concentrated mainly in west China, i.e., west of the Yinchuan-Kunming axis, especially within the western crustal body, but also seen in the northern crustal body, where they rest on the western wing of the North Xinjiang-Hinggan arc. Their structural lines have an en echelon or parallel arrangement, created by NE-directed compressive forces derived from a set of dextral, N-S-directed regional horizontal shear forces. They comprise long-line-shaped tight folds and associated faults, successively younger southwestward, occurring in the geosynclinal(and pre-geosynclinal in places) structural layers during the Jinning, Caledonian, Hercynian, Indosinian, Yanshanian and Himalayan orogenies, being a constituent part of the Tethys tectonic system.

The accompanying strong magmatic activity has produced a great amount of basic and ultrabasic rocks. In Mesozoic-Cenozoic times (chiefly Cenozoic in some places) diwa structures were superimposed on the pre-existing structural lines to form diwa fault-fold belts, with the exception of some areas which maintain their geosynclinal character. The diwa structures manifest themselves mainly in the diwa and platform structural layers, but they have also affected and reworked the pre-existing structural layers. The prominent diwa tectonism is block-faulting, but linear folding is also significant.

In the Kunlun, Bayanhar, Gandise and Himalayas, the NW tectonic systems form an inclined S-shape, with the central segment trending E-W and the northwestern and southeastern segments trending N-S.

NW tectonic systems also exist locally in east China. A NW-trending structural line composed of slightly metamorphosed rocks in the Lower Paleozoic geosynclinal structural layer of east Hunan and west Jiangxi extends for several tens of meters from the eastern edge of the Chaling diwa to central Jiangxi.

The above-mentioned five tectonic systems often intersect, interference, overlap or intertwine with each other. However, with regard to whole Chinese territory, the principal representatives of various tectonic systems show a regular change in their distribution and have a regular arrangement, forming a quite tidy pattern. The most characteristic feature is the feather-symmetric relationship between the tectonic systems lying east and west of the Yinchuan-Kunming axis represented by the N-S Helan Mountain-Daxue Mountain tectonic system in the N-S diwa region (Fig. 4.2). In this relationship the Yinchuan-Kunming axis is a feather axis. On each side of the north end of the axis, i.e., in the eastern and western segments of the northern crustal body, stand respectively the eastern and western wings of the North Xinjiang-Hinggan arc-shaped tectonic system. Here the direction and morphology of structural lines show a clear feather-like symmetry. South of the E-W Southern Tianshan foot-Yinshan tectonic system, structural lines within the eastern and western crustal bodies also have a feather-like symmetric arrangement: those east of the feather axis belong to the NE tectonic systems, but those west of it to the NW tectonic systems. These two gigantic tectonic systems, running in contrary directions and having an en echelon or parallel arrangement, meet at the Yinchuan-Kunming feather axis and oblique from the N-S tectonic system at an acute angle, which, pointing southward, diminishes as it moves further south. And the N-S tectonic system within the Yinchuan-Kunming feather axis serves as a transition zone between the structural lines developed widely in the eastern and western crustal bodies. Here several sets of structural lines are found intersected and interpenetrated. In the central segment, for example, two tectonic systems gather together to form some minor southward-convex arc-shaped structures, e.g., the Jintang arc. In the central segment and at the north end of the northern segment, the N-S tectonic system deflects to NE due to the predominance of the NE tectonic systems, forming a S-shaped tortuous tectonic zone. Furthermore, the principal fractures belonging to several tectonic systems respectively intersect along this tectonic zone and combine into the Yinchuan-Kunming zone of zigzag fractures (caused by X-type cracks).

2. Origination of the five tectonic systems and their stress analysis

There are different interpretations about the origination of the five tectonic systems in Chinese territory and about the mechanism of their interrelation and joining into a definite pattern. Based on the above points, we give the following analysis (Fig. 4.2).

As tectonic systems occurring in different parts of China differ from each other in the time of their origination and in the process of their evolution, and the structural layers involved and affected also differ from each other, stresses responsible for them were complex and varied from time to time and from place to place. But generally speaking, it may be thought that prior to Cenozoic time there existed two pairs of regional horizontal compressive forces, N-S- and E-W-directed and affecting large areas. They fought against each other: sometimes they were equally strong, and interfered with each other; sometimes they were not equally strong and occupied alternately the dominant position. As a result, a great and complex regional stress field varying from time to time was established.

N-S-directed horizontal compressive forces have produced gigantic E-W tectonic systems. In the process of their action, horizontal differential movements affecting large areas occurred in different areas. They were responsible for two pairs of horizontal couples of forces acting symmetrically with respect to the central China's longitude taken as a demarcation line and compounding into a pair of southward-directed horizontal bending forces. In the northern crustal body this pair of bending forces has given rise to the gigantic southward-convex North Xinjiang-Hinggan arc-shaped structure, which extends from the Altay and Northern Tianshan on the west to the Da Hinggan Mountains on the east.

South of the northern crustal body, within the eastern and western crustal bodies, the N-S-directed horizontal differential movements also caused local bending forces in some areas, which have produced middle and minor southward-convex arc-shaped structures. But no gigantic arc-shaped structures reflecting bending forces of a wide scope have appeared. On the contrary, these two crustal bodies are occupied by a gigantic N-S tectonic system and on both sides by NE and NW tectonic system. The reason is that E-W-directed horizontal compressive forces probably manifested themselves particularly apparently south of the Southern Tianshan-Yinshan tectonic system. Due to their interference, in the periphery of the top of the North Xinjiang-Hinggan arc, where E-W-directed tension should have been strongest, a N-S tectonic system representing a compression took shape. Furthermore, the southward-directed horizontal bending forces no longer kept their original character, but were turned into two sets of couples of regional N-S-directed forces acting symmetrically with respect to the N-S tectonic system and moving in parallel but opposite directions: dextral (clockwise) on the west and sinistral (counterclockwise) on the east side. From these two sets of couples of forces, two pairs of regional compressive forces, NE-SW- and NW-SE-directed, have been derived. The concrete expression is that the Asian crustal body came into a head-on compression separately with the Indian and Pacific crustal bodies, thus forming NW tectonic systems in the west and NE tectonic systems in the east. In the process of action of these two pairs of regional horizontal compressive forces, differential movements again took place locally in different areas, producing couples of horizontal forces of a limited scope to constitute local horizontal bending or torque forces. As a result, NW-SE-trending structures with an inclined S-shape were formed in the western crustal body, and northwestward- westward-convex

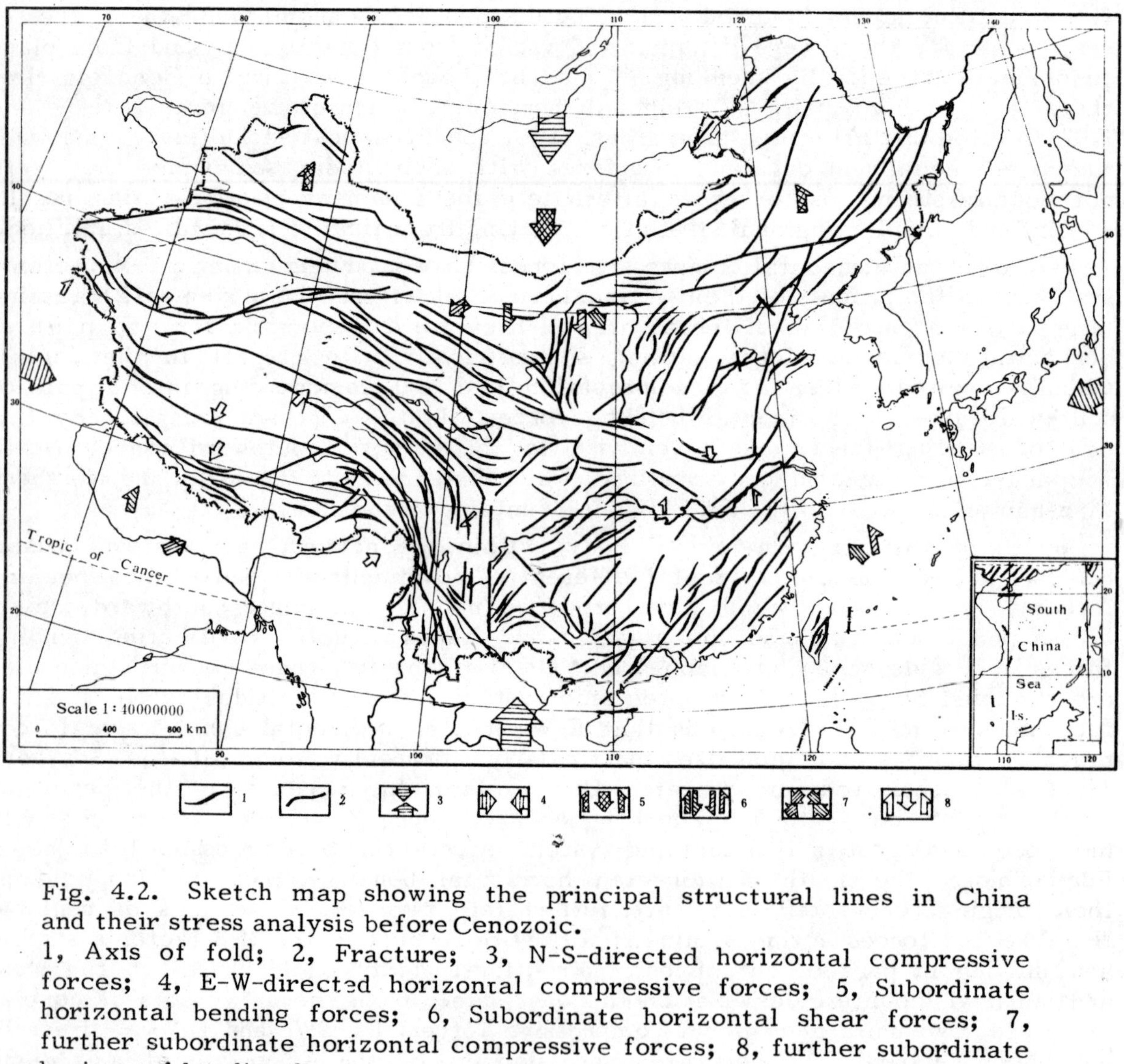

Fig. 4.2. Sketch map showing the principal structural lines in China and their stress analysis before Cenozoic.
1, Axis of fold; 2, Fracture; 3, N-S-directed horizontal compressive forces; 4, E-W-directed horizontal compressive forces; 5, Subordinate horizontal bending forces; 6, Subordinate horizontal shear forces; 7, further subordinate horizontal compressive forces; 8, further subordinate horizontal bending forces.

arc-shaped structures and sometimes counter-S-shaped and vortical structures appeared in the eastern crustal body.

E-W-directed compressive forces have produced the apparent gigantic N-S tectonic system at the demarcation line between the two pairs of horizontal shear forces in east and west China as well as minor and sporadically distributed N-S tectonic systems in other areas. It is worth to note that these compressive forces also in another way affected a large area. In east China, for example, they compelled NE-trending fractures, originally left-lateral when formed, to become right-lateral, with the result that the NE tectonic systems have apparently changed their trend. Moreover, the magnitude of the E-W-directed horizontal compressive forces increases from north to south, so that the angle between the NW and NE tectonic systems on the one hand and the N-S tectonic system on the other points southward, and diminishes as it moves further south. That is also why the northeastward-convex arc-shaped structures within the western crustal body and the northwestward-convex ones within the eastern are well-developed and more bent all in the south.

It should also be pointed out that in most cases the NE structures trend close to east-west or north-south, i.e., turn to ENE or NNE. Some geologists explained this turn as a result of decrease in the angle between A axis in the strain ellipsoid and the action force in the process of rotational strain. This explanation has furthermore led to another inference: the ENE structural lines are old structures formed in the Paleozoic, while the NNE structural lines are young ones created in the Cenozoic. As a matter of fact, however, the NNE structural lines were already formed extensively in north China as early as the Luliang orogeny, controlling the trend of geosynclines and the orientation of fold axes, and also appeared in the south-central segment of Xuefeng Mountain and in the Donggang Mountain ore field in east Hunan during the Caledonian orogeny (see Chapter 2). Therefore, the another possible explanation is: the NE structures trend close to east-west when the N-S-directed compressive forces apparently participate in the process of action of N-S-directed sinistral horizontal shear forces, or close to north-south when the E-W-directed compressive forces do so.

Finally, the principal reason why the N-S tectonic system (and the N-S diwa region controlled by it) has a westward offset in its southern part and swerves to NE in its central part (deforming the Bayanhar geosynclinal region into a triangle) is that in south China the more active compressive force is derived from east, because the Pacific crustal body pushes northwestward at a greater intensity than the Indian crustal body does northeastward.

It should also be pointed out that the above-mentioned stress analysis is suitable only for the major periods in the China's crustal development. Under the combination of complex internal and external factors, the stress field often varied, to a certain extent, from time to time and from place to place. For example, a striking phenomenon is that since the Cenozoic some compression or compression-shear fractures in east China have been converted into tension or tension-shear ones. This indicates that during this period the evolution of the crust in East Asia and the Pacific ocean was subjected to a tensile stress pulling the crust apart in a SE direction as well as to its internal factors.

V. A dynamic model for the development of the crust

In regard to the causes of the progressive development of crustal structure via transformation between mobile and "stable" regions, the theory of progression [13] holds that the Earth's materials not only repel but also attract one another. These are two aspects of the basic contradiction existing simultaneously in the evolution of the earth. In order to explain such a phenomenon an hypothetical dynamic model has been proposed.

On the one hand, lower-asthenospheric material (including heavy) decreases in density due to temperature increase. It creeps from deep to shallow zones, resulting in the outward diffusion of heat energy (including radioactive). Consequently, this part of the Earth contracts step by step due to heat loss, but on the contrary, the crust at the corresponding site expands because of being heated. At the same time, heat flow increases and causes strong crustal movements, predominantly lateral (chiefly compressional). On the other hand, upper-asthenospheric material (especially heavy) creeps from shallow to deep zones due to gravitational differentiation, resulting in accumulation of heat energy. Consequently, this part of the Earth expands step by step due to heat accumulation, but on the contrary, the crust at the corresponding site relatively contracts because of scatter of heat energy. At the same time, heat flow decreases and weakens the crustal movements there. Moreover, vertical movements predominate over lateral ones. The former may be caused by gravitational differentiation in the asthenosphere, vertical expression of expansion and contraction in the crust or within the Earth, isostatic adjustment, and alternation between horizontal and vertical movements. The latter are dominantly tensional. It can be assumed that the crust develops as an unity of the two opposites of contradiction, reciprocal repellence and attraction of the Earth's materials.

The above-mentioned two aspects of the contradiction within the Earth exert an influence in opposition to each other on the development of crustal structure and the tectonic character of a region. The reciprocal repellence of the Earth's materials and the outward diffusion of heat energy necessarily result in intense tectonism, magmatism and metamorphism, and thus appear as an intra-crust mobile factor for crustal development. The reciprocal attraction of the Earth's materials and the accumulation of heat energy, on the contrary, weakens tectonism, magmatism and metamorphism and, therefore, behave as an intra-crust stable factor for crustal development.

As already mentioned, repellence and attraction exist within the Earth simultaneously; therefore, the intra-crust mobile and stable factors for crustal development also exist simultaneously. But as we know, of the two contradictory aspects, one must be principal and the other secondary. The principal aspect is the one playing the leading role in the contradiction. The nature of a thing is determined mainly by the principal aspect of a contradiction, the aspect which has gained the dominant position. But this situation is not static; the principal and non-principal aspects of a contradiction transform themselves into each other and the nature of the thing changes accordingly. So far as the contradiction within the Earth is concerned, when repellence occupies a dominant position, the process of internal contraction due to heat diffusion is just one pregnant with the gradual intensification of attraction. This creates a condition for the principal aspect of the contradiction to change its position and for the repellence preponderance to negate itself. On the contrary, when attraction occupies a dominant position, the process of internal expansion due to heat accumulation is just one pregnant with the gradual intensification of repellence. This also creates a condition for the principal aspect of the contradiction to change its position once more and for the attraction preponderance to negate itself (negation of negation). If just an intra-crust mobile factor caused by the reciprocal repellence of the Earth's materials and the diffusion of heat energy holds a governing position in a certain area during a certain period and becomes the principal aspect of the contradiction, horizontal and dominantly compressional crustal movements prevail in the greater part of the period. These movements cause strong folding and fracturing, magmatic intrusion and extrusion, metamorphism, seismicity and sharp tectono-topographic contrast, thus forming a mobile region. In contrast, in another period of the same region or in another region of the same period, if just an intra-crust stable factor caused by the reciprocal attraction of the Earth's materials and the accumulation of heat energy holds a governing position and becomes the principal aspect of the contradiction, vertical crustal movements prevail in the greater part of the period. At this time, tectono-topographic contrast becomes small. Folding, fracturing, magmatism, metamorphism and seismicity turn weak. Heat flow decreases. The crust cools and is consolidated. Previously formed fractures and cracks are cemented due to filling by magma. And the region is converted into a (relatively) "stable" one. The subsequent development is on the analogy to this. The development of the earth's crust probably is just thus motivated by the struggle between the two opposites, the reciprocal repellence (diffusion) and attraction (coacervation) of the Earth's materials. Along with the transformation of the two contradictory aspects of the contradiction, the principal and the non-principal, mobile and "stable" regions transform themselves into each other. This is called the theory of mantle creep and heat energy diffusion-coacervation alternation.

In the history of transformation between mobile and "stable" regions in China, generally speaking the Proterozoic was a period when the mobile factor held a dominant position in the main part of the territory and became the principal aspect of the contradiction; therefore, geosynclinal regions spread widely. Sinian to Lower Paleozoic time was a period when the mobile factor gradually weakened and the stable one progressively intensified; therefore, geosynclinal reigons died out one after another and gradually decreased in area, while platform regions originated in succession and progressively increased in area. Upper Paleozoic to Lower Mesozoic time was a period when the stable factor held a dominant position and became the principal aspect of the contradiction; therefore, platform regions appeared in a great amount and spread widely over the greater part of the territory, whereas geosynclinal regions remained or originated only in a few places. Beginning in the Indosinian orogeny in the Late Triassic, the mobile factor intensified again. As a result, most platform regions died out step by step, giving place one after another to newborn mobile regions. These newly formed mobile regions have nothing in common with the previous ones, because the crust in China at this time differed markedly from that in the Proterozoic having more sialic structural layers, a more complex construction and a more thick and rigid basement. They are not the simple repetition of geosynclinal regions, but new-type mobile regions having more structural layers and featured by block-faulting--diwa regions. After the middle Mesozoic, the mobile factor held a dominant position; therefore diwa regions spread widely and only three small platforms remained. At the same time, mobile regions formed in Taiwan Province and in the Yarlung Zangbo

River-Himalayas are geosynclinal in type, because these areas have different internal conditions, less sialic structural layers and a relatively simple construction, being at an earlier stage of crustal development. Furthermore, the whole world is similar in crustal development to China. However, as the process of struggle between repellence and attraction of the Earth's materials varied from place to place and the principal and non-principal aspects of the contradiction are not universal, tectonic elements coming into being during the same period but in different areas are not identical.

In addition, the above-mentioned struggle between the two contradictory aspects within the Earth and their transformation into one another eventually lead to a variation in the Earth's rotative velocity. As a result of contraction within the Earth occurring after the reciprocal repellence of the Earth's materials and the outward diffusion of heat energy, the Earth's rotative velocity, according to the law of conservation of the angular momentum, gradually increases. On the contrary, as a result of expansion within the Earth occurring after the reciprocal attraction of the Earth's materials and the accumulation of heat energy, the Earth's rotative velocity gradually decreases. The variation in the Earth's rotative velocity, in turn, can stimulate the struggle between the two contradictory aspects within the Earth and their transformation into one another in a certain area. The reason is that during a period of increase in the Earth's rotative velocity, centrifugal forces become strong and, in combination with gravitation, may increase the horizontal component, thus reinforcing lateral movements in this period. Meanwhile, the outward migration of heavy materials due to magmatic intrusion and extrusion accompanying intense crustal movements may bring about outward diffusion of the Earth's mass, thus creating favourable conditions for a decrease in the Earth's rotative velocity and for strengthening attraction and causing the principal aspect of the contradiction within the Earth to change position. Afterwards, during a period of decrease in the Earth's rotative velocity, on the contrary, lateral movement become weak and magmatic activity insignificant. As a result, the Earth's mass is made to move to the center, thus creating a favourable condition for an increase in the Earth's rotative velocity and for strengthening repellence and causing the principal aspect of the contradiction within the Earth to change position.

As for the causes of directional development of crustal structure, it is assumed that the principal one is the horizontal differential creep of mantle material. The above-mentioned process of creep of mantle material causing transformation between mobile and "stable" regions must result in differential creep in different directions, including horizontal one. This would be responsible for the relative movement, including junction(collision) or separation or offset, of crustal blocks of different size and give rise to stress fields of different range varying from time to time and from place to place. Thus the above-mentioned tectonic systems and the peculiarities in their distribution came into being.

Apart from the above-mentioned basic internal cause, there are secondary causes which may have influence upon directional development. (1) The variation in the Earth's rotative velocity resulted from the above-mentioned struggle between the two contradictory aspects within the Earth gives rise to two sets of regional lateral compressive stresses, longitudinal and latitudinal, in the crust and to a change in their magnitude, and intensifies or weakens the relative displacements between some parts of the crust or some crustal bodies in definite directions.

This is a major factor determining the directional development of crustal structure in a large area. (2) The unevenness of tectonic development, disparity in expansion and contraction, and differences in internal and boundary conditions in different parts of the crust determine and produce corresponding directions of movement and stress fields. This can determine or influence upon the directions of structures and their differences between different crustal bodies or within them.

Finally, the attraction of other heavenly bodies may also cause or influence upon differential movement between different parts and bodies of the crust.

CHAPTER 5
RELATIONS OF TECTONICS TO MINERAL DEPOSITS IN CHINA

I. Tectonics and mineralizations

The relationships between tectonics and mineralizations may be summarized as follows.

1. Minerogenic specialization of tectonic elements

Firstly, each tectonic element, varying in sedimentary, magmatic and metamorphic formations and changing in structural and petrochemical features due to the differences in the history of their geologic development, can offer, besides common conditions of ore formation, its own speçific ones. This leads to the "minerogenic specialization" of tectonic elements. That is to say, tectonic elements of different character and types yield different mineral kinds or deposit types as well as common mineral kinds or deposit types, or differ from each other in key mineral kinds. These mineral are called specialized minerals. Here we take the key mineral kinds as example. In geosynclinal regions, strong magmatism marked by extensive development of basic and ultrabasic rocks brings about variegated rock types ranging in composition from ultrabasic to acid. Moreover, magmatic rocks of geosynclinal type are petrochemically characterized by a high content of iron-group elements. Chromite, nickel sulfides and other mineral assemblages, therefore, are particularly rich. The reason why more than 60% iron, 62% nickel and cobalt in the world are concentrated in pre-Cambrian times is the wide distribution and prevalence of geosynclinal (and pre-geosynclinal) regions at that time. In platform regions, small-scale basic magmatism predominates and yields mainly deposits of iron, copper and nickel associated with basic rocks. But determined by tectono-geomorphic features, exogenic mineralizations appear to be very important. Various deposits of iron, manganese, phosphor, aluminium, coal, oil shale, oil and gas, therefore, have a wide distribution. In diwa regions, as concerns the Cathysian type, magmatic activity is also magnificant and produces acid to basic as well as alkaline rocks, especially a large amount of ultraacid to intermediate intrusive and effusive rocks. These rocks are petrochemically characterized by the high contents of non-ferrous, rare and disseminated metals and radioactive elements; elements (such as Li, Be and B) with low atomic number but anomalous amounts in the crust also have a relative concentration. The above-mentioned characteristics are most apparent for the granitoids intruded in the maximum-mobility period of diwa development (Yanshanian age). Therefore, tin, bismuth, molybdenum, copper, lead, zinc, boron and lithium form industrial concentrations, and mineral deposits with associated cadmium, indium, germanium, niobium, thorium, uranium, beryllium, selenium, yttrium and other elements are very important. The reason why 80% tungsten in the world is concentrated in the Mesozoic, more than 85% molybdenum in the Mesozoic-Cenozoic, and 50% tin at the end of the Mesozoic is probably the rapid expanding of diwa regions at that time, which covered a large area and the maximum-mobility period of which corresponded mostly to the end of the Mesozoic.

Diwa basins contain important sedimentary deposits of oil, gas, oil shale, coal, gypsum and salt.

Secondly, in different tectonic elements, the same mineral kind has a different source of material to form its deposit under different conditions for magmatic activity. Here we cite iron deposits as example. (1) in different tectonic elements magmatic rocks of different composition occur with different intensity: In geosynclinal regions basic and ultrabasic rocks occur particularly abundantly; in platform regions basic rocks predominate but occur on a small scale; in diwa regions, as concerns the Cathysian type, acid rocks occur on a far larger scale than basic ones. Iron is more richly contained in basic and ultrabasic rocks (Table 5.1). (2) Acid rocks of different tectonic types vary in iron content: acid rocks of geosynclinal type contain more iron than those of diwa type. Table 4.2 is an example. It is clear that geosynclinal regions (including died paleogeosynclinal ones) have the best source of material to make up iron deposits and the most favourable conditions for magmatic activity. In fact, as proved by comparison between the three tectonic elements studied in detail, the largest iron deposits are frequently found in geosynclinal regions and, furthermore, among them the most important are iron deposits linked with submarine basic volcanic extrusion in the early period of geosynclinal development. Next come diwa regions, because diwa magmatic activity has generally passed through several pre-existing structural layers, including geosynclinal and pre-geosynclinal ones, and may absorb a great amount of iron from iron-bearing beds and from great and rich iron deposits formed previously in them. Moreover, some rocks (e.g., carbonate) in these pre-existing structural layers may serve as favourable country rocks which provide part of iron in the process of contact metasomatism with an invading magma to form iron deposits.

Table 5.1. The average iron content of various magmatic rocks in China as compared with its crustal abundance

Rock type	Ultrabasic rocks	Basic rocks	Intermediate rocks	Acid rocks	Alkaline rocks
Average iron content, %	9.06	8.90	5.07	2.36	3.73

The clarke of iron, 4.8

Finally, on account of the differences in the conditions of ore formation in different tectonic elements, even the same mineral kind may often be of different deposit type. We cite iron deposits again as example. Iron deposits formed at the pre-geosynclinal stage are represented by metasedimentary deposits (later reorganized with introduction of new material). Of iron deposits of geosynclinal type the most important are metasedimentary and volcano-metasedimentary iron ore deposits. Amongst iron deposits of platform type, sedimentary and residual iron deposits are widespread; in addition, there are volcanic and hydrothermal iron deposits. Among iron deposits of diwa type, contact-metasomatic and hydrothermal-skarn iron deposits are most common and volcanic iron deposits most promising. Moreover, in areas of distribution of diwa volcanics (e.g., in a zone of distribution of basalt along the Guangdong coast) there are iron deposits formed by weathering of these volcanics.

The principal characteristic features of endogenic and exogenic ore deposits of different tectonic types will be discussed below.

2. Inheritance of mineral deposits and progression of metallogenic processes

Any tectonic element not only possesses minerogenic specialization of its own, but also can inherit mineral deposits of its predecessors. Thus the inheritance of mineral deposits is brought into being. In other words, in any tectonic element we can find remains of ore assemblages of various tectonic types formed at the previous stages of crustal development as well as ore assemblages peculiar to itself. The former are called inherited mineral deposits. In platform regions, for example, we can find, generally speaking, ore assemblages of geosynclinal and pre-geosynclinal types inherited from the predecessors as well as ore assemblages of platform type. Similarly, in diwa regions we can find ore assemblages of platform, geosynclinal and pre-geosynclinal types, inherited totally by them, as well as ore assemblages of diwa type. In most cases, the more lately a tectonic element emerges in a development sequence, the more tectonic types of ore assemblages can be found and, therefore, the mroe comprehensive and complex the ores become, provided that the situation of absence and hypoplasia of the ores caused by some reasons do not exist. Thus, the progression of metallogenic processes is brought into being. For instance, the diwa region is the latest among the three tectonic elements studied in detail. Consequently, ores in it are more complex and varied than in both geosynclinal and platform regions. It is the richest in mineral deposits among the presently known tectonic elements. The diwa region has a wide distribution in Chinese territory (especially in its eastern part)and, therefore, abounds with mineral deposits of all kinds.

3. Multistage and uneven character of metallogenic processes

The progression of metallogenic processes results in the multistage formation of ores in a certain area (especially in a tectonic region with a complex construction). This is identical with the multistage character of crustal development. For instance, in the southeast of China at least three metallogenic stages (not including the pre-geosynclinal stage) have been recognized and studied in detail. The first metallogenic stage is the geosynclinal metallogenic stage (Proterozoic to early Paleozoic). Endogenic mineral deposits formed during this stage by strong and complex magmatic activity are: metamorphosed volcano-sedimentary iron deposits carried by geosynclinal marine effusives of Jinning-Caledonian age and deposits of tungsten, antimony, gold, lead, zinc and arsenic associated with such intrusive rock as granite. Exogenic mineral deposits comprise metamorphosed sedimentary deposits of iron, manganese and phosphorus formed in the process of the geosynclinal development. The second stage is the platform metallogenic stage (late Paleozoic to Middle Triassic). During this stage, broad marine transgressions and regressions took place, both source and sedimentation areas had a wide distribution and, therefore, exogenic mineral deposits formed in the process of sedimentation such as sedimentary iron ore, bauxite and coal are plentiful. The third stage is the diwa metallogenic stage (Late Triassic to present). Endogenic mineral deposits, formed during this stage again by strong and varied magmatic activity, include numerous non-ferrous and rare metals, such as tungsten, tin, bismuth, molybdenum, copper, lead, zinc, gold, antimony, mercury and iron carried by diwa granite and other igneous rocks. Exogenic mineral deposits formed under favourable conditions of climate and associated with widespread diwa basins include coal, oil shale, petroleum, gas and salt.

On the other hand, because of the unevenness of crustal development, different regions may differ from each other in tectonic character or type during the same interval of time, or tectonic elements of the same character and type presently live at different periods of their development. In addition, there may be other differences in the conditions of ore formation. As a result, different regions differ from each other also in the number of metallogenic stages and in the features of ore formation, thus bringing about metallogenic unevenness. For example, being of different tectonic character and type, the Taiwan geosynclinal region, the Sichuan platform region and the SE diwa region differ markedly in the number of metallogenic stages and in the degree of complexity of ore assemblage types. Differences in the features of ores also exist between the North China and the NE diwa regions because of their different history of geologic development, and between the NE and the Tianshan diwa regions because of their different type and different present stage of development. Even the diwa regions of the same Cathysian type also differ from each other in ores due to the differences in intensity of magmatism and volcanism.

4. Regularities in the spatial and temporal distribution of mineral deposits

On account of the above-mentioned many-sided reasons, the spatial and temporal distribution of mineral deposits in various parts of China show common regularities to some extent as well as their own specific features. Firstly, in point of the temporal distribution, sedimetary and volcanogenic ores of various tectonic types within an area are found to exist in a corresponding contemporaneous structural layer; magmatic ores, in pre-existing structural layers. Thus, a definite structural layer is a temporal structural position where specific ores or deposit types are located. In diwa regions, for example, sedimentary and volcanogenic mineral deposits formed in the process of the development of diwa basins are found to exist in the diwa structural layer, while mineral deposits associated with intrusion of diwa magmas can be found in the diwa structural layer and in part of older structural layers intruded by the magmas.

Secondly, in point of the spatial distribution, the above-mentioned mineral deposits can appear at any place where a corresponding structural layer exsits, or in any area where intrusive rocks contemporaneous with the structural layer exist. Thus, a definite tectonic region is a spatial structural position where specific ores are located. In geosynclinal regions, for example, mineral deposits formed in the early (proper geosynclinal) period, including exogenic and endogenic ones, are located mostly in fold belts and deep fracture zones, but sedimentary deposits formed in the late (fold belt) period are restricted to piedmont and intermontane depressions. In diwa regions, diwa-type endogenic mineral deposits exist both on geodome uplifts and in diwa subsidences, while diwa-type sedimentary deposits (except oil and gas which can migrate) are confined mainly to diwa basins. As regards mineral deposits inherited from pre-diwa stages, they exist within the limits of geodomes composed of the corresponding and older structural layers as well as under diwa-type sediments in diwa basins, and so on.

Finally, regularities in the distribution of tectonic regions can result in regularities in the distribution of mineral deposits.

The above-mentioned regularities in the temporal and spatial distribution of mineral deposits can serve as one of the guides to the search for and prognosis of minerals.

5. Control over and reworking of mineral deposits by tectonic elements

(1) Control over mineral deposits by structural features of various tectonic elements

Tectonic elements of different character and types differ in structural features. Mineral deposits of different tectonic type, therefore, are generally controlled by the corresponding or associated structural features. For example, chromite deposits associated with ultrabasic rocks are generally controlled by fractures closely related to various tectonic elements, especially to the earlier period of a geosynclinal stage, and their attitude and trend. The Yarlung Zangbo River deep fracture zone may be cited as example. The depositional features and trend of geosynclinal-type sedimentary deposits, e.g., geosynclinal-type deposits of iron in Hunan, are controlled by the shape and trend of geosynclinal troughs and affected after their formation by the type, attitude and trend of geosynclinal -type folds and faults. Gold-bearing quartz veins of gold deposits formed in the geosynclinal stage by the Caledonian igneous intrusion in some gold ore fields of Guangxi are controlled by thrusts at the reversed limbs of geosynclinal-type reversed isoclinal folds. Diwa-type endogenic deposits, e.g., deposits of tungsten and tin in the Nanling Mountains, are controlled by block faults, folds and domes of open type (including structures formed as a result of reworking of pre-diwa strucutral layers by diwa tectonism).

Practice has proved that an analysis of structural features of various tectonic elements plays a certain role in metallotectonic researches. Therefore, historical analysis in combination with stress analysis will contribute to the prospecting for and exploration of mineral deposits.

(2) Control over and influence upon subsequent mineral deposits by pre-existing tectonic elements

Sedimentary, magmatic and metamorphic formations and structural features of pre-existing tectonic elements can more or less control and influence the mineralizations of subsequent tectonic elements in the same area. Their role may be summarized as follows.

1) Metallotectonic control. In platform and diwa regions, for example, the type (deep or ordinary fracture), character (tension, compression or shearing)

and attitude of fractures existing in the basement can exert control over and influence upon mineralizations occurring in the platform and even the diwa stage. For endogenic mineral deposits, as these inherited fractures control the location, area and trend of intrusions as well as the character of magmas (ultrabasic, basic, etc.), they indirectly determine the kinds and deposit types of related ores and regularities in their distribution. For exogenic mineral deposits, these ancient fractures, continuously developing, exert control over (or influence upon) the location of sedimentary ore-bearing basins and mineral deposits laid down in them, over the limits of their distribution, over the process of their development and metallogenic peculiarities, and over the formation of ore beds and their subsequent deformations.

2) Conditions of country rocks. For example, deposits of tungsten and tin formed during the Mesozoic diwa stage in southeast China are mostly hydrothermal fissure filling veins, if the geosynclinal structural layer characterized by flysch-like clastic sediments serves as a country rock, but are dominantly contact-metasomatic deposits, if the platform structural layer characterized by extensive carbonate rock serves as a country rock. If mother rocks are intruded into a structural layer composed of clastic and carbonate rocks, both of the above-mentioned two deposit types can exist. Such deposits have been known, for example, in the Nanling Mountains. Contact-metasomatic iron deposits in Hebei Province, some iron (and copper) deposits in the lower reaches of the Changjiang (Yangtze) River and other contact-metasomatic deposits have also been lithologically controlled by Ordovician or Triassic limestones within the platform structural layer. The reason why skarn iron deposit is one of the important deposit types in diwa regions is that it is generally endowed with such historical conditions.

(3) Reworking of pre-existing mineral deposits by subsequent tectonic elements

Tectonic elements appearing after the formation of mineral deposits can rework them from all quarters. The result of reworking may be profitable or harmful. The most common reworking is as follows.

1) Reworking by structural features. In diwa regions, for example, mineral deposits formed during the platform, geosynclinal and older stages may be reworked, after entering into the diwa stage, by diwa tectonism, and no longer keep their original structural features. Mineral deposits in the basement may generally be reworked by block-faulting. Mineral deposits in the platform structural layer may be reworked by both folding (chiefly of open type) and faulting (frequently block-faulting). As a result of reworking by structural features, pre-existing mineral deposits sometimes may be metamorphosed into rich ores. Reworked by diwa tectonism, platform sedimentary deposits, in particular, have no longer kept their originally flat, broad, gentle and areally distributed platform-type structural features and ore beds, but appear to be very complex and sometimes to be overthrust by older strata or broken into many fragments, locally metamorphosed (especially dislocationally-metamorphosed) and enriched. We may cite iron deposits of Ningxiang type and coal beds of Permian age in south China as their examples.

2) Reworking by magmatic activity. This may be illustrated by sedimentary iron deposits of two tectonic types in south China reworked by diwa magmatic activity.

The first type is iron deposits of Sinian age formed at the geosynclinal stage.

In a certain area of Hunan Province, the iron deposits are represented by hematite and magnetite-hematite; the latter contains a very low proportion of magnetite amounting to 10% (caused probably by regional metamorphism at the geosynclinal stage). But in another area of Hunan Province, where iron deposits of this type have experienced the influence of diwa-age granitoids, the proportion of magnetite has markedly increased, its grains have coarsened and the iron content correspondingly increased. Iron deposits of the same type in a certain area of Guangxi Zhuang Autonomous Region have also been enriched by diwa magmatic activity: where the deposits are cut across by quartz veins, the iron content has generally increased. Iron deposits in a certain area of Guangdong Province were originally metasedimentary ones of geosynclinal type, but have been affected by diwa-age granite intrusions to a significant extent, which has played an important role in increasing the rion content.

The other type is sedimentary iron deposits of Devonian age formed at the platform stage. Affected by diwa igneous activity taking place in Mesozoic time, some iron deposits in a certain area of Hunan Province underwent metamorphism and reduction resulting in partial turning of hematite into magnetite. Such reworking has relatively enriched the deposits and increased the iron content. But iron ore beds in other places not reworked this way have a low iron content. Reworking of pre-existing mineral deposits by subsequent magmatic activity sometimes gives rise to new ore deposits (compound ore deposits) due to combination of ore-forming materials from two or many stages.

6. Polygenetic formation of mineral deposits

In view of the above-mentioned close relations between mineral deposits and tectonics, many mineral deposits (especially in diwa regions) are believed to be multistage, polygenetic and derived from more than one sources of material. The author called them polygenetic compound ore deposits (see below). For example, metasedimentary iron deposits in Liaoning Province originally were sedimentary iron-poor deposits. Later, they passed through some tectonic stages and underwent intense tectonism, metamorphism and strong magmatism in the geosynclinal and the diwa stage, with introduction of ore-forming material. Thus the deposits show hydrothermal minerogenic features as we see them today. The iron deposits in a certain area of Inner Mongolia originally was a sedimentary deposit, but was affected, after its formation, by magmas with a high content of alkalis, thus changing into a polygenetic compound ore deposit rich in other elements. The iron deposits in a certain area of Guangdong Province originally was a sedimentary deposit of early Paleozoic age accumulated in the early period of the geosynclinal development. It was reworked by metamorphic processes occurring in the late period of the geosynclinal development, and again affected by magmatic processes operating in the Mesozoic diwa stage. The primary ore of the siderite deposit in a certain area of Hubei Province was formed in sedimentary processes taking place in the late Paleozoic platform stage, but the great part of the deposit originated from magmatic processes occurring in the Mesozoic diwa stage. In most cases it can be seen that a deposit not only shows many features characteristic of a sedimentary or metasedimentary deposits, but also displays evidence of an endogenic deposit. That is why the problem of its genesis has hotly been debated for years and can not be solved. Therefore, from the point of view of "progressive mineralization", one must make a comprehensive tectonic analysis of geological data when he studies the genesis of mineral deposits.

II. Tectonics and endogenic mineral deposits of China[1]

Tectonics is one of the important conditions of endogenic mineralizations. Magmatic activity and associated mineralization are conditioned by tectonic regimes at different tectonic stages, while migration, dispersion and concentration of ore-forming elements, by structural features, country rocks and geochemical factors of different tectonic types. The former is a time factor of the formation of endogenic mineral deposits, whereas the latter provides space conditions for the formation of endogenic mineral deposits. Here we give a brief description of the problem.

1. Endogenic mineralizations in China and the temporal and spatial distribution of mineral deposits formed by them

In the process of the multistage development of the earth's crust, endogenic mineralizations evolve along with crustal movements and magmatic activities taking place in various geological periods. The whole geologic time in Chinese territory can be divided into eight main tectono-magmatic-metallogenic epochs, namely, Taishan, Wutai, Luliang, Jinning, Caledonian, Hercynian (subdivisible into three sub-epochs), Pacific (including Indosinian, early and late Yanshanian sub-epochs), and Himalayan.

Because of the unevenness of crustal development, every tectono-magmatic-metallogenic epoch varies from region to region in tectonic character, magmatic activity, mineral kinds and deposit types. A summary of the characteristics of the main tectono-magmatic-metallogenic epochs and associated minerals in various tectonic regions of China is given in Table 5.2.

Generally speaking, in the history of tectono-magmatic-metallogenic development in China, the Archeozoic era was mainly an epoch of magmatic mineralizations of pre-geosynclinal old tectonic elements. The Proterozoic to Paleozoic era was an epoch of predominantly geosynclinal-type magmatic mineralizations; in the late period of this epoch platform-type magmatic mineralizations appeared. The Mesozoic to Cenozoic, especially the Mesozoic, was an epoch of predominantly diwa-type magmatic mineralizations, but at the same time geosynclinal-type magmatic mineralizations still occurred. Of all the tectono-magmatic-metallogenic epochs the most important is the Pacific epoch. The principal endogenic deposits of non-ferrous and rare metals in China were formed mostly in this epoch and produced mainly in diwa regions. Next come the Caledonian and Hercynian epochs, with both geosynclinal- and platform-type magmatic mineralizations. Mineral deposits were also formed in other epochs and each epoch offers its own metallogenic characteristics.

The unevenness of crustal development and the inequality of tectonic stages in different regions have also led to the uneven spatial distribution of endogenic ores. But they still show a regular trend in their evolution and differentiation in Chinese territory, because they are conditioned by the distribution of tectonic regions and by the regularities in their development.

---

1. Liu Yixuan.

Table 5.2. Principal tectono-magmatic-metallogenic epochs of various tectonic regions in China and related endogenic mineral deposits

Epoch			Igneous rock types and related mineral deposits		
Tectonic Region			Geosynclinal region	Platform region	Diwa region
Cenozoic	Himalayan	Late (Q)	Basalt and andesite. Sulphur.	Basalt	Basalt and andesite. Cobalt and iron.
		Early (R)	Ultrabasic rocks, quartz diorite, granite and plagioclase granite; basalt, andesite, tuff, agglomerate, quartz porphyry. Chromium, gold, copper, iron, lead, zinc, mercury, silver, sulphur.	Basalt	Biotite granite, muscovite biotite granite, basalt, andesite, agglomerate, tuff. Copper, lead, zinc, gold, barite.
Mesozoic	Yanshanian	Late ($K-E_1$)	Peridotite, granite, granodiorite, pegmatite, hypabyssal porphyries; quartz porphyry, dacite, andesite, tuff, basalt. Chromium, nickel, platinum, copper,	Granitoids and acid to intermediate lavas. Iron, molybdenum, copper,	Biotite granite, granodiorite, diorite, monzonite, gabbro, peridotite, alkaline intrusive rocks, kimberlite, acid to intermediate volcanic and alkaline effusive rock. Antimony, mercury, gold, tungsten, tin, molybdenum, copper, iron, lead, zinc, manganese, chromium, nickel, vanadium, tantalum, rare elements,

Mesozoic	Yanshanian	Late ($K-E_1$)	iron, antimony, lead, zinc, tungsten, tin, rare metals, sulphur, barite, talc, jasper, lizardite, rock quartz, piezo-electric quartz.	lead, zinc, gold.	sulphur, arsenic, pyrophyllite, alunite, fluorite, barite, rock quartz, piezo-electric quartz, diamond.
		Early (J)	Granite, granodiorite, diorite, alkaline rocks; tuff, liparite, tholeiite. Iron, copper, molybdenum, lead, zinc, lithium, beryllium, mica, arsenomarcasite.	Small alkaline bodies, granitoids. Copper, lead, zinc, iron.	Biotite granite, granodiorite, basic-ultrabasic rocks, alkaline intrusive rocks, acid to intermediate and alkaline volcanic rocks, monzonite, quartz porphyry. Tungsten, tin, bismuth, molybdenum, iron, copper, lead, zinc, beryllium, niobium, tantalum, rare elements, chromium, nickel, titanium, platinum, asbestos, pyrophyllite, alunite.
		Undivided (J-E)			Alaskite, plagioclase granite, granite, granite porphyry, basic rocks. Molybdenum, copper, lead, zinc, gold, vanadium - titanomagnetite, fluorite, rock quartz.
	Indosinian	(T)	Granitoids, gneissose granite, basic effusive rocks. Iron, copper, chromium, nickel, sulphur, mica, asbestos.		Biotite granite, quartz diorite, granite porphyry, gabbro. Iron, rare metals (niobium, tantalum, beryllium, etc.), copper, chromium, nickel, platinum, lead, zinc, gold, mica, asbestos.
		Late (P)	Basic-ultrabasic rocks, granite,	Basalt, gabbro,	Granite, alaskite, basic-ultrabasic rocks, intermediate-basic volcanic

Late Paleozoic	Hercynian	Late (P)	quartz syenite, basalt, andesite. Chromium, nickel, diamond, non-ferrous and rare metals.	Copper, iron and cobalt.	rocks, basalt. Iron, rare elements, tungsten, tin, molybdenum, lead, zinc, chromium, copper, nickel, vanadium, titanium, beryllium, diamond, asbestos, plogopite, rock quartz.
		Middle (C)	Granitoids, tuff, andesite, intermediate to basic volcanic rock. Rare metals.		Biotite granite, granodiorite, basic and ultrabasic rocks. Iron, vanadium, titanium, copper, lead, zinc, chromium, nickel, magnesite, mica, rock quartz (piezo-electric quartz).
		Early (D)	Diabase, andesite, quartz porphyry, felsic rocks, tuff, agglomerate, basalt. Chromium, asbestos, lizardite.		Granite, granodiorite, basic-ultrabasic rocks, intermediate to basic volcanic rocks. Chromium, copper, nickel, gold.
Early Paleozoic	Caledonian	Late (S)	Granite, granodiorite, quartz diorite, intermediate to basic intrusive and effusive rocks. Chromium, asbestos, lizardite.		Biotite granite, granodiorite. Gold, lead, zinc, copper, iron, tungsten, tin, antimony, rare elements.
		Early(Cm-O)	Granite, acid-basic effusive rocks. Copper, gold, sulphur.		Granite, diorite, pegmatite, basic-ultrabasic rocks, intermediate to basic volcanic rocks. Chromium, nickel, platinum, iron, copper asbestos.

Sinian	Jinning	Pegmatite veins, biotite hornblende gneiss, magmatic complex, acid-basic effusive rocks and spilite kerato-phyre. Tungsten, tin.	Alkaline rocks. Asbestos, graphite.	Granite, diorite, pegmatite, basic-ultrabasic rocks, intermediate to basic volcanic rocks. Copper, nickel, iron, cobalt, gold, lead, zinc, tin, tungsten, antimony.
Early Proterozoic	Luliang		Metapegma-tite gra-nite, gran-ite porphy-ry. Lead, zinc.	Granite, granodiorite, granite porphyry, basic-ultrabasic rocks. Gold, copper, molybdenum, iron, lead, zinc, antimony, tungsten, tin, nickel, chromium, asbestors.
Late Archeozoic	Wutai			Granite, granodiorite, granite porphyry, pegmatite, basic-ultrabasic rocks. Copper, chromium, nickel, iron, titanium, gold, lead, zinc, rare elements, talc, asbestos, magnesite.
Early Archeozoic	Taishan		Granite gneiss, metavol-canics.	Granite, granodiorite, pegmatite. granulite, migmatite-granite, basic-ultrabasic and effusive rocks. Iron, copper. zinc, nickel, chromium, gold, rare elements, boron, asbestos, corundum, sillimanite, mica.

(1) Characteristics of endogenic mineralizations in east China

Diwa mineralizations occurring in the Pacific (especially Yanshanian) magmatic-metallogenic epoch are the most prominent. The important deposits of tungsten, tin, bismuth, molybdenum, copper, lead, zinc, antimony, mercury, iron and rare metals in China were all formed in this epoch. Next come geosynclinal-type mineral deposits of the Caledonian, Jinning and Luliang epochs which are less prominent than the former. The reason is that diwa regions of Cathysian type characterizd by intense magmatic activity had a wide development in east China. Diwa-type magmatic mineralizations lasted from the Late Triassic to the Paleogene or later. At first they evolved from weak to strong, and the sphere of their influence progressively increased. In the Middle and Late Jurassic and Cretaceous, along with the stepping of the diwa regions into the maximum-mobility period, the mineralizations reached their climax. But this evolution tended to delay from south to north and from west to east. For example, magmatic mineralizations in the SE diwa region began in the Late Triassic; in the North China diwa region, in the Early Jurassic; and in the NE diwa region, in the Early-Middle Jurassic. Their climax was also delayed from south to north. From the Tertiary on, magmatic activity in east China gradually weakened and decreased in areal extent. It generally began with acid or acid-to-intermediate, culminated with acid and ultraacid, and ended with basic rocks. These rocks often contain many alkali metals. Volcanism becomes stronger and stronger as it moves towards the Pacific coast. In areas of volcanic activity, extrusion and intrusion generally accompany each other, reflecting the same source of magma. Geosynclinal-type magmatic mineralizations in this area tended to be delayed from north to south, with the exception of the NE diwa region (Hercynian age), in the following sequence: the North China diwa region and the Yishan (Ordos) platform region (Luliang age), the larger part of the Central China diwa region, Yunnan, Guizhou, and Sichuan (Jinning age), the SE diwa region and the eastern Qinling Mountains (Caledonian age). The above-mentioned features of crustal development are associated with the peculiarities of the diwa and geosynclinal evolution systems in this area (see Chapter 4).

Diwa-type endogenic deposits in east China linked with volcanic rocks and found in the diwa structural layer were widely distributed in this area; those associated with intrusive rocks exist in all structural layers. They are controlled mainly by the following tectonic systems: the eastern wing of the North Xinjiang-Hinggan arc, E-W Yinshan tectonic system, Qinling-Huaiyang arcuate tectonic system, E-W Nanling tectonic system, and NE tectonic systems widely distributed in the whole area. Geosynclinal and pre-geosynclinal endogenic deposits are seen in the basement exposed or concealed in this area. They are also controlled by the above-mentioned tectonic systems, but the concrete conditions are not alike.

(2) Characteristics of endogenic mineralizations in west China

In this area geosynclinal mineralizations occupy an important position. They were recorded already in the Proterozoic, and took place successively in the Luliang, Jinning, Caledonian, Hercynian, Pacific, and Himalayan orogenies. With the appearance of the initial-mobility and especially the maximum-mobility period of geosynclinal development, a great many endogenic mineral deposits formed. Such magmatic metallogenic activity has not come to a close yet. The most widespread are geosynclinal-type mineral deposits of the Hercynian tectono-magmatic-metallogenic epoch. Besides, there are also geosynclinal-type mineral deposits of the Caledonian and the Pacific epoch. This is the principal area of distribution of geosynclinal-type endogenic deposits in China, and important deposits of chromium, nickel, cobalt, vanadium, titanium, copper, iron, lead, zinc and gold are distributed mainly in this area. The reason is that the geosynclinal stage in this area lasted long and ended late. Eugeosynclines marked by particularly intense magmatic activities have a wide development, and the geosynclinal structural layer outcrops extensively.

Geosynclinal-type endogenic mineral deposits in west China are distributed in the extensively outcropping geosynclinal structural layer and concentrated to form some geosynclinal-folded metallogenic belts (partly affected by diwa fault-fold belts). Such belts have a parallel arrangement, a progressive decrease in age and a trend chiefly in a WNW or NW direction, and bend partly like an "S" shape. These are: the Altay, Tianshan, Altun-Qilian, Kunlun-western Qinling Mountains, the Bayanhar and Yarlung Zangbo River-Himalayan geosynclinal regions. Among them the Altay and the northern Tianshan belong to the western wing of the North Xinjiang-Hinggan arc, the southern Tianshan to the E-W tectonic systems and the others to the NW tectonic systems. The age of these geosynclinal-folded metallogenic belts shifts as does the age of the geosynclinal-type magmatic mineralizations in east China, i.e., it is delayed also roughly from north to south, in the following sequence: Qilian Mountain (Caledonian age), the Kunlun Mountains (Hercynian age), the Bayanhar Mountains (Indosinian age), the Gandise Range (Yanshanian age) and the Yarlung Zangbo River-Himalayas (Himalayan age), with the exception of the Altay and Tianshan (Hercynian age). Diwa magmatic activity reached its climax probably in the Himalayan orogeny and was controlled by structural lines similar to geosynclinal-type ones but expressed by diwa fault-fold belts.

The N-S diwa region lying between east and west China is a transition zone. Mineral deposits formed by geosynclinal mineralizations of Luliang, Jinning, Caledonian, Hercynian and Pacific ages, by platform mineralizations of post-Luliang, post-Jinning, post-Caledonian, post-Hercynian ages, and by diwa mineralizations of Pacific and Himalayan ages occur in this zone, where deposits of nickel, iron, copper, chromium, lead, zinc, vanadium, titanium, tin and tungsten coexist. Ore-controlling structural lines are first the N-S and next the NW and NE tectonic systems intersected by the former.

2. A brief description of the principal magmatic-metallogenic formations of various tectonic elements in China

As stated above, different tectonic elements differ in tectono-magmatic mineralizations and yield different magmatic-metallogenic formations. Each of these formations offers its own peculiarities. The principal features of the magmatic-metallogenic formations of these different tectonic types in China are listed in Table 5.3.

Emphasis is put on the characteristics of some common magmatic-metallogenic formations in various tectonic elements as follows.[1]

---

1. By magmatic-metallogenic formations we mean a complex of igneous rocks-ores (or elements) genetically related to each other and formed at a definite tectonic stage and under a definite tectono-magmatic and geochemical condition.

Table 5.3. Metallogenic characteristics of geosynclinal, platform and diwa regions in China

Region	Ore-related magmatism	Structural control and distribution pattern	Country rocks	Metallogenic formations and deposit type	Specialized mineral associations	Magmatic-metallogenic epochs
Diwa	Intense. Chiefly alkaline and acid-to-intermediate rocks in the early, acid-to-intermediate as well as basic and meta-alkaline rocks in the middle, and basic effusive rocks in the late period.	Control by arch-faults in the early, by folds in the middle, and by fractures in the late period. Rocks occur in a short belt.	For intrusive ore formation: on positive units such as geodome: metamorphic, sedimentary and volcanic rocks in X, I, or II; in negative units such as diwa: sandstone, conglomerate, and volcanics in III. For volcanic ore formation: volcanics and volcanogenic sediments in III.	Acid rock polymetallic formation, calc-alkaline diorite-quartz diorite Fe-Cu formation in the early, acid granite W-Sn-Bi-Mo formation, acid-to-intermediate rock Fe-Cu formation, continental volcanic rock Fe-Cu formation and porphyry Cu formation in the middle, and basic volcanic rock sulfide formation in the late period. In X, I and II— chiefly hydrothermal, skarn and pegmatite types; in III—	Au, Cu, Pb, Zn, Fe, and rare elements in the early, W. Sn, Bi, Mo, Sb, Hg, Fe, Cu, Pb, Zn, and rare elements in the middle, and Cu, Pb, Zn, S, Au, Sb, alumstone, fluorite, pyrophyllite and barite in the late period.	In the eastern part the Pacific epoch, particularly the Yanshanian epoch, occupies first place; the Himalayan epoch comes second. In the western part the Himalayan epoch occupies first place; the late Pacific epoch comes second. In the N-S diwa region the most important is the Yanshanian epoch

Diwa				chiefly continental volcanite and hydrothermal type.		
Platform *	Weak. Occurs chiefly in the early or the late period and produces basic-ultrabasic and intermediate alkaline effusive and minor intrusions.	Control by inherited and marginal major deep fractures. Basic-ultrabasic rocks occur in a belt, but intermediate alkaline rocks in a ring.	For intrusive ore formation: on positive units such as swell--metamorphic and igneous rocks in X and I;in negative units such as basin--sandstone, shale, limestone and igneous rocks in II. For volcanic ore formation: volcanic and volcanogenic rocks in II.	Basalt Cu formation, basic-ultrabasic intrusive rock V-Ti-Fe formation, and alkaline rock Nb-Ta formation. In X and I: chiefly hydrothermal and magmatic types; in II: chiefly volcanic hydrothermal type,but magmatic type near major deep fractures.	V,Ti,Fe, Ni,Nb,Ta, apatite and talc associated with intrusive rocks;Cu, Fe,and rare elements associated with effusive rocks	The Hercynian epoch occupies first place, the Jinning epoch comes second, and the Pacific epoch predominates in marginal areas.
Geosynclinal **	Intense. Chiefly intermediate-to-basic submarine extrusion and basic-ultrabasic intrusion in the early,predominantly acid granite intrusion	Control by major deep fractures in the early, by folds in the middle, and by folds and faults in the late period.	For intrusive ore formation:on geanticlines--deeply metamorphosed rock in X such as schist and gneiss as well as igneous rocks in X and I; in geosynclines--slightly	Marine volcanic rock Fe-Cu formation, basic-ultrabasic rock Cu-Ni-Cr-Fe formation in the early, acid intrusive rock non-ferrous metal formation in the middle, and acid-to-intermediate	Fe,Cu,Cr, Ni,V,Ti, Pt and asbestos in the early,Au, Cu,Pb,Zn, W,Sn,Mo, Be,Nb and Ta in the middle,	In the western part the Hercynian and Caledonian epochs occupy first place; the Pacific and Himalayan epochs come second. In the eastern part the Luliang and Caledonian epochs

Geosynclinal **	in the middle, and mainly weak calc-alkaline intrusion and extrusion in the late period.	Rocks occur in a long belt.	metamorphosed rocks in I such as slate, phyllite and sandstone and igneous rocks. For volcanic ore formation: volcanic rocks and volcanogenic sediments.	volcanic rock polymetallic formation in the late period.	and Pb,Zn, Ag,Sb,Hg, As,Cu and Au in the late period.	occupy first place; the Jinning, Hercynian and Himalayan epochs come second. In the N-S diwa region the Jinning epoch occupies first place;the Caledonian,Hercynian and Indosinian epochs come second.

* Including paleoplatform regions.

** Including paleogeosynclinal regions.

(1) Magmatic-metallogenic formations of pre-geosynclinal type

Little is known about the magmatic-metallogenic peculiarities of old tectonic elements developed in Chinese territory at the pre-geosynclinal stage. According to available data, valuable mineral deposits formed at this stage are chiefly deposits of copper, nickel, gold, zinc and iron. We cite the following important magmatic-metallogenic formations as examples.

1) Copper-bearing metamorphic acid to intermediate volcanic and metamorphic granodiorite porphyry formation. The mineral deposits occur mainly as disseminations and stringers of copper ores. They are found in Archean rocks of the pre-geosynclinal structural layer (Jiangxian Group) in some zones of the North China diwa region and confined to metamorphic granodiorite porphyry between schist and metamorphic acid to intermediate volcanics or to the contact with the enclosing volcanics. The deposits have been controlled by fractures. Ore minerals are chalcopyrite, bornite, chalcocite, molybdenite, etc., which occur as disseminated grains on a large scale. Archean rocks, volcanics and subvolcanics have a wide distribution in the North China diwa region. Copper deposits of this type can be found in areas with similar geologic conditions.

2) Copper- and zinc-bearing marine volcanic granulite formation. This formation is seen in strata of probably Archean age, equivalent to the Anshan Group, in areas of outcropping of the pre-geosynclinal structural layer in the North China diwa region. Copper and zinc ores occur in rocks composed of biotite plagioclase gneiss, biotite granulite and amphibolite. The mineralization is genetically related to pegmatite. Copper is rich, but zinc is richer than copper. They are accompanied by gold, silver, cobalt, gallium, cadmium, selenium and rhenium, as is seen in a certain Cu-Zn deposit in Liaoning Province. The Archean granulite formation has a wide distribution in north and northeast China, and mineral deposits of this type can be found in areas with similar geologic conditions.

3) Copper- and nickel-bearing basic gabbro and gabbro-norite formation and gold-bearing metamorphic acid granite formation. The former is magmatogenic and contains copper and nickel with associated vein-like copper, iron and gold mineralizations. The latter contains a great amount of gold-bearing quartz veins and copper mineralization. Both the formations are seen in the Taishan Group of Archean age in Shandong Province. Mineral deposits of these two types will be found in the pre-geosynclinal structural layer of north, northeast and southeast China.

(2) Magmatic-metallogenic formations of geosynclinal type

Magmatic-metallogenic formations of geosynclinal type in Chinese territory came into being mainly in the early period or at the beginning of the late period of geosynclinal development. In the early period occurred chiefly iron and copper polymetallic sulfide ores and chromium, nickel and platinum deposits. They are associated with submarine intermediate-basic extrusion and basic-ultrabasic intrusions accompanying the activity of deep fractures in the process of strong differential vertical movements between geosynclines and geanticlines. In the late period there principally formed non-ferrous metals and polymetallic deposits. They are associated with acid-intermediate magmatic intrusions accompanying the closure of geosynclines and the folding of sediments. Among them iron and copper deposits associated with intermediate-basic marine volcanics, chromium, platinum and nickel deposits associated with basic-ultrabasic (especially magnesium) rocks, and primary gold veins are the most typical.

1) Magmatic-metallogenic formations in the geosynclinal period. The most important magmatic-metallogenic formation of geosynclinal type in this period is marine volcanogenic iron-copper formation. It can be subdivided into two principal subformations.

a. Copper- and iron-bearing spilite-keratophyre subformation, including copper-polymetallic and copper-iron deposits. It develops in eugeosynclines. Orebodies occur in volcanic rocks of the geosynclinal structural sublayer, associated with a spilite-keratophyre formation made up by marine natrium-rich intermediate alkaline-basic volcanic extrusions. Ore-bearing country rocks are metamorphic quartz-keratophyre, tuffs, volcanogenic terrigenous rocks and basic volcanics. Ore can be divided into two types: One of them contains pyrite and chalcopyrite as well as lead, zinc and tin, the associated elements being gold, silver, selenium, cadmium, germanium, nickel, etc. Copper pyrite of Baiyin type is its example. The other contains magnetite and chalcopyrite, and sometimes concurrently bears sedimentary mineralizations (see the third paragraph of this chapter). The magmatic-metallogenic subformation is known in Gansu, Shaanxi, Sichuan, Yunnan, Henan, Zhejiang, etc. The age of the mineralizations is chiefly Proterozoic to early Paleozoic. A certain copper deposit in Gansu Province and a certain copper deposit in Yunnan Province are their examples. Such deposits may be found in areas with similar geologic conditions in deep fracture zones of the northern Qilian, northern Qinling, Dabie, Longmen Mountains and Sichuan, Yunnan and west Zhejiang Provinces.

b. Iron- and copper-bearing greenschist subformation. It develops in eugeosynclinal areas of intermediate-basic and acid-intermediate volcanogeno-sedimentary rocks. The source of ore-forming material is mainly related to these volcanic effusives. The ore-bearing rocks underwent slight to moderate metamorphism. They are chiefly metamorphic intermediate-basic and acid-intermediate volcanics, metadiabase, gabbro-diabase, albite-diabase, albitite, amphibole-schist, plagioclase-schist, etc. Iron deposits of magnetite type occur mostly in intermediate-basic volcanic rocks and are accompanied by copper, vanadium, titanium, tin, zinc and other elements. Copper deposits occur mainly in plagioclase-schist, granodiorite and greenschist. The main ore minerals are pyrite, chalcopyrite and molybdenite in association with gold, silver, sulphur, selenium and rhenium, as is seen in a certain copper deposit in Inner Mongolia. This metallogenic subformation is known in Yunnan, Sichuan, Xinjiang and Inner Mongolia. The age of the mineralizations is Silurian, Carboniferous and Permian respectively. Ore deposits of this type may be found in areas with similar geologic conditions, especially on the borders of deep fractures.

2) Magmatic-metallogenic formations in the geosynclinal period-fold (belt) period. These formations are represented by a copper-, nickel-, chromium- and iron-bearing basic and ultrabasic formation, which occurred mostly in the geosynclinal period, but appeared also at the beginning of the fold (belt) peirod. It is a magmatogenic deposit, controlled by major deep fractures, and subdivisible into two subformaitons.

a. Copper- and nickel-bearing ferriferous basic and ultrabasic subformation. It is related to submarine basic and ultrabasic extrusive-intrusive complexes originating in the geosynclinal period. These complexes consist mainly of pillow spilite-keratophyre and basalt accompanied by gabbro and diabase intruded along bedding planes and norite, peridotite, pyroxenite and peridotite-norite cutting diagonally bedding planes. Ore minerals are magnetite, pentlandite, chalcopyrite and nickel- and cobalt-bearing pyrite and sperrylite. Ore bodies occur generally in the upper part of the bottom of plutonic bodies or in veins. Examples are copper and nickel deposits in some areas of Sichuan, Yunnan and Gansu Provinces. In south Sichuan and central Yunnan Provinces, in the Yinshan, Qilian, Qinling, Yanshan Mountains and other areas with similar geologic conditions, mineral deposits of this type are expected to be found in zones of distribution of ferriferous basic and ultrabasic complexes controlled by major deep fractures.

b. Chromium-bearing magnesian basic and ultrabasic subformation. It occurs mostly in major deep fracture zones (especially in those serving as the boundary of major tectonic regions), has a zonal distribution and shows a linear arrangement. According to the relation of its generation to periods of geosynclinal development, the basic and ultrabasic rocks bodies may be roughly divided into two kinds.

(a) Rock bodies formed in the geosynclinal period and seen in intermediate-basic submarine effusives and basic and ultrabasic intrusives. They are mostly magnesian, rich in chromium ores and known along the Yarlung Zangbo River. Parts of the Caledonian and Hercynian rock bodies occurring along major deep fracture zones in the Tianshan, Yinshan, Altun, northern Qilian, western Qinling, Translongmen, Kunlun, Helan Mountains may contain valuable ores.

(b) Rock bodies formed in the fold (belt) period and seen in intrusions along fractures in fold belts. They are mostly ferriferous. When basic magma is well differentiated, there may occur magnesian differentiates containing chromium, platinum and nickel. Such ore deposits are expected to be found in Shaanxi, Gansu, Qinghai, Xinjiang, Sichuan, Yunnan, in rock bodies of this kind intruded along fracture zones in geosynclinal fold belts of Hercynian, Caledonian, Jinning and Pacific ages.

3) Magmatic-metallogenic formations in the fold (belt) period.

a. Iron- and copper-bearing metadiabase-gabbro formation producing hydrothermal-metasomatic iron deposits and bedded stringers and disseminations of copper. Iron and copper deposits formed in Jinning age are their examples.

b. Vanadium-bearing gabbro-plagioclasite titanomagnetite formation producing late-magmatic deposits. The iron deposit in a certain place of Hebei Province, formed in Luliang age, is its example.

c. Lead- and zinc-bearing acid granite formation.

d. Tungsten-, tin-, molybdenum- and beryllium-bearing granite-granodiorite formation producing high-temperature hydrothermal tungsten deposits, hydrothermal sulfide-tin deposits, and skarn deposits of tungsten and molybdenum.

e. Polymetallic ore-bearing acid to intermediate volcanic rock formation producing sulfide deposits of lead, zinc, gold, silver and copper, which are seen in the five active geosynclinal regions and the paleogeosynclinal fold belts in west China.

f. Gold-bearing acid rock formation yielding gold-bearing quartz veins. It is the most important industrial type of gold deposits in China and one of the most principal characteristic features of endogenic metallizations in geosynclinal regions. Gold veins in Hubei, Hunan, Jiangxi and north Xinjiang and gold deposits in northeast China are the products of granitic intrusion in paleogeosynclinal regions. The Jinguashi and Ruifang gold deposits in Taiwan Porvince are examples occurring in an active geosynclinal region. Gold-bearing quartz veins strictly controlled by fractures have a complex morphology and appear in groups. Country rocks are mostly schists and slates, and contain a small amount of sulfides. Silicification, pyritization and chloritization are the principal alterations, and gold content varies greatly.

(3) Magmatic-metallogenic formations of platform type

These formations were formed in the early and late periods of platform development in the process of long-term vertical crustal movements affecting a large area. They originated during the extrusion of platform-type magmas under the action of major deep-seated fractures. The following is some typical examples of ore-bearing formations occurring in the residual-stability period of platform development.

1) Iron- and copper-bearing basalt formation. This formation is known in the southwestern part of the Sichuan platform region as well as in Sichuan, Yunnan and Guizhou Provinces. Mineral deposits occur in the Permian Emeishan basalt of the platform structural layer and formed within the variegated tuff breccia and gray compact massive basalt. They are hydrothermal bedded iron and copper deposits. Ore minerals consist chiefly of chalcosine and less of chalcomiclite, chalcopyrite and tetraedrite, with associated native copper, magnetite, titanioferrite and hematite. In recent years large iron deposits have been discoverd. The basaltic volcanism has been controlled by N-S-trending major deep-seated fractures and related to the rejuvenation of the fractures during the Hercynian movement in the residual-stability period of platform development. In many areas of China basic lavas poured out along fissures in the initial- and residual-stability periods of platform development. Such geologic conditions are favourable for the search for ore deposits of this type.

2) Titanium- and iron-bearing basic and ultrabasic rock formation. It is situated chiefly in Sichuan and Yunnan Provinces and related to basic and ultrabasic rocks such as gabbro, norite and peridotite emplaced in the residual-stability period of the platform stage. The rocks are characterized by rich TiO_2 and relatively high content of alkali metals. Ore deposits were formed by magmatic segregation and controlled by high-angle tensile-shear fractures, the Tangdan major fracture and the Anning River deep fracture. The age of emplacement of these ore deposits is approximately the same as the Permian basalt in the adjacent area, indicating that internal relations must exist between them. Mineral deposits of this type may be of considerable extent and can be found in areas of basic and ultrabasic rocks controlled by major deep fractures in Sichuan and Yunnan Provinces.

(4) Magmatic-metallogenic formations of diwa type

Diwa regions have a wide distribution in China. Magmatic-metallogenic formations of diwa type are well-developed and rich in mineral deposits with varied types. Mineralizations took place in all of three periods of diwa development. So far as east China is concerned, the initial-mobility period saw ore deposits of gold, copper, lead, zinc and rare elements associated with granodiorite, granite and muscovite-biotite granite and ore deposits of iron and copper associated with calc-alkaline diorite and quartz diorite to accompany arcogenesis, faulting and folding. The maximum-mobility period was marked by ore deposits of tungsten, tin, bismuth and molybdenum associated with biotite granite, ore deposits of iron and copper associated with granodiorite-diorite-granite, ore deposits of iron and copper associated with calc-alkaline intermediate to basic and acid to intermediate continental volcanics, and porphyry copper-molybdenum deposits to accompany faulting and folding. The residual-mobility period was characterized by deposits of copper, lead, zinc, gold, silver, alumstone and barite to accompany faulting. Generally speaking, the most typical are non-ferrous, rare and disseminated metals associated with acid and intermediate rocks, which reached a climax in the maximum-mobility period. Below we cite some important examples with brief descriptions.

1) Magmatic-metallogenic formations in the initial- and maximum-mobility periods.

a. Tungsten-, tin-, bismuth- and molybdenum-bearing biotite granite formation.

It is chiefly a pneumatolytic hypothermal deposit of wolframite-quartz vein type. The most famous deposit was formed in the Nanling Mountains during the early Yanshanian orogeny and is known as ore deposit of Nanling type. It is also distributed in small amounts in north China. In the Nanling Mountains, it is widely developed in biotite granitic intrusion and its country rocks. Of the five magmatic phases of Yanshanian age occurring during the maximum-mobility period of diwa development in south China, the first, second and fourth, especially the second, appear to be the principal metallogenic stages. Country rocks are chiefly pre-Devonian metamorphosed sandstones and shales within the geosynclinal structural layer and partly pre-existing granites or Paleozoic strata within the platform structural layer. The dominant minerals are wolframite, powellite, cassiterite, molybdenite, bismuth glance and native bismuth associated with beryllium, lithium, niobium, tantalum and various disseminated elements. Ore bodies are characterized by a dense concentration and a deep extension. Ore-controlling structures trend N 70° - 80° E, but the strike of ore veins is generally related to NNE-or-NNW-trending fractures.

This formation also includes ore deposits of cassiterite-quartz vein type, cassiterite-scheelite (or wolframite)-quartz vein type and bismuthine-native bismuth (containing also tungsten, tin, and molybdenum)-quartz vein type. Ore deposits of these types sometimes contain many rare, rare-earth and disseminated elements, thus forming ore deposits of various subtypes. The metallogenic formation occurs in east China, in the N-S and West Yunnan diwa regions, and in the SE diwa region as the most promising area.

b. Tin-, tungsten-, molybdenum- and beryllium-bearing granite-granodiorite formation.

If forms skarn deposits of tin, tungsten, molybdenum and beryllium occurring as a single ore deposit named skarn (tin) deposit of Gejin type. It is related to granite, granodiorite, granite porphyry and quartz porphyry emplaced in the early Yanshanian orogeny during the maximum-mobility period of diwa development, and occurs at the contact between intrusives and Paleozoic limestones within the platform structural layer. Ore-controlling structures trend NNE or NNW and lie generally at the edges of diwa basins adjacent to geodome uplifts. The metallogenic formation has a great ore potential in the Nanling, Yanshan and eastern Qinling Mountains, in east Yunnan and south Liaoning Provinces and in the middle and lower reaches of the Changjiang (Yangtze) River.

c. Iron- and copper-bearing granodiorite-diorite-quartz diorite formation represented by skarn deposits of iron and copper.

Skarn deposits of iron. These deposits occur chiefly in areas of carbonate rocks (mostly in the platform structural layer, seldom in the geosynclinal, pre-geosynclinal and diwa structural layers), associated with intermediate or weakly acid small rock bodies, such as diorite, quartz-monzonite, granodiorite and its porphyry. Ore minerals are magnetite, hematite, pyrite, and chalcopyrite. Iron ore bodies frequently occur in groups in contact zones between intrusives and country rocks. Among them those associated with diorite occasionally contain copper, cobalt and sulphur as seen in the Yanshan-Liaoning area and in the middle and lower reaches of the Changjiang (Yangtze) River. Those associated with granites have a more complex mineralogical composition: tin, lead, zinc, etc., as seen in Inner Mongolia, north Hebei and south China. There may be formed compound ore deposits resulted from repeated activities of hot solution, e.g., iron ore deposits in a certain area of Hubei Province. Along E-W tectonic zones and NNE or NE-trending fracture zones of the North China, Central China, SE and NE diwa regions, iron ore deposits are expected to be found in areas with similar geologic conditions.

Skarn deposits of copper. These deposits spread most widely in the middle and lower reaches of the Changjiang (Yangtze) River and occur often as rich ones. Ore-forming mother rocks are diwa-type granodiorite, quartz diorite, monzonite and diorite intruded during the Yanshanian orogeny and occurring as hypabyssal stocks and small intrusives. The dominant ore mineral is chalcopyrite which is located at the contact between these intrusives and carbonate rocks belonging to the platform structural layer. Controlled obviously by fractures, these copper ore deposits generally occur in groups at the edges of geodome uplifts, especially at the intersections of sub-latitudinal with NNE-trending fracture zones. Along fracture zones of the middle and lower reaches of the Changjiang (Yangtze) River, in the Nanling and eastern Qinling Mountains and in the Yanshan-Liaoning area, copper deposits of this type may be of value in areas with similar geologic conditions (Figs. 5.1 and 5.2).

d. Iron- and copper-bearing continental volcanic rock formation.

This formation is one of the most important types of rich iron and copper deposits in China, distributed chiefly in continental volcanogenic basins of Jurassic and Cretaceous ages along the east coast, and controlled by more than two sets of fractures. It may be divided into two subformations.

(a) Iron-bearing continental intermediate to basic and acid to intermediate (intermediate alkaline) volcanic rock subformation. It was formed in the maximum-mobility period of diwa development and associated with continental intermediate to basic and acid to intermediate andesite, trachyte, trachyandesite and alkaline to intermediate subvolcanic rocks of the Upper Jurassic Jiande Group. The ore usually has a high content of iron. The metallogenic subformation occurs: i. in volcanic rocks to form magnetite deposits associated with diorite porphyry bodies, with trachyte, tuff and andesite as their country rocks, e.g., a certain iron deposit in Anhui Province; ii. at the contact between augite-diorite porphyrite and biotite andesite to form magnetite-hematite-siderite deposits, e.g., a certain iron deposit in Jiangsu Province; and iii. in lavas and tuffs of trachyandesite-andesite composition to form magnetite-hematite-pyrite (pyrrhotite)

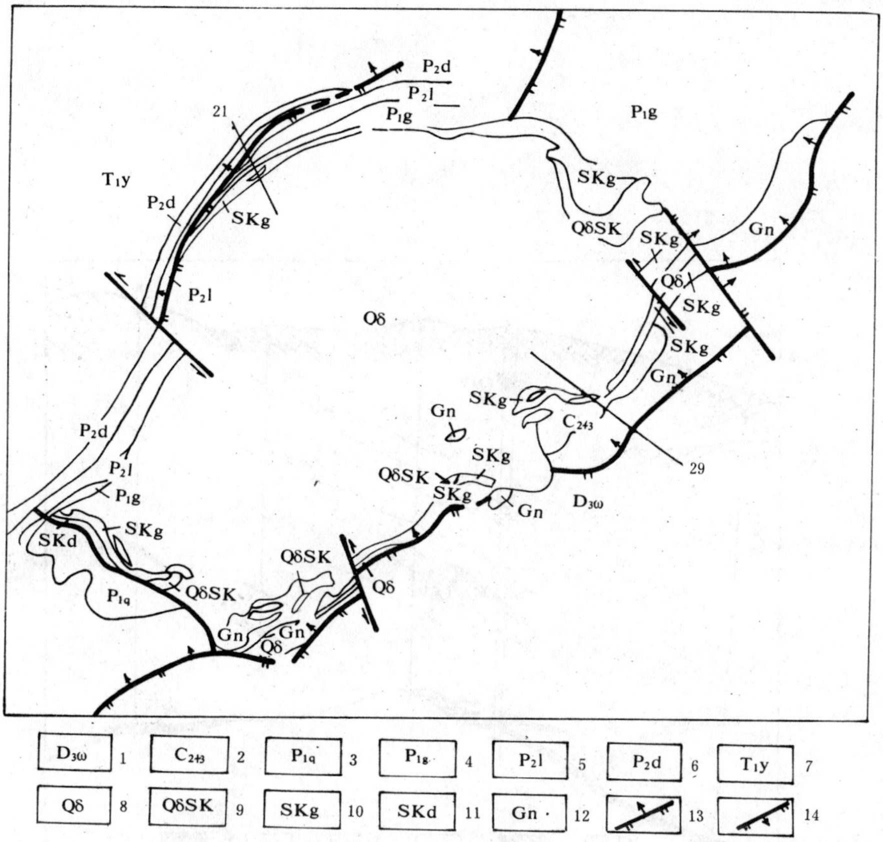

Fig. 5.1. Simplified structural map of a skarn-type copper deposit in the lower Changjiang area (after 812 Geological Team, 1974).

1, Quartzite of the Wutong Group; 2, Limestone and dolomitic limestone; 3, Siliceous and carbonaceous shales of the Qixia Formation; 4, Siliceous limestone, chert and manganese-bearing shale of the Gufeng Formation; 5, Siliceous and carbonaceous shales of the Longtan Formation; 6, Siliceous shale and limestone of the Dalong Formation; 7, Striped marble; 8, Quartz diorite of Yanshanian age; 9, Skarnized quartz diorite; 10, Garnet skarn; 11, Diopside skarn; 12; Gossan; 13, Reverse fault; 14, Normal fault.

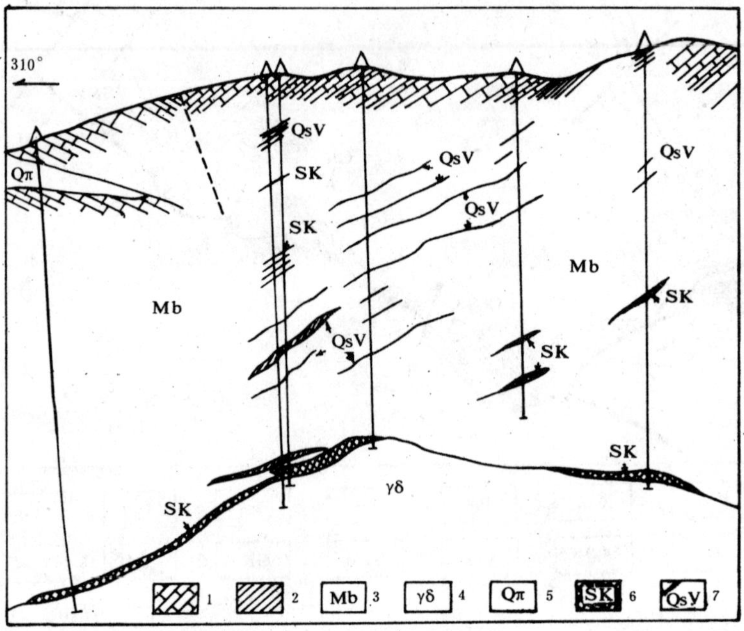

Fig. 5.2. Geologic cross section through a polymetallic deposit in the Nanling area (after 206 Geological Team).
1, Limestone; 2, Shale; 3, Marble; 4, Granodiorite; 5, Quartz porphyry; 6, Skarn-type ore deposit; 7, Quartz sulphide vein.

deposits, e.g., some iron deposits in Anhui and Jiangsu Provinces (Figs. 5.3, 5.4 and 5.5).

The subformation is genetically related to volcanic or subvolcanic rocks with an intermediate petrochemical composition and displays features of multistage mineralizations and repeated replacement processes. Volcanic rocks at the mineralized zones have generally bent into gentle folds and small uplifts, and ore deposits are localized at the edges of volcanic terrains controlled by fractures or within diwa basins. In the SE, Central China, NE and North China diwa regions, volcanogenic diwa basins, controlled by fractures, have a wide distribution and a great ore potential.

(b) Copper-bearing continental acid to intermediate volcanic and subvolcanic rock subformation. It is known in the middle and lower reaches of the Changjiang (Yangtze) River as a hydrothermal-metasomatic copper pyrite deposit. It occurs at the contact between calcareous rocks and continental acid to intermediate volcanic and subvolcanic rocks or hypabyssal small intrusive bodies. Ore deposits have been controlled by NE- and NNE- trending fractures, e.g., copper deposits in Zhejiang and Anhui Provinces.

e. Copper- and molybdenum-bearing granodiorite porphyry-alkali granite porphyry and calc-alkaline volcanic rock formation.

This formation produces chiefly porphyry copper and molybdenum deposits, an important copper deposit type in China. It is widespread in diwa regions of Cathysian type in east China, associated with calc-alkaline continental volcanic, subvolcanic or hypabyssal intrusive rocks of Yanshanian age, and known as porphyry copper-molybdenum deposit of Cathysian type. It may be divided into two types. One of them occurs in calc-alkaline continental volcanic rocks within the diwa structural layer. The period of the mineralization almost coincides with the age of volcanism. Ore deposits occur immediately around the necks of andesite or dacite, trachyte and subvolcanic rock. The mineralization also occurred in the surrounding volcanic, subvolcanic and sedimentary rocks. Ore fields generally lie in diwa basins or at their edges, where continental volcanism has taken place. An typical example is a certain porphyry copper deposit in Jiangsu Province. The other type is intrusive porphyry (or volcano-plutonic) copper deposit situated in various structural layers of a diwa region. In space it does not necessarily occur within the diwa structural layer, nor has it direct relation to diwa-type volcanism. But in time the mineralization-related porphyry may be regarded as comagmatic with the diwa volcanic rocks at the place in question. The porphyry is their hypabyssal-abyssal igneous rock, e.g., granodiorite porphyry, or a product of the intermittent period after a main volcanic event, e.g., some porphyry copper deposits in Jiangxi Province (Fig. 5.6).

Porphyry copper-molybdenum deposits of Cathysian type in various diwa regions of China, especially in those of east China, have a great ore potential. Such deposits are expected to be found along major deep fractures, especially at the intersections of the very notable NNE-trending deep fractures with fractures trending in other directions in east China.

f. Rare element-bearing acid granite formation.

It is one of the magmatic-metallogenic formations characteristic of diwa regions of Cathysian type. Diwa granite in these regions is petrochemically characterized

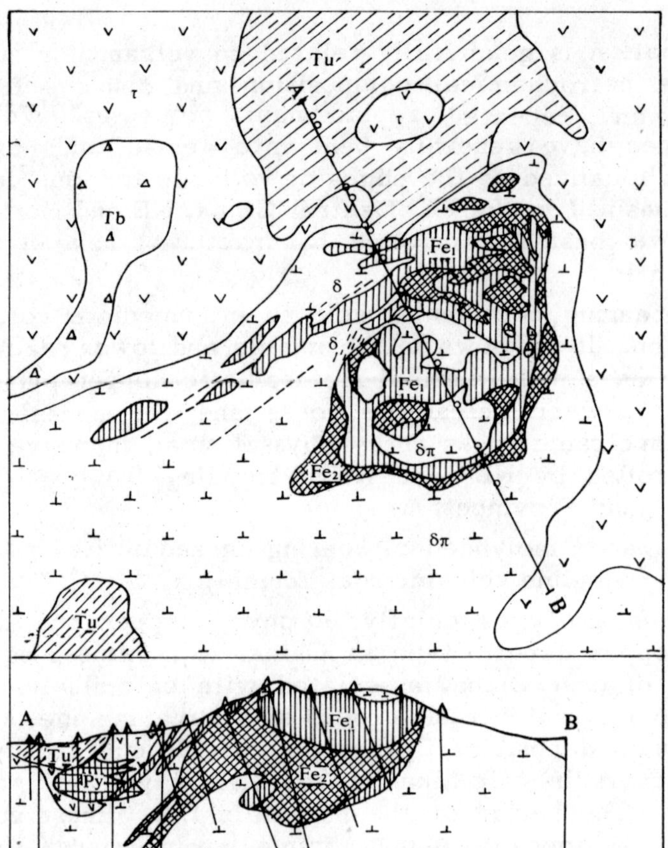

Fig. 5.3. Iron ore bodies controlled by a subvolcanic rock body structure in Aoshan, Anhui Province (after 322 Geological Team).
1, Tuff; 2, Tuff breccia; 3, Trachyte; 4, Diorite porphyry; 5, Rich magnetite; 6, Poor magnetite; 7, Pyrite; 8, Foliated diorite.

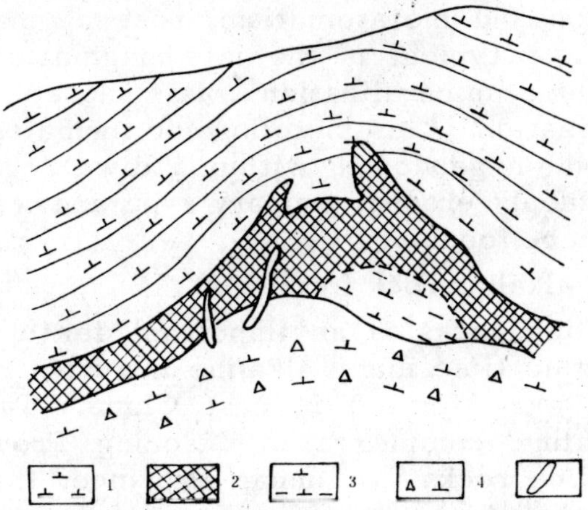

Fig. 5.4. Cupola -like ore body at the Zhangjiatang iron mine, Anhui Province (after 1122 Geological Group).
1, Andesite; 2, Hematite ore body; 3, Ore-bearing trachyandesite; 4, Breccia andesite; 5, Pyrite lode.

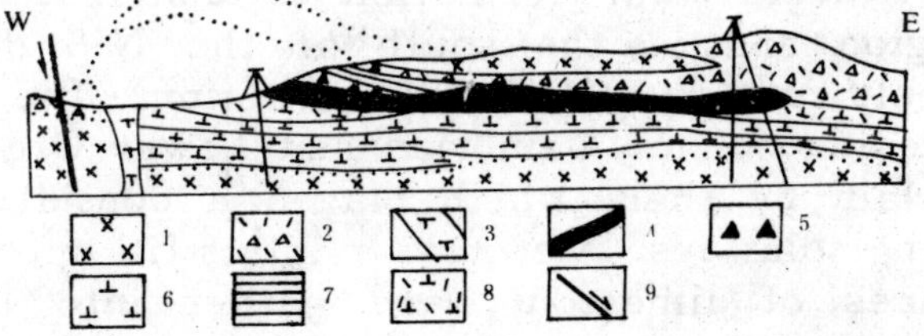

Fig. 5.5. Example of iron ore control by stratification of volcanic rock (after 1122 Geological Group).
1, Diorite; 2, Tuff; 3, Syenite porphyry; 4, Ferriferous tuff; 5, Agglomerate; 6, Andesite; 7, Tuffaceous shale; 8, Trachyte; 9, Fault.

by high contents of SiO_2 and alkali metals (potassium and sodium). Rare element deposits associated with diwa granite are of several types: magmatic, late-magmatic differentiation and metasomatism, post-magmatic and weathering crust. Among them the most typical is the late-magmatic differentiation and metasomatism type. The mineralization is closely associated with k-feldspathization and albitization. This metallogenic formation frequently occurs in granitic arches developed on geodomes within a diwa region. If the granitic arches have not yet been deeply eroded, primary type and weathering crust type deposits of rare elements can be found.

g. Rare element-bearing alkaline rock formation.

The occurrence of alkaline rocks is an important feature of magmatism in diwa regions. During Mesozoic time many alkaline intrusive rocks and associated deposits of rare elements were formed in east China. Typical examples are nepheline syenite and alkaline complexes in Liaoning Province, alkali-syenite in Shandong Province, alkaline rocks in Yunnan Province, the Ordos complex in Inner Mongolia, alkaline rocks at the northern foot of Dabie Mountain, alkaline complexes in Hubei Province and syenite in Guangdong Province. These alkaline rocks have yielded deposits of rare elements. Ore deposits associated with alkaline rocks may be found in diwa regions of Cathysian type in east China.

h. Antimony- and mercury-bearing epithermal formation.

Hydrothermal deposits of antimonite-quartz type appear to be characteristic of endogenic mineralizations in east China's diwa regions, especially in the Central China and SE diwa regions. The deposits have two forms of occurrence, stratiform and vein-like. The dominant ore mineral is antimonite, which occurs typically in the Xikuangshan Mine, Hunan Province. Here note should be taken of the fact that there are lamprophyre dikes in the mine. Further investigations are needed to clarify whether or not the antimony deposit is asociated genetically or spatially with the dikes.

The mercury-bearing hydrothermal formation is distributed chiefly in the Yunnan-Guizhou diwa region and in the south of the N-S diwa region. The Hunan-Guizhou mercury belt runs in a NNE-SSW direction. The mercury deposits are confined to a definite horizon of Cambrian and Lower Ordovician age. The horizon is generally overlain by shale which played a considerable role in the concentration of ore-bearing solutions. Mercury veinlets fill up, with replacement, fissures and bedding planes of limestone, and gather into layers, veins and disseminations. The dominant ore mineral is cinnabar and the alteration includes silicification and dolomitization.

i. Diamond-bearing kimberlite formation.

Kimberlite commonly occurs in diwa regions and in a small amount in other tectonic elements. It appears to have been derived from the upper mantle and brought about by reactivation of deep fractures in diwa regions. Its appearance suggests that diwa-type tectonism is a reflection of material movement in the earth's interior. East China is one of the principal areas of distribution of kimberlite, which occurs mainly in deep fractures tectonically very active since Mesozoic time (Fig. 5.7). In the light of tectonic conditions of formation of kimberlite, we can find some new areas of distribution of kimberlite in other parts of China.

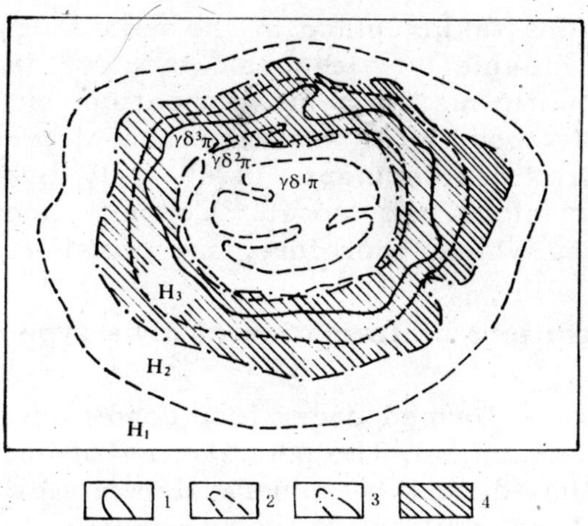

Fig. 5.6. Horizontal section of a porphyry copper ore body in Jiangxi Province (after Jiangxi Geological Institute).
1, Boundary of granodiorite porphyry body; 2, Boundary of alteration zone; 3, Boundary of ore body; 4, Copper-molybdenum complex.

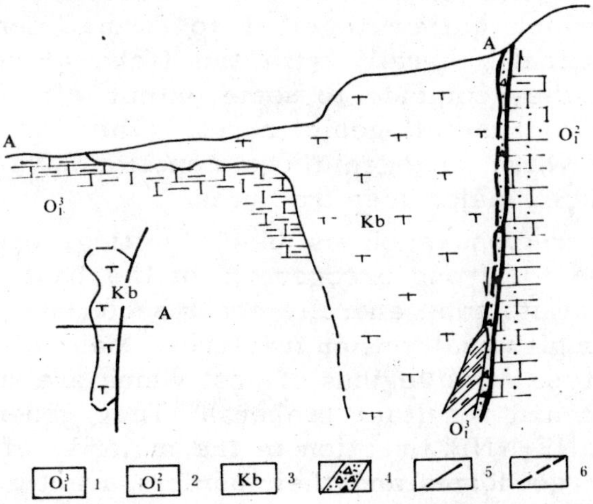

Fig. 5.7. Diagrammatic illustration of morphology of the No. 20 rock body from a kimberlite rock body swarm in east China (after Shandong Geological Bureau).
1, Thin argillaceous limestone; 2, Dolomitic limestone; 3, Kimberlite; 4, Crush zone; 5, Fault; 6, Inferred boundary.

2) Magmatic-metallogenic formations in the residual-mobility period.

Endogenic mineralizations taking place in the residual-mobility period of diwa development are associated chiefly with basic effusive (predominantly basaltic) rocks. The known magmatic-metallogenic formation is copper-bearing basalt formation which generally occurs in intermontane depressions formed in the fault-fold belt period of diwa development. It is widely distributed in diwa regions of east China, especially along the coast. Copper mineralizations have been reported from these places which, therefore, are promising areas for the search for copper.

3. An analysis of the conditions of formation of diwa-type endogenic ore deposits in China

Endogenic ore deposits are formed under four conditions: structural, magmatic rock, country rock, and geochemical. The peculiarities of endogenic mineralizations in diwa regions indicate that diwa-type mineral deposits differ from geosynclinal- and platform-type ones in the conditions of their formation.

(1) Structural conditions

Being the main ways of diwa tectonism, block-faulting and arcogenesis-fracturing are believed to be the most important structural features of diwa regions. Their appearance is due to the fact that diwa regions have inherited the consolidated and rigid paleogeosynclinal structural layer (and the overlying platform one) and developed on them as a foundation. Therefore, ore control by block-faulting is particularly apparent (Figs. 5.8 and 5.9).

The linear alignment of the known endogenic deposits in Cathysian-type diwa regions of east China, which gather together to form a series of metallogenic belts (Fig. 5.10), is identical generally with the trend of regional fractures or fault-folds. As a result, they coincide to some extent with each other in space, forming fracture (fault-fold)-metallogenic zones. This is different from that in geosynclinal regions, where endogenic ore deposits have been controlled by folds within a fold belt and by major deep fractures.

Metallotects in diwa regions such as block-faulting, open folds and swells originate and develop due to strong arcogenesis of the basement. In the process of rising of a geodome, at its top generally occur secondary tension forces which give rise to a series of longitudinal tension fractures. Many of the fractures widely distributed in Cathysian-type diwa regions of east China are such tension fractrues (high-angle normal faults and resultant grabens). They generally run parallel to the axes of archs (in a NE-NNE direction in the majority of areas). Originating and developing mostly on geodomes and their borders, fractures of this kind serve as canals for the rising magmas and have been filled up with them. The areas of distribution of newly formed contemporaneous sediments lying between geodomes are diwa subsidences. As a result, rock bodies of various forms were created. They were extensively developed in east China like a string of beads or a chain, generated at the top and on the borders of geodomes and separated by diwa basins arranged in a short belt. Moreover, igneous rocks generally coexist closely with effusive ones, constituting a series of volcano-plutonic cycles. This characteristic feature is different from that of the early period of geosynclinal development, during which major deep fractures extend for several hundred to more than one thousand kilometers and have been filled up intermittently with basic and ultrabasic rocks.

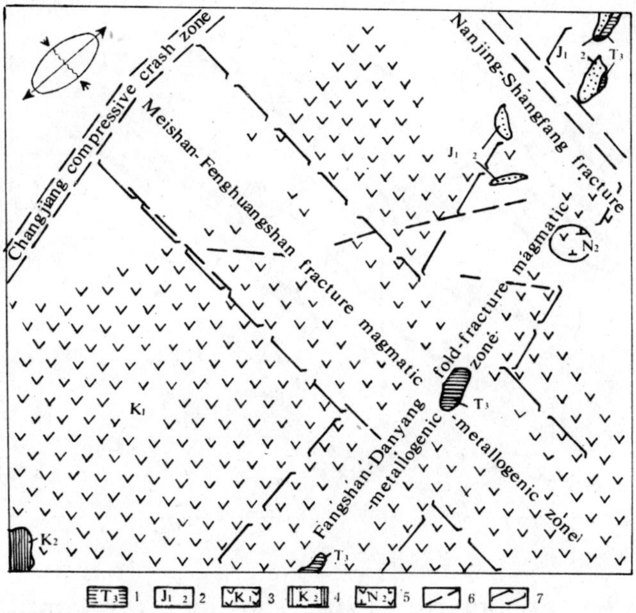

Fig. 5.8. Structures controlling volcanic-hosted iron deposits and their stress analysis (after 807 Geological Team).
1- Huangmaqing Formation; 2- Xiangshan Group; 3- Longwangshan, Yunheshan and Dawangsha Formations; 4- Niangniangshan Formation; 5- Dongxuanhuan and Fangshan Formations; 6- Fracture; 7-Approximate boundary of tectono-magmatic metallogenesis.

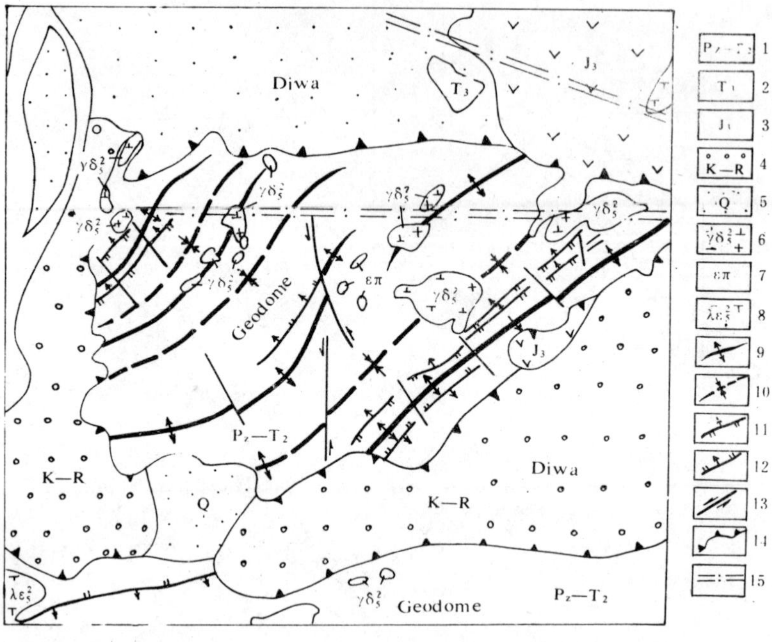

Fig. 5.9. Schematic map showing ore-controlling structures in a copper field, Central China diwa region. Mineral deposits are located in the platform structural layer (Paleozoic to Middle Triassic) folded and faulted in a NE direction, but controlled by an EW-trending basement fracture, thus found at the intersections of two sets of fractures (after Wang Daojing).

1- Paleozoic-Middle Triassic (platform structural layer); 2- Huangmaqing Formation; 3- Volcanics; 4- Red bed; 5- Quaternary sediments (2-5 is diwa structural layer); 6- Granodiorite; 7- Syenite porphyry; 8- Quartz syenite (6-8 is products of Yanshanian diwa magmatism); 9- Axis of anticline; 10- Axis of syncline; 11- Reverse fault; 12- Normal fault; 13- Strike-slip fault; 14- Boundary of geodome; 15- Concealed basement fracture (inferred from aeromagnetic data).

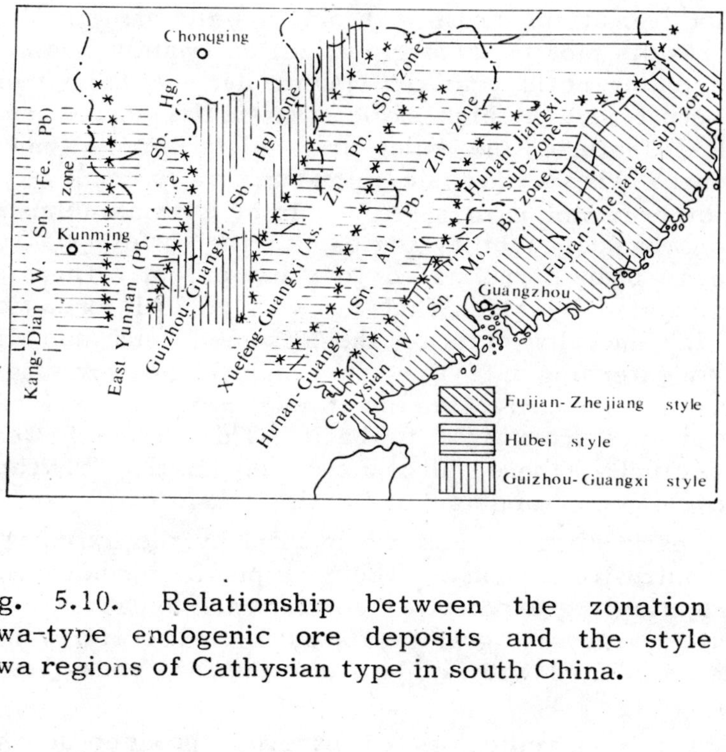

Fig. 5.10. Relationship between the zonation of diwa-type endogenic ore deposits and the style of diwa regions of Cathysian type in south China.

(2) Magmatic rock conditions

Magmatic mineralizations in diwa regions evolve differently from those in geosynclinal regions. They are characterized by the particularly extensive development of non-ferrous, rare and disseminated metal deposits of many genetic types in the earlier period, and by Cr, Fe, Ni and Co mineralizations in the later period. This change is due to the fact that magmatic processes in diwa regions proceed from acid to basic.

The principal reasons why diwa-type magmas develop from acid to basic may be two. One of them is petrogeochemical. As seen from Tables 4.1, 4.2. 4.3, 4.4 and 4.5, geosynclinal-type granite can be ascribed mostly to the gabbroic series, and its ore-forming material is characterized by "basaltphile" elements, such as Fe, Mn, V, Ti, Cr, Cu, Zn and the metals of the platinum group, while diwa-type granite can be assigned mostly to the granitic series, and its ore-forming material is characterized by "granitephile" elements, such as Sn, W, Be, Li, Bb and Ta. The former is probably generated from the differentiation of gabbroic (basaltic) magma in oceanic (or transition) crust with a thin sialic structural layer, whereas the latter probably results mostly from granitization (mainly assimilation of sialic material related to palingenetic granitic magma derived from remelting of the sial) in continental crust with a thick sialic structural layer. Therefore, showing different geochemical features and being in a different tectonic environment, diwa-type granite generally lacks the basic "initial magma" preceding geosynclinal-type granite. The other reason is tectonic. After the maximum-mobility period i.e., in the residual-mobility period, all the structural layers above the basement have already been consolidated to some extent. Therefore, structural disturbances are dominated by block-faulting with rigid deformation and tension (shear) character. It generally cuts to the basement and sometimes produces major deep fractures extending into the upper mantle and drawing out basic and ultrabasic magmas. Consequently, in the whole process of diwa development magmatic activity evolves from acid to basic. This order, however, is merely a general law. Actually there also exist magmatic "rhythms" in every tectono-magmatic cycle comprising acid and basic rocks.

Magmatic processes in diwa regions are marked by the extensive development of acid extrusive-intrusive rocks. Their typical formations include the liparite-dacite formation, granodiorite formation, hypabyssal granite-quartz porphyry formation, and abyssal granite formation. Therefore, non-ferrous and rare metal deposits associated with acid to intermediate rocks are most common in diwa regions.

The granitic body in diwa regions is a synthetical product of repeated igneous activities, that is the compound rock body. For this reason, it is not right to say about the "ore deadline" of such a rock body. The compound rock body formed by repeated igneous activities shows mineralizations of various types superimposed upon each other. As a result, metallogenic phenomena, more complex than in any other known tectonic region, have been created in diwa regions. The extremely typical case is that in a not very large sector of an area, different deposit types, from pegmatite in the early to meso-epithermal and even telethermal in the late period, may coexist. This fact reflects the polygenesis of endogenic mineral deposits in diwa regions.

(3) Country rock conditions

The features of endogenic mineralizations in diwa regions are intimately related to the country rock condition. Geosynclinal and platform structural layers constitute the foundation of a diwa region, while endogenic mineralizations of the diwa stage occur mainly in the maximum-mobility period. Consequently, endogenic mineral deposits can be localized not only within the diwa (including intrusive and volcanic rocks) but also within the geosynclinal and platform structural layers. Such a country rock condition, along with the influence of tectonic and magmatic processes in the diwa stage, has led to a series of complex metallogenic features unknown in other tectonic elements. Ore-forming country rocks on geodome uplifts are dominantly geosynclinal-type sandstones and shales as well as deeply metamorphosed rocks of the crystalline basement, while those in diwa subsidences are mainly platform-type carbonate rocks. These two rock types show a definite regularity in their spatial distribution, which may be taken as one of the guides to the search for mineral deposits of various types.

On geodome uplifts, geosynclinal and pre-geosynclinal metamorphic rocks predominate and have been folded and consolidated several times. Therefore, in the process of diwa tectonism, structural disturbances are characterized by the extensive development of fractures of different size. As a result, all deposit types (high- to low-temperature) of various mineral kinds are expressed as veins. The development of fractures gives rise to fine, low-pressure canals with a high-oxidation condition, thus contributing to the high concentration of oxyphile elements such as W, Sn, Be, Mo and Ta in pneumatolytic hot solutions.

In diwa subsidences, carbonate rocks, chemically highly active, generally have a wide distribution (except when they are covered by diwa sediments or volcanics). As a consequence, sharn and hydrothermal metasomatic mineral deposits can attain an extensive development and occur chiefly within the diwa subsidences, forming a series of metallogenic provinces of typical skarn or metasomatic deposits, as in the Yanshan-Liaoning area, the middle and lower reaches of the Changjiang (Yangtze) River and the Jiangxi-Hunan-Guangxi area. In other tectonic regions, e.g., geosynclinal (or paleogeosynclinal), such a country rock condition exists only on geanticlines (or paleogeanticlines). Skarn metasomatic deposits occur in acid volcanic rocks as well as in limestones. Examples are some polymetallic deposits in northern Qilian Mountain (formed in the geosynclinal stage).

(4) Geochemical conditions

It is generally believed that crustal material is derived from the mantle and that the sial results from the differentiation of it. In the process of the differentiation, elements (and compounds) with a low melting point, a small specific gravity, a great ionic radius, a great atom volume and a strong volatility (e.g., alkali metals, silica and volatile components) constantly migrate upwards, while elements with a high melting point, a great specific gravity, a small ionic radius and a small atom volume (e.g., Mg, Cu, Pt, Fe, Co and Ni) progressively move downwards. This process cannot be completed once, but goes on advancing along with tectonic development, and the differentiation grows more and more perfect. Therefore, the development of the earth's crust has resulted in an increase in the thickness and areal extent of the sial (granitic layer) and in the quantity of acid to intermediate and alkaline rocks and gaseous elements in the material composition of the earth's crust. This has led to the geochemical differences between different tectonic stages (geosynclinal, platform and diwa). Consequently, during the diwa stage, element associations of basic and ultrabasic magmas are poorly developed, but those of acid and alkaline magmas well-developed. That is why endogenic mineralizations taking place in the diwa stage are exclusively associated with granitoids and mineral deposits of volatile elements, such as antimony, mercury and fluorite, attain an extensive development.

Diwa-type magmatic rocks are geochemically characterized by high abundances of Si, K, Na, Li, Rb, Cs, Be, Nb, Ta, Th, U, W, Sn, Mo, Cu, Pb and Zn and low abundances of Mg, Fe, Cu, Co, Ni, Ti and V. This is particularly true for diwa-type granitoids. Most characteristic of the chemical features of diwa-type magmatic rocks is the high abundance of alkali metals (potassium and sodium), especially potassium. Therefore, in the endogenic metallogenic process during the diwa stage, ore deposits of Li, Be, W, Sn, Bi, Mo, Cu, Pb and Zn attain a special development.

Diwa-type granite is rich in volatile elements, so that alteration is well-developed. Mineralization is intimately related to alteration, e.g., Nb, Ta and Be to K-feldspathization and albitization, W and Zn to greisenization, and B and Be to skarnization.

Geochemical conditions of country rocks also show specific features in diwa-type endogenic mineralization.

As mentioned earlier, exposed on geodome uplifts is mainly flysch clastic rock of geosynclinal facies. It is geochemically characterized by the predominance of Si and Al and the abundance of Fe, Mn, B, F, Sn and Be. When diwa-type granitoids are intruded, Si and Al add to ore-forming solution and the pH value increases, which makes ore-forming elements to appear chiefly as oxides, e.g., cassiterite. The introduction of Fe and Mn greatly contributes to tungsten mineralization resulting in wolframite deposits. Moreover, in tungsten deposits, Be, Sn and F exist in a great number, forming W-Be and W-Sn deposits. Thus it can be concluded that wolframite mineralization may be regarded as one of the typical and indicative mineralizations of granitoids on geodome uplifts (areas of distribution of flysch clastic rock of geosynclinal facies).

In diwa subsidences, exposed are mainly carbonate rocks of platform facies, except when covered by diwa-type sediments or volcanics. They are rich in Ca, Mg and Fe, but poor in Si and Al. When diwa-type acid and intermediate rocks intrude, a vast amount of skarn deposits occur. Mineral resources include Be, W, Sn, Mo, Fe and Cu. Here we take skarn iron deposits as example. On the one hand, country rocks are rich in iron, thus providing a source of iron; on the other hand, they are poor in Si and Al, thus avoiding the dispersion of iron carried by magmas. Moreover, Mg and Ca, especially Mg, can play a role of a precipitating agent in the formation of magnetite. The existence of Ca contributes to the concentration of molybdenum and tungsten. Consequently, skarn molybdenum and tungsten deposits can also be formed.

From the above-mentioned analysis of the various conditions of formation of endogenic ore deposits in diwa regions, it can be concluded that diwa mineralization is a new-type mineralization. As diwa regions in China have a wide distribution, the mineralization suggests that the domain of the search for minerals in China is very extensive. It should also be pointed out that the above-mentioned four conditions are interrelated and that they act on the diwa-type mineralization comprehensively and indivisibly.

III. Tectonics and exogenic mineral deposits of China[1]

Relations of tectonics to exogenic mineral deposits are expressed in that different sedimentary formations contain different mineral deposits. Therefore, exogenic mineral deposits formed under different tectonic conditions and at different stages of crustal development offer their own mineral kinds and deposit types. The migration and concentration of ore-forming material and the attitude and distribution of the formed mineral deposits were all conditioned by tectono-topographic contrast, paleogeographical features and sedimentary processes. Here we give them a brief description from two points.

1. Temporal and spatial distribution of exogenic mineral deposits in China

On the one hand, because of the multistage character of crustal development, sedimentary deposits of different tectonic types may coexist in the same region; those of the same mineral kind and of the same tectonic type may occur in several horizons of the same region. On the other hand, caused by the unevenness of crustal development, sedimentary deposits of the same mineral kind and the same tectonic type in different regions may appear at different times and in different horizons. This has resulted in a certain sequence of sedimentary deposits of various tectonic types and various mineral kinds, their distribution features and temporal and spatial regularities.

In Archean time there existed a pre-geosynclinal tectonic element in China. Exogenic mineral deposits formed at that time include large metasedimentary iron deposits as well as phosphorus and manganese ores within the pre-geosynclinal structural layer (crystalline basement). They are represented by Anshan-type iron deposits and Donghai-type phosphorus deposits in north and northeast China.

In Proterozoic time the larger part of China already entered into the geosynclinal stage and geosynclinal regions developed most widely. As a result, geosynclinal-type exogenic deposits occurred most extensively. The known ores are iron, manganese and phosphorus. They exist in the geosynclinal structural layer, e.g., in Proterozoic metamorphosed rocks widely exposed in north, northeast, central and northwest China. Among them the most famous are sedimentary iron and manganese deposits in a certain area of north China.

In Sinian time, central, south and northeast China, Yunnan, Guizhou and Sichuan were still in the geosynclinal stage, during which exogenic deposits of iron, manganese and phosphorus were formed. Typical examples are Sinian phosphorus deposits in south China. At that time north China already entered into the platform stage, during which such platform-type exogenic deposits as residual and sedimentary iron deposits and sedimentary manganese deposits were formed.

In early Paleozoic time, platform regions in China were rapidly expanding, but geosynclinal regions still covered a considerable area. Exogenic minerals formed at that time include phosphorus, iron, manganese, coal and salt. To platform type belong iron and phosphorus deposits (Lower Proterozoic) in north China and phosphorus deposits (Early Cambrian) in Yunnan, Guizhou, Sichuan and the Tarim basin. To geosynclinal type belong volcano-sedimentary iron and manganese deposits (Cambrian to Ordovician) in northwest China and phosphorus deposits (Early Cambrian) in the Tianshan and Beishan Mountains.

---

1. Chen Jiachao and Guan Yinwen.

In late Paleozoic to early Mesozoic time, the large part of east China and many areas of northwest China already entered into the platform stage. Therefore, platform-type exogenic deposits were rich and varied at that time. However, geosynclinal regions continued their development. Mineral kinds generated at late Paleozoic to early Mesozoic time include iron, manganese, copper, aluminium, phosphorus, coal, petroleum, gas, oil shale and salt. Among them to platform type belong: iron deposits (Devonian) of shallow-marine facies in southeast China, iron deposits (Carboniferous to Triassic) of continental facies and alternately marine and continental facies in north and south China, manganese (Lower Devonian, Middle Carboniferous, Middle Permian and Middle Triassic) and phosphorus deposits (Middle Permian) in south China, bauxite deposits (Carboniferous and Permian) in north China, Yunnan and Guizhou, coal fields (Carboniferous and Permian) in north China, petroleum and natural gas in southeast and north China, Yunnan, Guizhou, Yinshan (Ordos) and Sichuan, sandstone-type copper (Middle Permian and Upper Triassic) and salt deposits (Middle Triassic) in Yunnan, Guizhou, Sichuan and south China. To geosynclinal type belong: volcano-sedimentary manganese (Upper Carboniferous) in the central Tianshan Mountains and coal fields (Carboniferous and Permian) in northwest and northeast China.

In middle to late Mesozoic time the larger part of China already entered into the diwa stage. At that time exogenic deposits occurred widely and occupied a prominent position. In addition, some platform and geosynclinal regions remained. Important diwa-type exogenic deposits are: coal fields (Lower and Middle Jurassic) and iron deposits (Jurassic) in north China, coal fields (Upper Jurassic), oil shale (Jurassic to Cretaceous), and petroleum in northeast China, coal fields (Upper Triassic to Lower Jurassic), oil shale (Jurassic to Cretaceous), salt (Mesozoic) and sandstone-type copper deposits (Jurassic to Cretaceous) in south China, and volcano-sedimentary iron deposits (Jurassic to Cretaceous) widely distributed along the east coast. Geosynclinal-type exogenic deposits discovered in the Kunlun, Bayanhar, Yarlung Zangbo River-Himalayan and Gandise regions include Mesozoic coal and iron. Platform-type exogenic deposits found in the Songliao, Yishan (Ordos), Sichuan and north Xinjiang regions include Mesozoic coal, petroleum, gas, salt and iron (Sichuan).

In Cenozoic time diwa- and geosynclinal-type exogenic deposits are the most important. Mineral kinds include petroleum, gas, coal, salt, aluminium, iron, manganese, copper and phosphorus. To geosynclinal type belong: Tertiary petroleum and natural gas on Taiwan and Diaoyu Island, Tertiary coal fields in the southern Bayanhar Mountains, and part of Quaternary salt deposits in Xizang. To diwa type belong: oil shale in Liaoning and Guangdong Provinces, Tertiary coal fields, iron and salt deposits in south China, Tertiary gibbsite deposits in Fujian and Guangdong Provinces, Tertiary salt deposits in northwest China, and petroleum in many provinces. In addition, there were deposited phosphorites (bird's droppings) at that time in the Xisha , Nansha, Zhongsha and Dongsha Islands and in adjacent areas of South China Sea.

In regard to the temporal distribution of exogenic mineral deposits, the Archean in China was an epoch of exogenic mineralizations in pre-geosynclinal tectonic elements. The Early Proterozoic to Sinian was an epoch of predominantly geosynclinal-type exogenic mineralizations, but at the end of this period platform-type exogenic mineralizations also appeared. The Paleozoic was an epoch of predominantly platform-type exogenic mineralizations, and towards the late Paleozoic platform-type exogenic deposits became widespread. But geosynclinal-type exogenic mineralizations also took place at that time. The Mesozoic to Cenozoic was an epoch of predominantly diwa-type exogenic mineralizations; at the same time there also occurred geosynclinal-and platform-type exogenic mineralizations. Concerning the spatial distribution of exogenic mineral deposits, diwa- and platform-type exogenic deposits occur most widelyin east China, but pre-geosynclinal-type ones are also important. Generally speaking, exogenic mineral deposits in east China have been affected, reworked and controlled chiefly by diwa-type open folds and block-faulting structures and occur in short belts. In west China, geosynclinal-type exogenic mineral deposits are common, but platform- and diwa-type ones also present. Exogenic mineral deposits in west China have been controlled and affected chiefly by geosynclinal-type tight folds and major deep fractures and occur in long belts.

2. Examples of ore-bearing sedimentary formations of various tectonic types in China

A sedimentary formation is generated in a definite tectonic setting. Each tectonic element or each development stage offers its own sedimentary formations. Exogenic mineral deposits are generally confined to definite sedimenary formations which in this case constitute ore-bearing formations. Below we list some important ore-bearing sedimentary formations of various known tectonic types in China and give a discussion on their characteristic features and temporal and spatial regularities.

(1) Pre-geosynclinal-type ore-bearing sedimentary formations

Ore-bearing sedimentary formations generated in pre-geosynclinal tectonic elements await further detail study. The most famous formations are as follows.

1) Iron-bearing formation.

It may be represented by Anshan-type iron deposits in the cherty iron formation-metasedimentary iron deposits. The formation consists of thick ferriferous quartzite and have a clear banded structure formed by the alternate layering of silica and magnetite. It occurs in the deeply metamorphosed Archean rock (Anshan Group) within the pre-geosynclinal structural layer and consists of, from below: lower migmatite bed, hornfels bed; lower iron band and biotite albite schist (the indicator horizon); upper iron band, chlorite and quartz schist; upper migmatite bed. Ore bodies have the form of a layer and sometimes grade into lens-shaped ore shoots formed by subsequent transformation and hydrothermal metasomatism.

Iron deposits of this type, of great dimensions, have a wide distribution in northeast and north China and in the Qinling Mountains, being one of the most important iron deposit types known in China.

2) Phosphorus-bearing formation.

It is an important phosphorus formation in China and is represented by Donghai-type phosphorus deposits. To our knowledge at present, it is widely distributed in Anhui, Hubei and Jiangsu Provinces. The ore is confined to the Archean Yuntai Group which consists of quartz-mica schist, siliceous marble, marble and phosphorite.

Similar horizons are also extensively distributed in north China. The Sanggan Group in north China is such a horizon. Its phosphorus-bearing part consists chiefly of deeply metamorphosed hornblende gneiss interbedded with marble. Ore beds occur at the bottom of the gneiss or marble.

(2) Geosynclinal-type ore-bearing sedimentary formations

1) Ore-bearing sedimentary formations in the geosynclinal period.

Ore-bearing sedimentary formations in the geosynclinal period generally stretch in a definite direction, and mineral deposits in them occur in long belts.

The formations are of marine facies with great thickness. The type and tenor of ores change rapidly, especially across ancient coastlines or basins. Each sedimentary rhythm involves two or three (sometimes more) components. Regressive sediments are often absent. Discontinuous rhythms prevail. Multi-order rhythms are conspicuous.

These ore-bearing sedimentary formations exist in all geosynclinal regions of China or in the geosynclinal structural layer of other tectonic elements. The geosynclinal stage in west China lasted long and ended late. Therefore, geosynclinal-type ore-bearing sedimentary formations there attain a more extensive development, occur in more stratigraphic horizons and have a more wide distribution than in east China.

a. Iron-bearing formation.

(a) Silica-iron subformation.

This subformation in China was commonly generated in ancient geosynclines, including those of Luliang, Jinning and Caledonian ages, and occurred mostly in the initial period of geosynclinal evolution. Its ore material was derived from sediments deposited on the borders of volcanoes (or from terrigenous sediments). Consequently, iron deposits are often associated with marine siliceous, arenaceous and argillaceous sediments. Iron beds are interlayered with siliceous shale and ferriferous quartzite.

Iron ore deposits formed at the geosynclinal stage in east China belong chiefly to this type. The Proterozoic in north China, the Sinian, Cambrian and Ordovician in south China, the Kunyang and Lancang Groups in Yunnan Province, the Lower Cambrian in the Beishan and Tianshan Mountains and other early-geosynclinal structural sublayers are expected to contain iron deposits of this type.

(b) Iron-bearing volcano-sedimentary subformation.

This subformation is seen at present mainly in west and central China. It is an ore-bearing formation produced by intense volcanism in the process of geosynclinal subsidence in these areas. Iron has been derived from submarine volcanic extrusion occurring nearby. Iron deposits are often interbedded with intermediate to basic volcanic rocks.

The subformation consists of numerous marine intermediate to basic lavas, pyroclastic rocks and tuffs as well as marine arenaceous, argillaceous, calcareous and siliceous sediments. The facies changes abruptly. Rhythmic sedimentation is very distinctive. The total thickness attains over 10,000 m. Iron ore is composed mainly of hematite and magnetite. Ore bodies comprise commonly many beds.

Qilian Mountain, Guangxi, Yunnan and Hubei Provinces abound with such iron deposits. In the geosynclinal structural layer of geosynclinal, platform and diwa regions, the above-mentioned iron deposits may be found in areas of sodium-rich submarine volcanic explosion.

b. Manganese-bearing formation.

The depositional environment of this formation bears a remarkable resemblance to that of the iron-bearing formation. Manganese is contained in many horizons and the area of its exposures is wide. In east China, the manganese-bearing formation occurs at the top of the Hutuo Group in Shanxi Province and in the black shales in the lower part of the Nantuo Tillite and in the upper part of the Nantuo Sandstone in south China. In the former region it is a siliceous manganese-bearing subformation composed of siliceous and clastic rocks, and the ore beds have changed into ferriferous quartzite. In the latter region it is a manganese-bearing carbonate subformation known as carbonate manganese deposit.

In west China the manganese-bearing formation is slightly different from that in east China: besides the siliceous manganese-bearing subformation (e.g., the Lower Cambrian of the Tianshan and Beishan Mountains) similar to that in east China, there is a volcano-sedimentary manganese-bearing formation directly related to volcanogenic sediments. The Lower Ordovician of Qilian Mountain and the Upper Carboniferous of the Central Tianshan Mountains consist of a great amount of marine intermediate to basic lavas and pyroclastic rocks with abrupt facies changes.

c. Phosphorus-bearing formation.

It is found in the geosynclinal structural layer of various regions in China. According to available geologic data, it is of great economic value in the Sinian and Lower Cambrian.

The Sinian geosynclinal-type phosphorus formation is present mainly in south China, but it also exists in Qilian Mountain. In the former case it occurs chiefly in the Upper Sinian Xifengsi Limestone, while in the latter case it is composed of siliceous slate, limestone, marble, schist and five phosphorus-bearing beds.

The distribution of the Early Cambrian geosynclinal-type phosphorus-bearing formation remains to be studied. The formation is known in northwest China where the number of ore beds changes considerably and both their top and bottom comprise metamorphic phyllites and siliceous rocks. Extending into the Tianshan Mountains, they grade into dolomitic limestone, ferriferous limestone and quartzose siliceous rocks of shallow-marine facies. In Zhejiang, Jiangxi and Hunan Provinces of east China, the shallow-marine black siliceous phosphorus-bearing formation was also deposited in Early Cambrian time. Part of it may be of geosynclinal type.

d. Magnesium-bearing formation.

In the geosynclinal stage sometimes occur thick dolomite beds which have a wide distribution and great dimensions, forming large dolomite deposits. Dolomite can change into magnesite under metamorphic processes, or itself is a primary dolomite-magnesite formation, as seen in Liaoning Province and north China.

2) Ore-bearing sedimentary formations in the geosynclinal and the fold belt period.

a. Coal-bearing formation.

It includes two coal-bearing subformations: one in the geosynclinal and other in the fold belt period.

(a) Coal-bearing subformation in the geosynclinal period.

Examples of this subformation are: the Carboniferous of the Altun, western Qinling and Kunlun Mountains; the Permian of the Taizi River valley in northeast China, of the Qamdo area in north Xizang, and of the Heqing area in Yunnan Province; and the Tertiary coal-bearing strata on Taiwan and Diaoyu Island. They are generally of shallow-marine, littoral and alternately marine and continental facies. Coal beds change rapidly in thickness and facies, and often alternate with sediments, reflecting an unstable state of sea basins at that time.

In addition, there are intercalations of black carbonaceous shale within the geosynclinal deposits of Lower Cambrian age in southeast China. It contains a significant content of carbon and may be taken as fuel named "rock coal". It is also worth notice that the lower part of the shale contains Nickel, molybdenum, vanadium and other metals.

(b) Coal-bearing subformation in the fold belt period.

It is a continental coal-bearing formation accumulated in piedmont or intermontane depressions during the fold belt period. Examples are the Yerqiang coal measures of Jurassic age in the Kunlun Mountains and coal-bearing rocks in the south of the Bayanhar geosynclinal region.

b. Oil- and gas-bearing formation.

It is developed both in the geosynclinal and the fold belt period and seen more frequently in young geosynclinal regions. For example, Tertiary rocks of the Taiwan geosynclinal region (including sea areas near Diaoyu Island) are rich in oil and natural gas produced in the geosynclinal and the fold belt period.

(3) Platform-type ore-bearing sedimentary formations

These formations include residual and ordinary sedimentary deposits.

Platform regions have small tectono-topographic contrast and large areas of erosion and sedimentation. Therefore, exogenic mineral deposits are particularly rich and varied, and ore-bearing horizons exist in large numbers and have a wide distribution.

Most of the platform regions appearing in the history of crustal development in China have already died out and have been converted into diwa regions. But the platform structural layer in many areas is still well preserved, thus providing favourable conditions for the summary and study of the metallogeny of the platform region.

Platform-type ore-bearing sedimentary formations have been generated in a tectonic setting with vertical movements affecting a large area and with a small tectono-topographic contrast, and in the process of alternating marine transgression and regression on a large scale, accompanied by the deposition of extensive marine, continental or alternately continental and marine sediments.

Sometimes marine regression lasted long for several periods (e.g., from Late Ordovician to Early Carboniferous in north China). Such a long break in sedimentation must have caused pre-existing strata to be deeply eroded and removed, providing a material source for sedimentary mineral deposits and creating favourable conditions for the formation of the deposits.

Platform-type ore-bearing sedimentary formations in many cases possess a small thickness, ranging generally from tens to hundreds of meters, and a planar distribution pattern. The lithology, facies and thickness change slightly and can be correlated easily over a vast area. The sedimentary rhythm generally involves both transgressive and regressive sequences, being of continual type. A single rhythm attains a great thickness and mineral deposits often occur in the regressive sequence of the rhythm.

Sedimentary mineral deposits are the important aspect of platform-type mineral deposits. Because of the particularly long duration of the platform stage and the wide distribution of platform conditions in east China, platform-type ore-bearing sedimentary formations appear to be well-developed and very important in this part of China.

1) Iron-bearing formation.

This formation includes two iron ore subformations: residual and sedimentary.

a. Iron-bearing subformation of ancient weathering crust type.

This subformation is derived from ancient geosynclinal- or pre-geosynclinal-type iron-bearing siliceous rocks, which have undergone weathering and leaching and, as a result, from which silica has been removed and iron relatively enriched. Iron deposits generally lie in weathering crust at the top of an iron-poor occurrence, passing gradually downwards into it. They are formed under the conditions of: i. a rich source of iron, ii. prolonged weathering under a favourable climate, and iii. a relatively stable tectonic environment providing a flat and broad land surface.

Iron deposit of this type has been formed, under a hot and damp climatic condition, by weathering and leaching of rock on an erosion surface with a gradually decreasing tectono -topographic contrast. Generally speaking, it can be formed in any development stage of any tectonic elements, but in the light of the above-mentioned ore formation conditions, it develops best in the initial-stability period of platform development. Such an iron deposit generally occurs on an unconformity between a geosynclinal (or pre-geosynclinal) and a platform structural layer. This was the most favourable time for the formation of iron deposits of ancient weathering crust type, because at that time crustal movements were dominated by gentle downwarping and upwarping affecting a large area, tectonic contrast was small, the whole paleogeosynclinal fold belt was reduced to a condition of peneplanation, and the climate was hot and damp.

There are indications that iron deposits of this type exist in northeast China. In this area, the Sinian (belonging to the platform structural layer) rests unconformably on the ferriferous quartzite of the Yingtaoyuan Formation of the Anshan Group (belonging to the pre-geosynclinal structural layer). The unconformity implies a break in sedimentation for a long span of time and shows conspicuous phenomena of weathering and leaching. At the bottom of the Sinian lies a basal conglomerate cemented by iron. Part of the conglomerate is derived from a metasedimentary iron-poor occurrence. On the unconformity there have been locally formed residual iron deposits varying in thickness. An example is the residual iron-rich deposit in Liaoning Province where it rests on the unconformity between the Sinian and the Yingtaoyuan Formation of the Anshan Group.

b. Oolitic ironstone subformation.

This formation is an iron deposit of neritic or littoral facies in transgressive sediments. Iron beds have a constant thickness, a high tenor and a planar distribution pattern. They are seen chiefly in finger bays surrounding the uplift (erosion) areas of a platform and have been found in north, east, central and south China as well as in Sichuan Province.

North China in the initial-stability period of its platform development saw the accumulation of iron deposits, which lie in the lower part of the Chuanlinggou Formation Shale of the Changcheng Group and rest unconformably on a platform structural layer composed of pre-Sinian rocks. Besides, there occur iron deposits in the lower part of the Tieling Formation Limestone and at the bottom of the Xiamaling Shale.

Iron deposits at the bottom of the Cambrian in Shanxi Province, whose ore beds rest unconformably on the Hutuo Group, have been formed in the transgressive sediments of neritic and littoral facies during the maximum-stability period of platform development. The ore-bearing formation has a small but constant thickness, and generally contains a hematite bed in its lower part. In Hebei Province there is also a hematite bed with a small thickness but a fixed stratigraphic position lying between the Lower and Middle Cambrian.

Middle to Upper Devonian sedimentary iron deposits in south China comprise oolitic hematite and have been formed in finger bays appearing in the maximum-stability period of platform development. They are related to the position and shape of ancient shorelines. The beginning of the platform stage and the marine transgression onto the platform was progressively delayed from southwest to northeast. As a result, iron beds stratigraphically lie at a higher and higher level as they move northeastward. For instance, they lie within the Middle Devonian in Sichuan and Hunan Provinces, and within the Middle and Upper Devonian in west Hubei, Jiangxi and Fujian Provinces. They are constant in thickness and occur between calcareous and clastic rocks. Iron ore is composed of oolitic hematite. The lower part is rich in remains of organisms indicating that the deposits are of neritic and littoral facies in transgressive sediments.

Middle Carboniferous sedimentary iron deposits in Shanxi Province represent the products of the beginning of transgressive sediments deposited during the residual-stability period of platform development. Both the iron-bearing formation and iron beds are constant in thickness. The thickness of ore bodies depends on relief. Iron ores occur in slag, mud-mass and oolitic forms, being the products of ancient weathering crust.

c. Siderite subformation.

In Sichuan Province, siderite deposits occurring in the Late Triassic to Early Jurassic Xiangxi coal measures are associated with coal. They are widespread in a planar distribution pattern all over the whole Sichuan platform region, being the products of the residual-stability period of the development of this platform region.

2) Manganese - bearing formation.

It is common in east China and falls into the following categories:

a. Manganese-bearing subformation in the initial-stability period. It is distributed chiefly in north China, of Sinian age and confined to the lower part of the Gaoyuzhuang Formation of the Changcheng Group, to the upper part of the Hongshui Formation of the Jixian Group or to the lower part of the Tieling Group.

Similar subformations also exist in the lower part of the Doushantuo Formation above the Sinian Nantuo Tillite in the Lishui River valley, Hunan Province, and contain manganese-bearing carbonate.

b. Manganese-bearing subformation in the maximum- and residual-stability periods. It is well-developed also in east China and is one of the principal manganese-bearing formations in China. The age of the ore formation is Early Paleozoic to Early Triassic. The main stratigraphic positions are as follows:

(a) Late-Devonian manganese deposits. They are distributed mainly in central and southwestern Guangxi Zhuang Autonomous Region and confined to the Upper Devonian Liujiang Formation composed of siliceous rocks, lenticular limestones and siliceous shales of neritic facies. Ore beds consist of oolitic or pisolitic carbonate.

(b) Middle Carboniferous sedimentary manganese deposits. They are distributed in Guangxi Zhuang Autonomous Region and Jiangxi Province and appear to be manganese-bearing carbonate rocks of neritic facies.

(c) Permian sedimentary manganese deposits. They are distributed chiefly in western and southwestern Guangxi Zhuang Autonomous Region, and are also exposed on the west side of Xuefeng Mountain, Hunan Province, and in Sichuan Province. In southwestern Guangxi Zhuang Autonomous Region, they occur chiefly in the upper part of the Permian Heshan Formation composed of argillaceous limestone, carbonate manganese ores and calcareous sandstones and shales of neritic facies. Ore beds generally have oolitic and microbedded structure. On the west side of Xuefeng Mountain, they occur in the Lower Permian Qianyang Formation and are often associated with bauxite. In Sichuan Province, they occur in the Upper Permian Longtan Formation and have a wide distribution and a fixed stratigraphic position.

Finally, there are Early Triassic sedimentary manganese deposits in east Yunnan Province accumulated during the late half of the residual-stability period of platform development. Both the manganese-bearing formation and the ore beds change abruptly.

3) Aluminium-bearing formation.

The important aluminium-bearing formation in China is generated mostly at the platform stage and that of economic value is concentrated in east China. Ore beds generally lie on the ancient erosion surface at the bottom of the aluminium-bearing formation and were formed at the beginning of a transgressive sequence. The bauxite beds in the Middle Carboniferous Benxi Formation of north China, in the Middle Carboniferous and Lower Permian of Guizhou Province, and in the Upper Permian Heshan Formation of west Guangxi Zhuang Autonomous Region and southeast Yunnan Province are all the principal bauxite-bearing strata.

The aluminium-bearing rock of the Upper Permian Heshan Formation in west Guangxi Zhuang Autonomous Region has a thickness of not more than 200 m. It rests disconformably on the Maokou Limestone and underlie, without any marked break, the siliceous rocks of the Dalung Formation. Ore beds lie on an ancient erosion surface of the Maokou Limestone at the bottom of hte aluminium-bearing rock and were controlled by relief. Overlying the ore beds are, in upward sequence, coal beds, argillaceous limestone and thin limestone. This suggests that bauxite was formed at the beginning of a marine transgression.

4) Phosphorus-bearing formation.

This formation is generated mostly in slowly subsiding basins, i.e., on open continental shelves, at the end of the initial- and the beginning of the maximum-stability period of platform development. It is characterized by insignificant lithologic and facies changes, continuous deposition of carbonate sediments and by a low content of coarse-grained clastic rocks. Ore bodies have been formed at the beginning of a marine transgression and at the end of a marine regression, i.e., during the period of a rapid change in the depth of sea water and the rate of sedimentation.

Platform-type phosphorus deposits of economic value in China are known to be confined to two stratigraphic positions.

a. The Sinian Doushantuo Formation. It is distributed in Daba Mountain, Hubei, Yunnan, Guizhou and Jiangxi Provinces. It has also been laid down at the bottom of the Huangyaguan Formation of the Changcheng Group, at the bottom of the Chuanlinggou Formation and in the middle part of the Xiamaling Formation of the Qingbaikou Group in north China.

b. The Lower Cambrian. It is distributed in Yunnan, Guizhou and Sichuan Provinces, and in Kalpin and Kuruktag, Xinjiang Uygur Autonomous Region. It has also been found at the bottom of the Mantou Formation in Hebei and Shanxi Provinces.

In addition, phosphorus deposits have also been generated in the Lower Cambrian of Hubei and north Hunan, in the Silurian of north Jiangsu, in the Middle and Upper Devonian of southwest Guangxi Zhuang Autonomous Region and Sichuan, in the Lower Permian Hufeng Bed of Jiangsu and Zhejiang Provinces, and in the Jialingjiang Group (middle part of the Middle Triassic) of north Guizhou Province.

5) Coal-bearing formation.

This formation is generated chiefly in the residual- and less in the maximum-stability period of platform development. The Taiyuang Group (Upper Carboniferous), the Shanxi Formation and the Lower Shihezi Formation (Lower Permian) of north and the Upper Permian of south China are the principal coal-bearing formations in China. They all appear in the maximum- and residual-stability periods of platform development.

The formation can be subdivided into two subformations according to sedimentary facies.

a. Alternately marine and continental and neritic coal-bearing subformation.

This subformation consists chiefly of alternately marine and continental sandstones and shales with intercalations of coal and limestone. It occurs in the Benxi Group (Middle Carboniferous) and the Taiyuan Group (Upper Carboniferous) of north China, in the Ceshui coal measures (Lower Carboniferous) of Hunan Procince, in the Zishan coal measures (Lower Carboniferous) of south Jiangxi and north Guangdong Provinces, in the Lower Permian (Liangsha and Tongkuangxi Formations and Ma'anshan and Qianyang coal measures) and Upper Permian Longtan Formation of Yunnan and Guizhou Provinces and Guangxi Zhuang Autonomous Region, and in the Carboniferous of Qilian Mountain. Among them the Taiyuan Group (Upper Carboniferous) Carboniferous of north and the Longtan Formation (Middle Carboniferous) of south China are the principal coal-bearing strata in China.

b. Continental coal-bearing subformation.

This subformation has a wide distribution, and coal beds contained in it are also of great importance. The Lower Permian Shanxi Formation (lower part of Lower Permian) and the Lower Shihezi Formation (middle part of Lower Permian) of north China, the Wayaobu coal measures (middle part of Upper Triassic) and the Yan'an Group (Lower Jurassic) of Shaanxi Province, the Shiguaigou coal measures of Daqing Mountain , Inner Mongolia, the Xiangxi Group (Lower Jurassic) of Sichuan Province, and the Shahezi coal measures (Middle to Upper Jurassic) of the Songliao Plain are all platform-type continental coal-bearing formations. Among them the most impostant are the Lower Permian Shanxi Formation and the Lower Shihezi Formation.

6) Oil- and gas-bearing and oil shale-bearing formation.

Platform-type oil and gas fields in China have been generated in two paleogeographical environments: one is shallow sea, bay and lagoon, the other is inland basin. At the present stage of our knowlege, the most important are Mesozoic-Cenozoic inland lake basins. Surrounded by the uplifts of diwa regions, the three residual platforms, the Songliao (Fig. 5.11), Sichuan and Yishan (Ordos) exist as relatively subsiding areas. Such a tectonic condition in combination with wave-like development of oscillatory movements can create a favourable environment for oil generation.

Platform-type oil shales have a wide distribution in China.

7) Salt-bearing formation.

Marine salt-bearing subformation. It is an important type, formed in lagoons during a marine regression caused by the rising of a platform. Examples are gypsum, rock salt and sylvite at the top of the Ordovician limestone in Shanxi Province, large sylvite deposits in the Permian of Shaanxi Province, and gypsum and sylvite in the Middle Triassic Jialing River Limestone of Daba Mountain, Sichuan Province. Platform-type gypsum and salt have also been deposited in Yunnan, Xizang and west Hubei.

Continental salt-bearing subformation. In the period of protracted rising of a platform, salt is usually deposited in inland lake basins. Examples are gypsum and salt deposits in the Shiquanfeng Formation (middle part of Middle Permian) and the Ermaying Formation (Lower to Middle Triassic) of Shanxi Province and in the Cretaceous of Sichuan Province and Turpan, Xinjiang Uygur Autonomous Region.

(4) Diwa-type sedimentary ore-bearing formations

Diwa-type sedimentary ore-bearing formations are generated under strong tectonism marked by the extensive development of block-faulting and arcogenesis-faulting. Diwa basins have an en echelon arrangement in a definite direction and in a definite system, and occur in groups. Mineral deposits in them also have such a distribution pattern. Most diwa basins in the diwa regions of east China, for example, trend chiefly in a NE direction and belong to the NE tectonic systems.

Diwa-type sedimentary ore-bearing formations are generally or predominantly continental in facies. They show a marked facies, lithologic and thickness lateral variation. Ore-bearing formations of the same age show great differences from each other in different diwa basins, and even in the same diwa basin, they also differ considerably from each other in different parts.

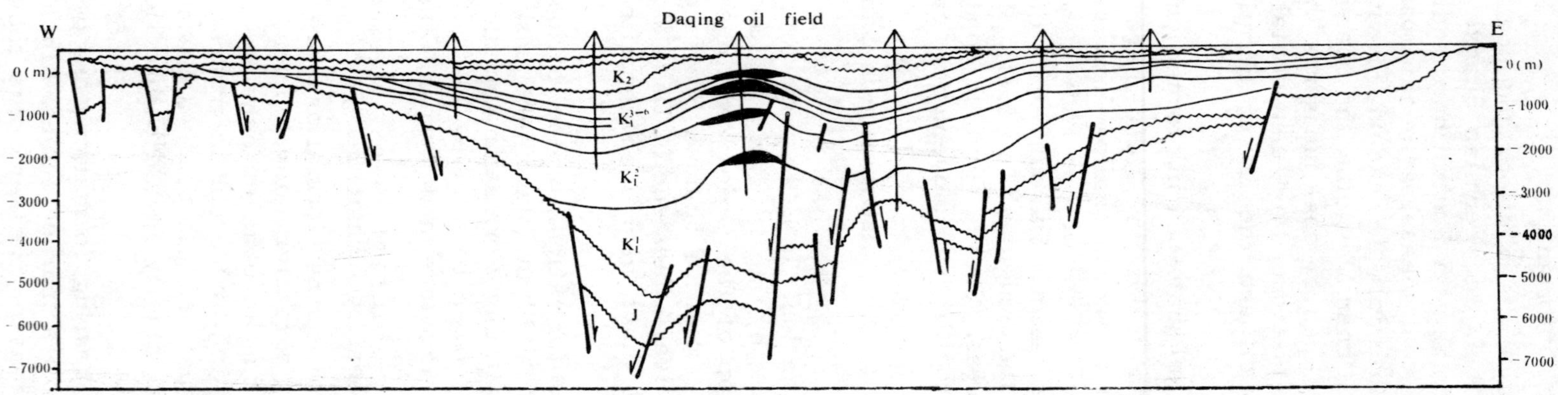

Fig. 5.11. Cross section through the Songliao Basin showing the Daqing oil fields.

Diwa basins generally have a large amplitude and a great rate of subsidence. As a result, ore-bearing sedimentary formations usually contain coarse-grained clastics with a great thickness attaining several thousand and sometimes even ten thousand meters. Ore beds in them can repeatedly appear in vertical section, totalling at maximum more than one hundred in number. Therefore, diwa-type sedimentary deposits generally comprise many beds. Diwa-type sedimentary ore-bearing formations show conspicuous rhythms. The rhythms are mostly continental flysch ones, of discontinuous type, and exist in large numbers. An individual rhythm has a great thickness measured in several or ten meters.

The above-mentioned principal characteristics make a distinction between diwa-type sedimentary ore-bearing formations and geosynclinal- and platform-type ones.

Each period of diwa development offers its own ore-bearing formations. So far as east China is concerned, the initial-mobility period is dominated by coal-bearing formations. After entering into the maximum-mobility period, especially in the diwa regions along the southeast coast of China, sedimentary processes are often accompanied by strong volcanic extrusion, thus creating diwa-type volcano-sedimentary ore-bearing formations. In the residual-mobility period of diwa development, oil-and coal-bearing formations occupy a dominant position.

1) Iron-bearing formation.

This formation may be represented by volcano-sedimentary iron-bearing formation related to volcanic extrusion. According to traditional theories, it appeared only in geosynclinal regions and no longer emerged after the Early Carboniferous. Countless facts have proved, however, that this formation also appeared in diwa regions, chiefly in Jurassic-Cretaceous time. As known from available data, it is distributed mainly in east China. Examples are the Jurassic and Cretaceous of Anhui, Jiangsu and Zhejiang Provinces. They consist generally of continental volcanic lavas, pyroclastic rocks, volcanic tuffs, pebbles and sandy-argillaceous sediments. Iron ore occurs as beds and lens in volcanogenic sediments. The characteristics of this volcanic rock iron deposit type, a new important iron deposit type in China, have already been described in the preceding chapter.

In addition, at the diwa stage iron also accumulated in lake (or sea) basins within a diwa subsidence. It has been suggested that it was derived from erosion areas on geodomes.

2) Coal-bearing formation.

Diwa-type coal-bearing formation is of great economic importance in China [94, 95,]. It is generated chiefly in the initial- and residual-mobility periods of diwa development, but sometimes also in the maximum-mobility period.

Coal-bearing formation generated in the initial-mobility period has a wide distribution. The Yipinglang coal measures (Upper Triassic to Lower Jurassic) of Yunnan and Guizhou Provinces, the Anyuan (Fig. 5.12) and Xiwan coal measures (Upper Triassic to Lower Jurassic) of southeast China, the Huating coal measures of Liupan Mountain, the Jurassic of Qilian Mountain, and the Upper Triassic to Lower Jurassic of northeast, north, central and east China were all generated in the initial-mobility period of diwa development.

Coal-bearing formation generated in the maximum-mobility period is known chiefly in the Late Jurassic to Early Cretaceous coal-bearing strata of northeast China and north Hebei. It also has a sporadic distribution in Yunnan and Guizhou Provinces.

The residual-mobility period is one of the most important coal-forming epochs in northeast China, during which some Tertiary coal fields were formed. Coal fields of this period composed of lignite have also been found in south, southwest and east China.

Coal-bearing formation of this type in China is mostly of continental facies with many coal beds. But an individual coal bed is thin. The lithology, facies and thickness change abruptly. The Anyuan coal measures of Jiangxi Province, for example, have a thickness of only 200 m in some places, but thicken up to 1,400 m at a short distance. Coal beds range from 4 to more than 80 in number. Branching and pinch-out are common.

3) Oil- and gas-bearing (including oil shale-bearing) formation.

Diwa-type oil- and gas-bearing formation consists of the following two subformations.

a. Oil shale-bearing subformation.

Diwa-type oil shale is generally associated with coal and similar to it in depositional features. As known at present, it is distributed mostly in diwa regions of east China, but has also been found in west China, e.g., in Qilian Mountain.

Oil shale-bearing sedimentary subformation occurs in every period of diwa development, but that of large size is seen chiefly in the early residual-mobility period (Tertiary).

Oil shale-bearing subformation generated in the initial-mobility period belongs to the same rock series as the above-described coal-bearing formation generated in the initial-mobility period. The Anyuan and Xiwan coal measures in south China, the Lower Lufeng Formation in Yunnan Province and the Manitemiao Group in Inner Mongolia are all the principal oil shale-bearing subformations deposited in this period.

Oil shale-bearing subformation generated in the maximum-mobility period is known as the Lower Hinggan volcanic rocks (Upper Jurassic) in northeast China and the Upper Jurassic in Hebei and Liaoning Provinces. Oil shale beds have also been found in Cretaceous strata within diwa basins in southeast and southwest China, west Liaoning and Liupan Mountain.

Oil shale-bearing subformation generated in the residual mobility period has a wide distribution in the Tertiary strata of east China. For example, oil shale in the Tertiary Fushun Formation of Liaoning Province (Fig. 5.13) and in the Tertiary Maoming Formation of Guangdong Province has a high grade and a great thickness.

b. Oil-and gas-bearing subformation.

Diwa regions provide favourable conditions for the extensive development of source, reservoir and sealing beds. Therefore, diwa-type oil- and gas-bearing sedimentary formations in China are of great importance and produce many large and rich oil and gas fields. As known at present, they are chiefly continental in origin. Among them the most important is the oil-bearing formation in the

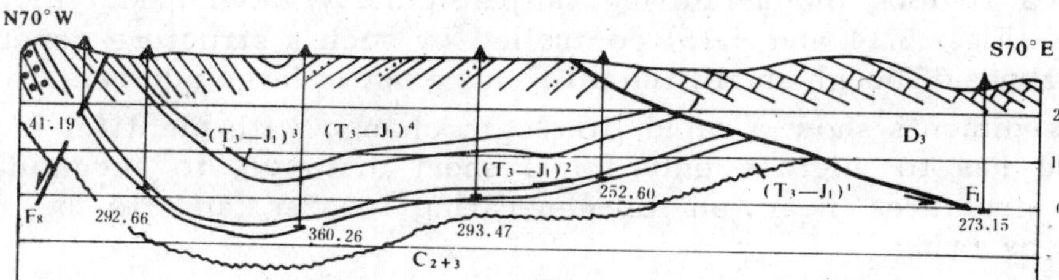

Fig. 5.12. Cross section through the Shixiajiang coal field in Hunan Province.

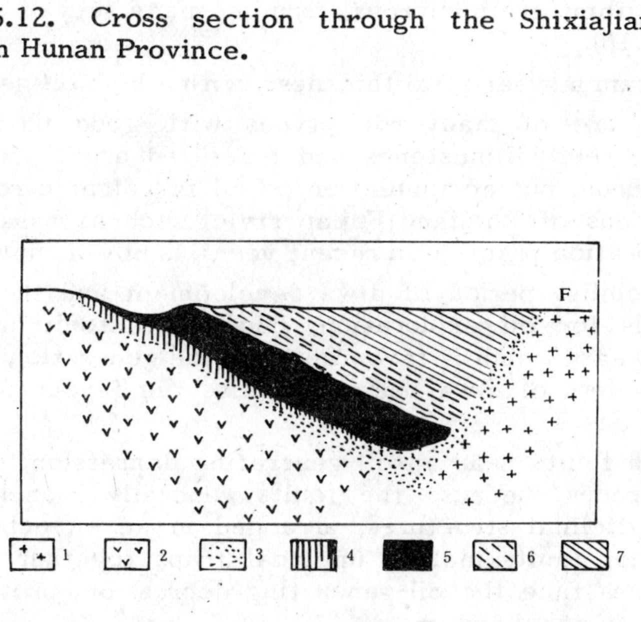

Fig. 5.13. Cross section through the Fushun diwa in Liaoning Province.
1- Gneiss; 2- Basalt; 3- Tuff; 4- Dark brown shale; 5- Main coal bed; 6- Oil shale intercalated with thin coal beds; 7- Green shale; F- Fault.

residual-mobility period. Large oil and gas fields discovered recently in east China, except those seen in the three residual platform regions, all belong to this type [40, 41, 42, 43].

So far as diwa-type oil fields in east China are concerned, their principal features (a-f) and the trend in the search for oil (g-j) can be discussed briefly as follows.

(a) In diwa regions block-faulting is particularly developed. Block-faulting type oil fields (Figs. 5.14 and 5.15) controlled by such a structure generally occur in groups and zones. Though small, the oil fields exist in large numbers.

(b) Diwa sediments show a rapid lithologic change with plentiful faults, which causes oil and has to migrate only for a short distance, to accumulate in the second-order structures near an oil-generating center and to be distributed concentrically as a ring.

(c) Producing horizons exist in large numbers, and oil fields in different horizons frequently get into connection with each other.

(d) Sedimentary rhythms are numerous, forming many sets of source, reservoir and sealing beds (Fig. 5.16).

(e) Source rocks attain a great total thickness, with a high oil generation index.

(f) Reservoir beds are of many rock types with good physical properties. Sandstones, limestones, shelly limestones and fractured argillites are the usually encountered reservoir beds, but accumulation of oil in volcanic rocks (extensively developed in diwa regions of Zhejiang-Fujian style) such as basalt, trachyte and tuff, as proved by exploration practice in recent years, is not unknown.

(g) The residual-mobility period of diwa development and the miodiwa region are the principal goals for oil exploration, because tectonic movement during this time is relatively weak, which is favourable for the generation and preservation of oil and gas. The borders of diwa basins are also the favourable zones for the accumulation of oil and gas.

(h) In major growth faults near an oil-generating depression, oil and gas pools frequently occur in groups, because the faults generally branch out noses and clamshell-shaped or anticlinal structures, arranged in an en echelon manner, as well as derivative minor faults making up local traps together with the faults. Many of these structures face the oil-generating depression, providing favourable sites for the accumulation of oil and gas.

(i) Low protrusions near an oil-generating depression are a new domain for the search of shallow secondary oil pools. Frequent crustal movements in diwa regions have crused the repeated migration of oil and gas. In case of existence of source rocks at depth, of canals (faults or unconformities) for oil migration and of traps at shallow places, shallow secondary oil pools can be generated.

(j) Pre-diwa structural layers in a diwa region are also the places for the search of oil and gas and merit our attention. Some oil and gas fields in them may have been inherited from the platform or geosynclinal stage; others, formed at the diwa stage, take the pre-diwa structural layers as reservoir beds to form secondary oil pools (Fig. 5.17).

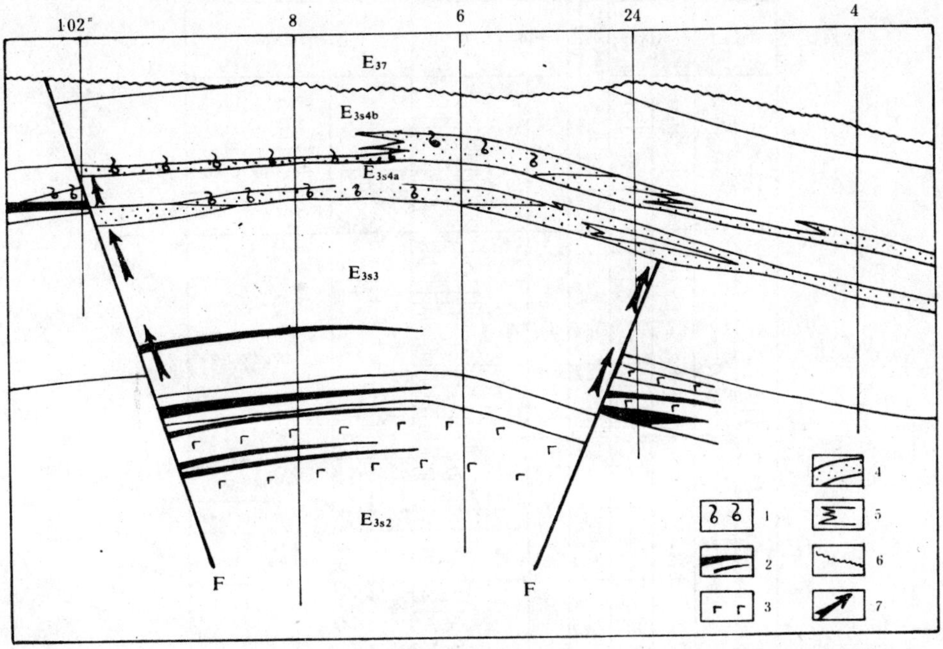

Fig. 5.14. Cross section through an oil pool in a certain diwa-type oil field.
1, Gas bed; 2, Oil bed; 3, Basalt; 4, Sand bed; 5, Mudstone Lens; 6, Unconformity; 7, Direction of migration of oil and gas.

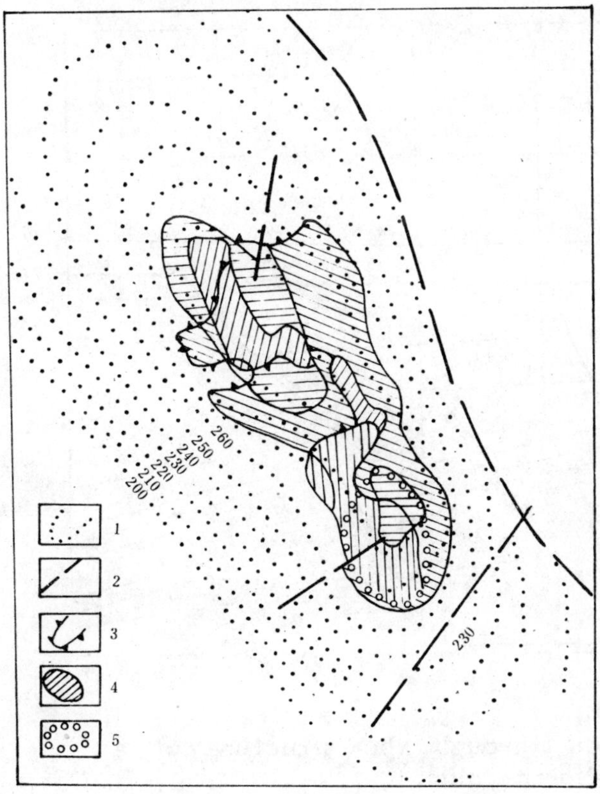

Fig. 5.15. Diwa oil pool of compound type.
1, Contour; 2, Fault; 3, Line of pinch-out; 4, Oil pool; 5, Gas roof.

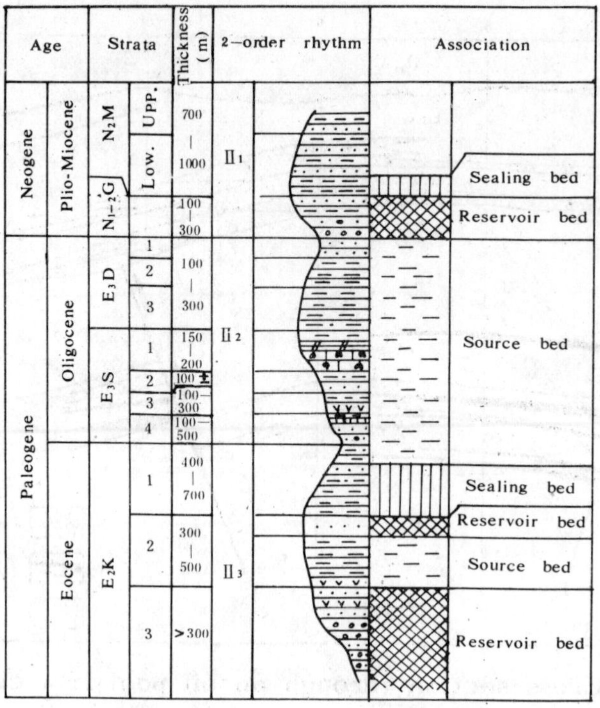

Fig. 5.16. Depositional rhythm in Tertiary sediments of a diwa basin, North China diwa region.

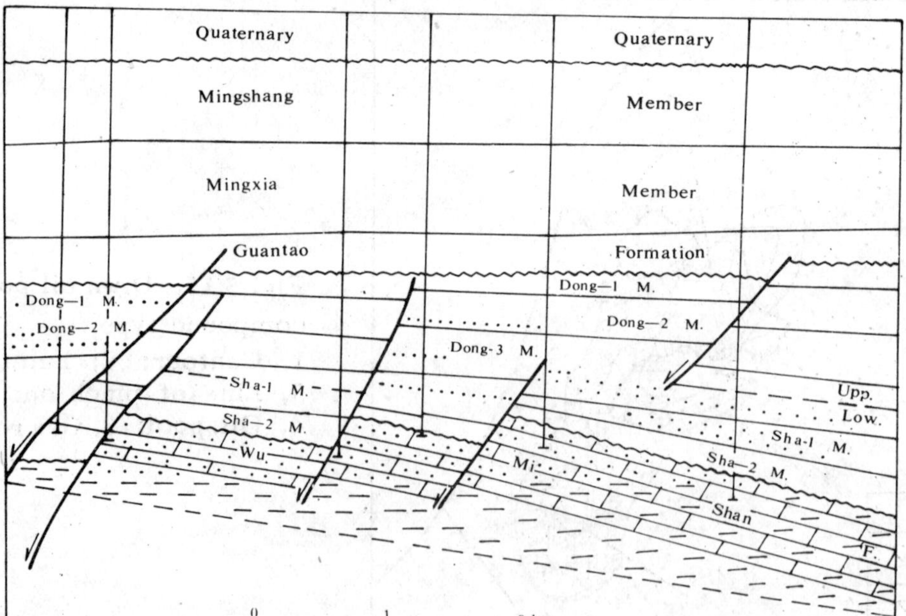

Fig. 5.17. Cross section through the structure of a buried-hill-type oil field, North China.

4) Copper-bearing formation.

This formation may be represented by copper-bearing sandstone (shale or conglomerate) in a copper-bearing clastic formation. It has been deposited in various periods of diwa development and in diwa basins of various regions, and is found, for example, in the Jurassic and Cretaceous of Yunnan Province, in the Cretaceous and Tertiary of south China and the Tarim basin, Xinjiang Uygur Autonomous Region. The formation is principally of fluvial and deltaic facies, with a great thickness amounting sometimes to 10,000 m, and shows clear rhythmic sedimentation of continental discontinuous type. The facies and lithology change abruptly. Ore-bearing horizons exist in great numbers; many of them are economically valuable. For example, Middle and Upper Jurassic to Upper Cretaceous red beds in central Yunnan Province, about 10,000 m thick, contain copper-bearing sandstones and shales everywhere, among which economically valuable ones are concentrated chiefly in the Matoushan Formation (lower part of Upper Cretaceous).

It is worth noting that in a few places the copper-bearing clastic formation is known to contain some rare elements. Various placers also merit our attention in the future.

5) Salt-bearing formation.

Diwa-type salt-bearing formation in China is chiefly continental in facies and has a wide distribution. It has been laid down in all diwa basins formed in various periods of diwa development, but more commonly it has been deposited in the maximum- and residual-mobility periods. Examples are Tertiary salt deposits in the NE, North and Central China diwa regions. Dawa-type salt-bearing formation is also distributed in northwest China. Salt deposits of this age, with many beds and a great total thickness, represent an important salt-forming epoch in China. Moreover, salt-bearing beds in the upper part of the Upper Triassic Yanchang Formation of Helan Mountain, northwest China, in the Lower Lufeng Formation (Middle Jurassic) of Yunnan Province, in the Cretaceous of Dahong Mountain, central China, in the Cretaceous of Anhui Province, in the upper part of the Jinggu Group (Cretaceous) of Yunnan Province, and in the Cretaceous of southeast China are also of this type (Fig. 5.18). It is reported that in the North Xizang diwa reigon a great amount of modern salt lake deposits are being formed.

Being at different stages of crustal development and under different tectonic conditions, sedimentary ore-bearing formations of different stages (e.g., geosynclinal stage, platform stage and diwa stage) differ from each other. The foregoing discussion has dealt only with the principal features of some sedimentary ore-bearing formations in space and in time. Further studies of the regularities governing their deposition are not only important for the development of metallogeny, but also of great practical significance in the exploration for mineral deposits.

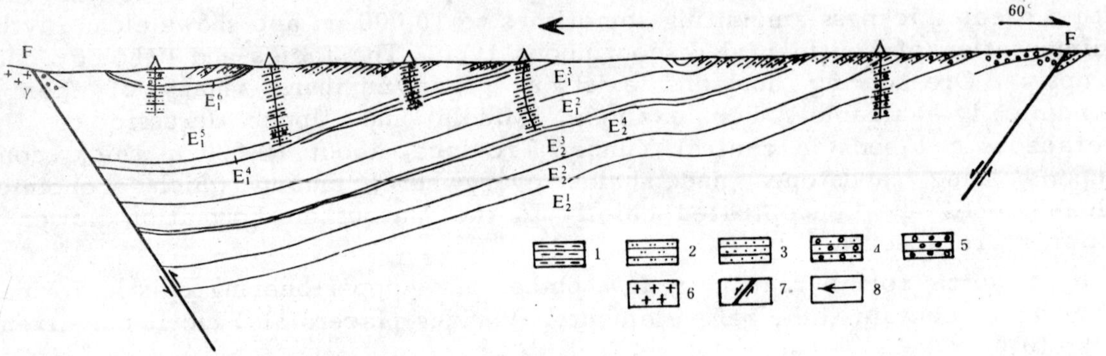

Fig. 5.18. Cross section through the ore-forming basin in Guangdong Province, where the Longtian gypsum deposits are located.
1, Mudstone with gypsum; 2, Siltstone; 3, Sandstone; 4, Sandstone and conglomerate; 5, Conglomerate; 6, Granite; 7, Fault; 8, Direction of migration of sedimentation center.

IV. Polygenetic compound ore deposits

1. Definition and Peculiarities

Various ore deposits in the crust as we see them today, particularly older ones, have underwent changes and reworking in varying degree after their formation. Polygenetic compound ore deposits as defined by the author (1975, 1978)[39] indicate those ore deposits which evidently possess many-sided genetic characteristics as a comprehensive result of more than one mineralizations. They can be ascribed to neither endogenic nor exogenic ore deposits. Only by singling them out as a third genetic type of ore deposits can we understand and expound the mechanism of their formation and the regularities governing their distribution in time and space and determine the direction for the search for them.

Polygenetic compound ore deposits are formed through many mineralization stages dervived from many sources, and belong to many genetic types.

Various mineralizations and ore-forming materials producing polygenetic compound ore deposits may occur either at the main stages (pre-geosynclinal, geosynclinal, platform, and diwa) of crustal evolution within the same ore-bearing area or at various substages (for example, early, middle and late periods of diwa development) within the main stage.

2. Principal models for metallogenesis

Polygenetic compound ore deposits vary greatly in many respects, depending on the concrete history of tectonic development and conditions of ore formation. Based on available data, they may be divided into three models according to their common genetic characteristics, i.e., superimposition-richening, reworking-richening, and reconstruction-enrichment. Each model may be subdivided into several modes of formation.

(1) Polygenetic compound ore deposits of superimposition-richening type

Polygenetic compound ore deposits of this type have underwent two or more tectonic-metallogenic stages. Magmatic, metamorphic and other ore-forming processes alone or jointly carry new ore materials to form a new ore deposit, being superimposed on an older deposit as a base formed at an earlier stage (generally with reworking). Ore deposits of this type have often become most complicated with a tenor higher than that for original ones. Such ore deposits are formed in the following ways.

1) Sedimentary + hydrothermal or/ and contact-metasomatic type. For example, the Bayanobo iron-rare earth-niobium deposit was formed by means of a close combination of two ore deposits (Figs. 5.19 and 5.20). One of them is a sedimentary iron ore deposit which occurred in a limestone-sandstone-mudstone (partly metamorphosed into dolomite, quartzite and slate) series of the Sinian Bayanobo Group and served as a base. The other is a rare earth-niobium deposit which resulted mostly from the Hercynian-Indosinian granitic hydrothermal (metasomatic) process. However, superimposition (and reworking) varies in different parts of this polygenetic compound ore deposit. At the West Mine, the primary features of the base deposit of sedimentary origin are well preserved and proved recently by drilling. But at the Chief and East Mines, more rare earth and niobium (probably and iron) have been introduced, and the base deposit has been subjected to stronger reworking, so that iron content has also been increased.

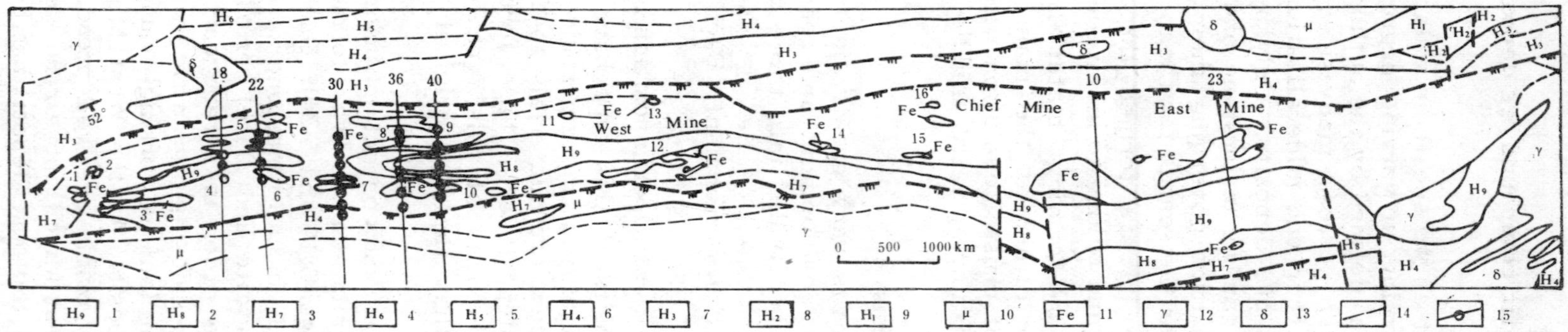

Fig. 5.19. Geologic map of the Bayanobo iron ore district, Inner Mongolia, China (after the Bayanobo Mine). 1, Quartzite and slate; 2, Argillaceous limestone; 3, Siliceous limestone alternating with quartzite; 4, Light-coloured quartzite; 5, Dark-coloured slate; 6, Dark-coloured quartzite; 7, Dark-coloured slate; 8, White quartzite; 9, Black slate and coarse-grained quartzite; 10, Migmatite; 11, Iron ore; 12, Granite; 13, Diorite; 14, Fault inferred from geophysical prospecting; 15, Profile and number of drill hole.

The Diantan iron deposit in Yunnan Province is another example. Its base deposit is a sedimentary iron ore deposit which occurs in the Carboniferous-Permian limestone and on which during the Yanshanian orogeny a hydrothermal contact metasomatic deposit produced by granitic magma was superimposed.

2) Metasedimentary + hydrothermal or / and contact metasomatic type. It may be represented by the Shilu iron ore deposit (Figs. 5.21 and 5.22). The base deposit is a metasedimentary deposit in the Paleozoic carbonate-clastic rocks of the Shilu Group which was reworked later by the hydrothermal metasomatic process of Mesozoic granitic intrusions and on which a skarn deposit was superimposed.

The Gongchangling iron ore deposit, Liaoning Province, is based on a metasedimentary iron ore deposit in the Anshan Group of Archean age which was enriched later by hydrothermal reworking and superimposition.

3) Volcanogeno-sedimentary or volcanogeno-metasedimentary + hydrothermal or / and contact-metasomatic type. The Daihongshan iron deposit, Yunnan Province, is an example.

(2) Polygenetic compound ore deposits of reworking-richening type

By this type we mean a polygenetic compound ore deposit formed by the individual or joint action of magmatic hydrothermal solutions, subsurface water solutions and tectonic processes which have strongly reworked a pre-existing deposit (sometimes with superimposition to some extent). The pre-existing deposit can be improved in quality by means of reworking (chiefly reworking, but also superimposition) or can migrate by plastic flow to a favourable stratigraphic or structural position, forming a more valuable, vein-like, stratiform, stratoid or chambered ore deposit. Such a compound ore deposit has the following three ways of formation.

1) Magmatic hydrothermal reworking-richening type. Under this type we understand that polygenetic compound ore deposit reworking of which is due to magmatic hydrothermal solutions.

a. Sedimentary + hydrothermal reworking type. The most distinct example is the Daye iron ore deposit (Fig. 5.53). Formerly this deposit was considered to be of simple skarn type. Recently, however, siderite has been discovered in the wall rocks. On the example of the Iron Hill Mine, in addition to 130 million tons of magnetite, there are 10 million tons of siderite. Siderite has been subjected to slight metasomatic alteration and sometimes occurs as veinlets. Its content is 30-40%, in some cases 20-30%, and rarely 45%. According to Zhou Shengsheng (1979), the Iron Hill magnetite deposit originally was a Triassic siderite deposit which was subjected to hydrothermal metasomatism of Yanshanian granodiorite and mostly changed into a magnetite deposit. The present-day siderite deposit is actually a relict after reworking.

The Devonian Ningxiang-type iron deposits in east Hunan had also been partially affected by Mesozoic magmatism leading to their reworking-richening and thus forming polygenetic compound ore deposits. As is calculated, of 29 Ningxiang-type iron deposits in Chaling and Youxian, 9 were reduced, magnetized and transformed partially from original hematite into magnetite with a relatively increased iron content. But in the vicinity of these deposits, the iron content of the unreworked (i.e., primary) ore is generally low.

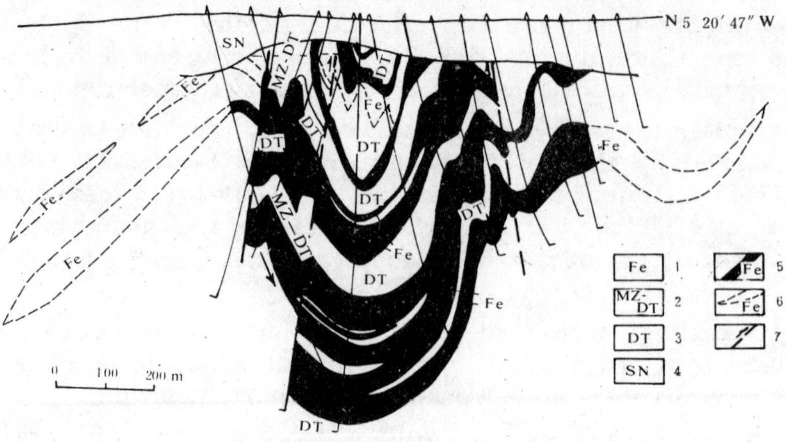

Fig. 5.20. Geologic-geophysical synthetic map along the No. 30 profile in the Bayanobo iron ore district, Inner Mongolia, China (after the Bayanobo Mine).
1, Iron ore; 2, Mineralized dolomite; 3, Dolomite; 4, Sedimentary cover; 5, Orebody controlled by drill holes; 6, Orebody inferred from geophysical prospecting; 7, Inferred fault.

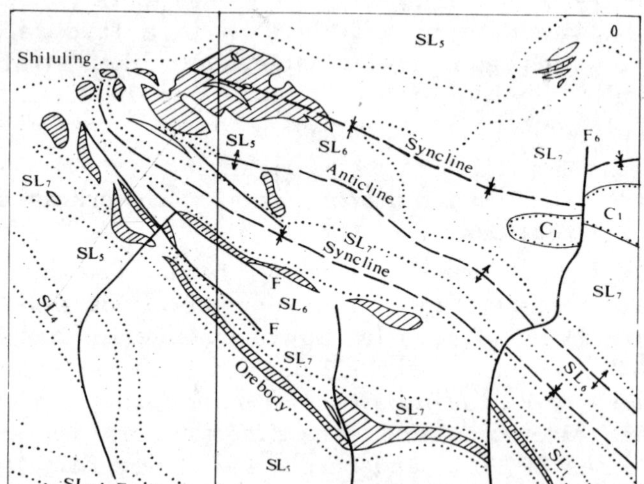

Fig. 5.21. Map of the Shilu iron ore district, Hainan Island, Guangdong Province, China (after 934 Prospecting Team).
1- Geologic boundary; 2- Fault; 3- Syncline and anticline; 4- Granite; 5- Iron orebody.

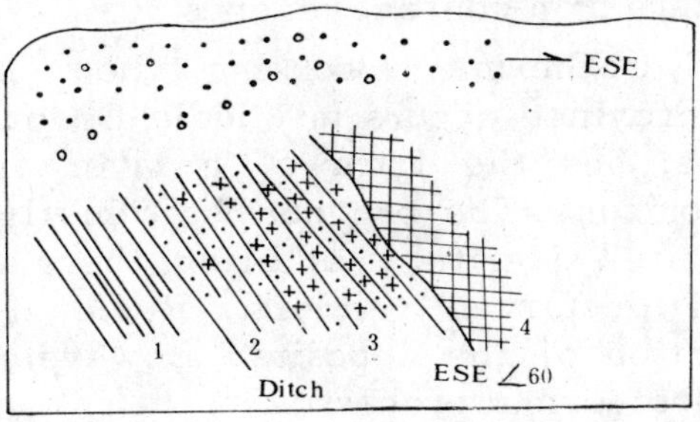

Fig. 5.22. Transitional relationship of the lowermost iron ore bed in SL_3 to the ore-bearing deposits in the Shilu iron ore district.
1, Ferriferous phyllite; 2, Ferriferous sandstone; 3, Siliceous iron ore; 4, Hematite.

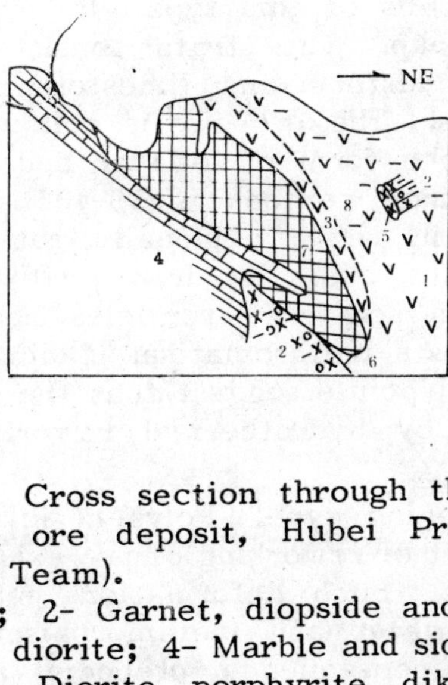

Fig. 5.23. Cross section through the Iron Hill Mine, Daye iron ore deposit, Hubei Province (after 609 Geological Team).
1- Diorite; 2- Garnet, diopside and andalusite skarn; 3- Altered diorite; 4- Marble and siderite; 5- Diopside skarn; 6- Diorite porphyrite dike; 7- Magnetite orebody; 8- Biotite-bearing diopside diorite.

Besides, the Sinian Xuanlong-type iron deposits in north China generally consist of hematite, but the latter was metamorphosed into magnetite at the contact with Yanshanian granite, where iron content reaches 50%.

b. Metasedimentary + hydrothermal reworking type. The Daxigou polymetallic siderite deposit in Shaanxi Province occurs in Middle Devonian iron-bearing clastic and clastic-carbonate rocks, has the form of a layer and is controlled by an ore-bearing stratigraphic horizon. The ore consists chiefly of siderite, magnetite in small amounts and magnetited siderite. In addition, there are associated elements: lead, zinc, etc. Magnetite occurs in veinlets. Regional metamorphism is one of the causes of magnetization of the deposits. However, there is probably also subsequent reworking by hydrothermal processes.

The Sinian Jiangkou-type iron deposits in Hunan Province generally consist of hematite in Lanyang and occur in the Sinian Jiangkou Formation. Though they consist of magnetite and hematite in Jiangkou and other places, the proportion of magnetite amounts only to 10%. The iron content of these metasedimentary ores generally ranges only from 26 to 30%, and grains are very small and difficult to dress. But at Guangdimiao, Qidong County, affected by Yanshanian granodiorite intrusions, the proportion of magnetite has been increased to 50% or more, and grains have coarsened and become easy to dress.

The Etouchang iron deposit in Yunnan Province occurs in the Sinian Kunyang Group. Its original deposit was of metasedimentary type and was also clearly enriched at the contact with granite.

c. Volcanogeno-sedimentary + hydrothermal reworking type. The Dangqu siderite deposit in Xizang (Tibet) is of this type. Its original deposit occurs in the mid-Jurassic Yanshiping Group. The strata consist of shallow-marine and continental-marine sandstone, mudstone and limestone, partially with gypsum and meta-alkaline acid interbeds. The orebody is composed mainly of siderite, occurs along bedding planes in the form of a layer and lens, and is concordant with the enclosing rocks. Its iron content is 30-40%. In addition, there is specularite which occurs often in veins, diagonally cuts the stratification and frequently intersects the siderite ore. Its iron content is more than 50%. Sometimes specularite occurs as scattered particles between the siderite ore and gangue minerals. There exists metasomatism leading to a slight alteration of siderite. The above-mentioned points suggest that the deposit is a polygenetic compound ore deposit formed by hydrothermal reworking and metasomatism of the original siderite deposit.

2) Structural reworking-richening type. Polygenetic compound ore deposits of this type are formed as a result of reworking of pre-existing deposits by folding, faulting and other deformations. Such deformations may be accompanied by hydrothermal solutions, heated water and metamorphism. This type may also be richened in ore grade. As a consequence, orebodies are often deformed and plastically flow to be concentrated in a favourable sructural position, forming an ore pillar or ore shoot. The common modes of formation include folding-reworking type and faulting-reworking type, which can appear individually or in association.

Polygenetic compound ore deposits of structural reworking-richening type originally may be sedimentary or metasedimentary, rarely volcanogeno-sedimentary or volcanogeno-metasedimentary. The Wufeng stone coal deposit of fissure type in Hunan Province may be cited as an example. Its original deposit occurs in Sinian-Cambrian rocks. During Mesozoic time faults cut the deposit and rendered it sufficiently plastic to permit it to flow to a favourable structural position. After such reworking, the ore was metamorphosed into anthraxolite stone coal, the heat output of which is 2-4 times as high as that of the primary ore, amounting to 3,500 cal/ gram at the maximum. Orebodies are characterized by fault breccia and show crosscutting relationships with the wall rocks, and are often accompanied by quartz veins. They are usually 0.4-0.5 m thick, but have become ore shoots with a thickness of 30 m when meet with a favourable structural position (i.e., a site of increased dip angle of a normal fault plane). Evidently, thickening is due to a change in the attitude of the fault plane. According to Tong Yuming (1980), the Huangcai anthraxolite deposit in central Hunan Province is a polygenetic compound ore deposit formed as a result of evolution of oil.

3) Underground water reworking-richening type. Polygenetic compound ore deposits of this type are formed as a result of reworking of pre-existing deposits by underground water solutions, particularly by underground water containing CO_2. One of examples is the Huangmei siderite deposit (Fig. 5.24). Its original deposit consists of siderite beds which occur in the mid-Carboniferous Huanglong Formation and the Lower permian Qixia Formation with fixed stratigraphic positions. The ore is dark-grey, its ooide is fine-grained (0.02-0.25 mm), and the average content is 33.83%. However, the ore has mostly changed into yellow one, its ooide has coarsened (0.4-0.8 mm), and the content has also been raised up to 39.25% on an average. Moreover, veinlets often stretch out from the original beds and fill various fissures. According to available data, this is a polygenetic compound ore deposit richened by circulation of underground water, which in the process of subsequent structural reworking has made the useful components remelted and reworked. The studies by Sun Jiafu (1979) have proved this view.

According to Liao Shifan, at the Caiyuanzi ore segment of the Tiekuang Mountain ore district, Guizhou Province, stratiform dark grey siderite of sedimentary origin above a 2,150 m elevation has been transformed into cream-coloured to form chambered and vein-like orebodies. 10% siderite at Guanyin Mountain, Guizhou Province, below a 1,200 m elevation has been transformed into cream-coloured, but the others are still dark grey; above a 1,900 m elevation, cream-coloured siderite has reached 30-40%. This is thought to be related to the action of underground water and CO_2 content in it.

(3) Polygenetic compound ore deposits of reconstruction-enrichment type

By this type we mean polygenetic compound ore deposits whose ore-forming materials originally were scattered in source beds, but later due to the action of undergroud water or hot aqueous solutions and magmatic hydrothermal or metamorhpic hot solutions it migrated to a favourable stratigraphic or structural position to form an ore deposit. The formation of the deposits may be accompanied by structural deformations. Their modes of formation are as follows.

1) Underground water reconstruction-enrichment type. Many deposits of uranium, lead, zinc, antimony, mercury and barite in Hunan, Guangxi and Guizhou Provinces belong to this type. The GErman metallogenist Hans Schneiderhohn put forward the concept of "regenerated mineral deposits" to interpret the origin of the above-mentioned ore deposits in Western Europe. Such ore deposits also may be ascribed to this type.

According to Yao Zhenkai (1980), a certain uranium deposit in China occurs in strata of various ages, from Sinian to Cenozoic, but the age of mineralization without any exception is late Mesozoic-Cenozoic (Fig.5.25). Meanwhile, the intensity of uranium mineralization is in direct proportion to the density of cracks. In the cracks and their close vicinity, the intensity of uranium mineralization has been distinctly increased. Moreover, controlled by faults and cracks, uranium minralization may cut across several horizons (Fig. 5.26). Such a strata-bound and structure-bound phenomenon cannot be explained, unless it results from the subsequent reconstruction-enrichment. It is clear that the formation of the deposit can be explained in terms of mobilization and redistribution of the primary contents of uranium in the source beds by circulation of hot aqueous solutions formed by dynamic metamorphism.

The Sidingchang lead-zinc deposit in Guangxi bears no relation to magmatic intrusives and was regarded by Soviet geologists as a telethermal ore deposit. It seems now that the deposit may be considered as a polygenetic compound ore deposit of underground water solution or hot solution reconstruction-enrichment type. Finally, the enriched part of the lead-zinc deposit in southwest Hnan Province is apparently controlled by fractures and genetically may be ascribed to this type.

2) Metamorphic-hydrothermal enrichment type. The Hongtoushan copper-zinc deposit in Liaoning Province occurs in copper-bearing metamorphic rocks (2,100 m.y.) composed of pre-Sinian interbedded biotite plagioclase-gneiss, sillimanite-gneiss and garnet-biotite-hornblende gneiss. The form of orebodies varies from prismatic, vein-like, stratoid to chambered. They are in jigsaw-like contact with the enclosing rocks and subjected to clear metasomatism. At the same time, however, they are confined to a definite stratigraphic position, occur in conformity with the schistosity of the enclosing rocks and only few veins diagonally cut the schistosity. Orebodies of various forms are closely joined, and along strike and dip there exist bulge, contraction, branching and compounding. Ore pillars often appear at the turn of strata, in the subordinate folds and at the intersections of various cracks. They are seen at the plunge of an inverted fold. The above-mentioned attitude is controlled obviously by interstratified cracks at the axis of the fold. According to available data, the deposit is intimately related to the mineral content of the enclosing rocks. Analyses show that the copper content of the biotite plagioclase-gneiss in the Ar_1S_2 bed is 200 ppm, and that from two pieces of biotite even reaches 300-500 ppm, i.e., 3-5 times higher than the clarke. However, when the original rocks are subjected to migmatization and transformed into striped biotite migmatite and biotite migmatite-gneiss, the copper (and zinc) content is regularly decreased. It is reasonable to infer that some of the enclosing rocks were rich in copper and zinc and served as a source of ore-forming materials, and that the subsequent hypogene hydrothermal process of migmatization resulted in the migration of ore-forming elements from the source to a new place.

3. Tectonic conditions of formation of polygenetic compound ore deposits

Tectonic environments of formation of polygenetic compound ore deposits of the above-mentioned various models are diverse and may be different either between ore deposits of various models or between various ore deposits of the same model. According to the tectonic characteristics of China, they may be briefly summarized as follows.

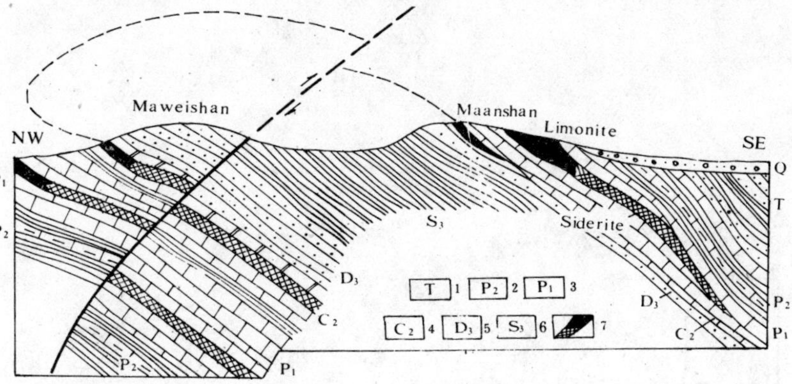

Fig. 5.24. Cross section across the Huangmei siderite deposit in Hubei Province (after Hubei Metalluro-Geological Exploration Corporation).
1, Sandstone and shale of the Puqi Group and limestone of the Daye Formation; 2, Shale and marl of the Longtan Formation; 3, Siliceous rock and flint limestone of the Maokou and Qixia Formations; 4, Limestone and dolomite of the Huanglong Formation; 5, Sandstone .. the Wutong Formation; 6, Sandstone and shale of the Shamas Formation; 7, Limonite and siderite.

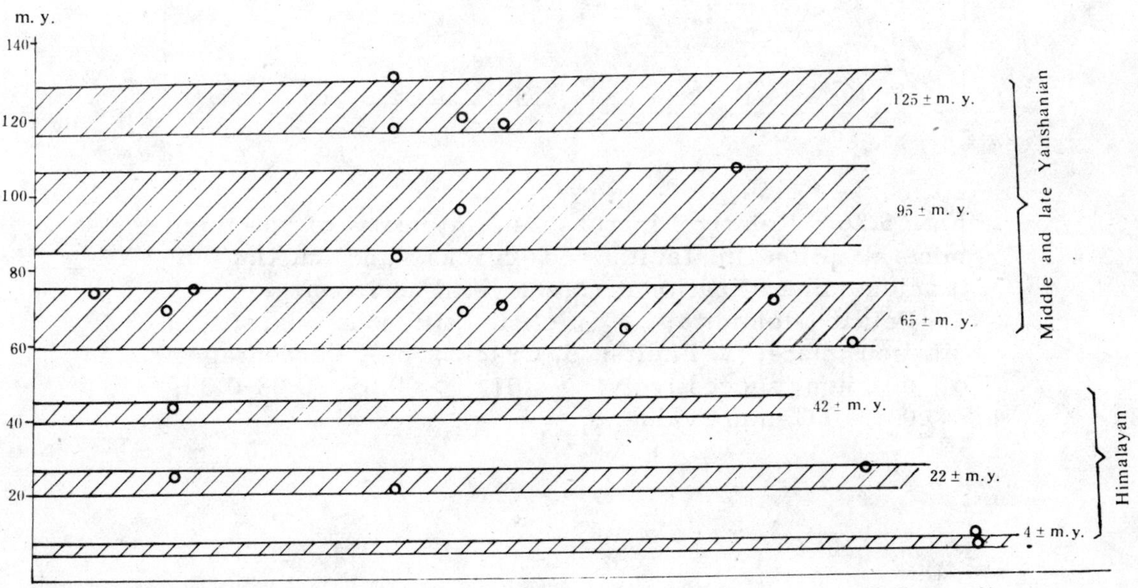

Fig. 5.25. Relationship between epoch of ore-bearing horizon and age of mineralization in some strata-bound uranium deposits.

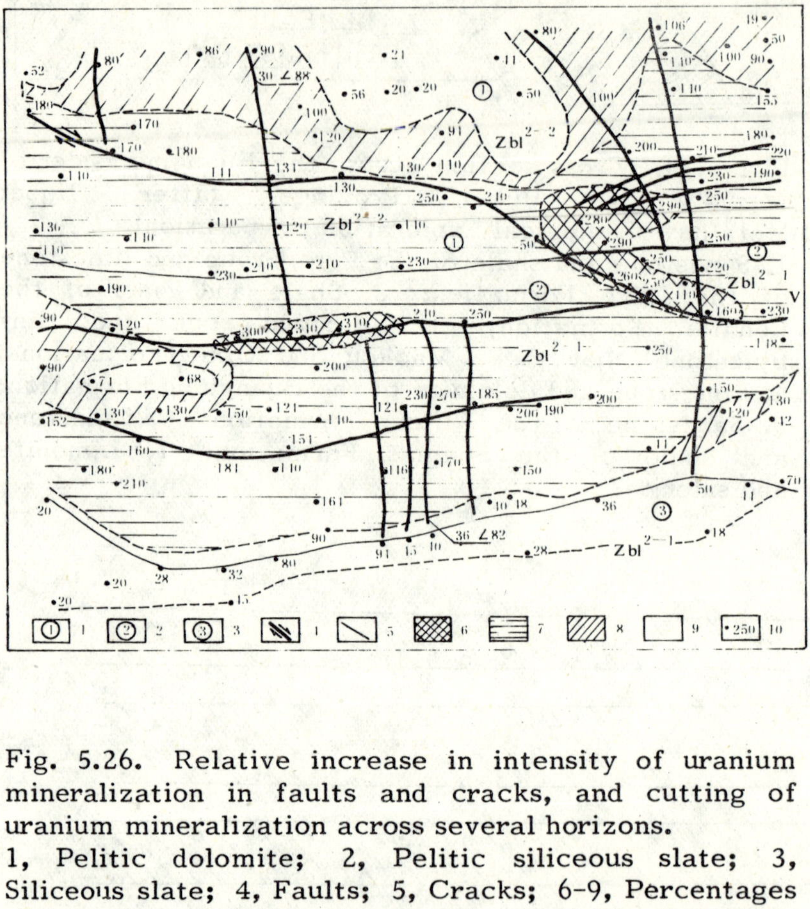

Fig. 5.26. Relative increase in intensity of uranium mineralization in faults and cracks, and cutting of uranium mineralization across several horizons.

1, Pelitic dolomite; 2, Pelitic siliceous slate; 3, Siliceous slate; 4, Faults; 5, Cracks; 6-9, Percentages of uranium successively: > 0.1, > 0.05, 0.03-0.05, < 0.03; 10, Gamma value.

(1) A geosynclinal (or pre-geosynclinal) ore deposit on which a diwa ore deposit is superimposed

Polygenetic compound ore deposit formed under such a tectonic condition is mostly seen in diwa regions. Its base deposit is generally a metasedimentary of volcanogeno-metasedimentary (caused by regional metamorphism in all cases) one formed at a geosynclinal or pre-geosynclinal stage. Later, during a diwa stage, because of the introduction (often accompanied by reworking) of ore-forming materials carried by magmatic processes (often accompanied by tectonic processes and contact or dislocation metamorphism), the base deposit is transformed into a new and more complicated deposit. Here we take iron deposits as example. Iron deposits of geosynclinal (or pre-geosynclinal) type may be of large dimensions. Though originally the iron content is only 20-30%, it can be considerably increased and the deposit can change into a big and rich one, if a deposit of diwa type is superimposed on it later.

The above-mentioned Shilu-type iron deposit is a striking example. This deposit is located in the south of the SE diwa region. Its original deposit occurs in the Shilu Group, which is characterized chiefly by a flyschoid formation and subjected to intense folding of tight type and regional metamorphism. The moderately-slightly metamorphosed sequence in the Shilu Group is dated at early-middle Cambrian, and was lithified, folded, uplifted and metamorphosed during the Caledonian orogeny. Later, the area entered into a stage of tectonic quiescence, the platform stage, which lasted to the Paleozoic or the beginning of the Mesozoic. Though the age of the Shilu Group is not yet determined, most geologists agree that it was deposited in a geosynclinal environment. Therefore, the evidence for the metasedimentary origin of the Shilu iron deposit is beyond doubt, and this deposit is just one of the deposits typical of geosynclinal regions. However, the evidence only shows one aspect of the genesis of the deposit, and refers to the first stage in the complex process of its formation, but not its entire process. After the mid-Mesozoic, the area entered into a diwa stage. Starting from the Indosinian orogeny (or from the end of the Hercynian orogeny, in the range of 200-250 m.y.), this new stage of crustal evolution was characterized by intense tectonic movement. Folds are chiefly of broad type. Block-faulting is well-developed and accompanied by a great mass of granodiorite intrusions (chiefly of Yanshanian age). According to Wu Qinsheng (1979), during this stage, under the bake of the intrusions, thermal metamorphism took place, andalusite-quartz schist, andalusite-chiastolite hornstone, tremolite-diopside rock and tremolite-diopside-bearing dolomite formed and, finally, local Sr isotope homogeneity occurred. At the same time, hydrothermal contact metasomatism also took place and the metasedimentary deposit formed during the first metallogenic stage was subjected to intense reworking and got additional ore-forming materials, forming compact and massive magnetite. As a result, the base deposit changed into a more complicated and richer deposit as we observe it today.

(2) A platform ore deposit on which a diwa ore deposit is superimposed

Polygenetic compound ore deposit of this type means what is based on a sedimentary deposit formed during a platform stage and enriched by a diwa ore deposit carried mostly by magmatic activity.

The above-mentioned Bayanobo deposit is (or partly is) one of the polygenetic compound ore deposits formed under such a tectonic condition. This deposit is located on the border of the North China diwa region. The ore-bearing Bayanobo Group formerly was considered to be of Ordovician age and its sedimentary formations to be of geosynclinal type, i.e., part of the Mongolian Hercynian paleogeosynclinal sediments. Recent investigations indicate, however, that the Bayanobo Group corresponds to the lower part of the Sinian with its absolute age being 1,530 m.y. It belongs to platform sediments and is equivalent to the Zhangcheng Group deposited during the first period of the platform stage in the North China diwa region. Its main part is unmetamorphosed. Locally occurring metamorphism is uneven in development, controlled clearly by the adjacent granite intrusion, and belongs to contact metamorphism. On the other hand, the sedimentary formations are characterized by a wide distribution of arkose, and folds are of tight type. The ore deposit is located on the northern border of the North China paleoplatform of Sinian age, in a paleosyneclise, the Bayanobo paleosyneclise. Ten odd km north of it lies the boundary with the Mongolian paleogeosyncline, the Yinshan south foot-Chifeng-Kaiyuan deep fracture. Being close to the geosynclinal region, the sediments in the syneclise possess the characteristics of a transition zone from geosyncline to platform, and folding movement occurred intensely and took a tight type. In this paleosyneclise, 500 km long in an E-W direction and 100 km wide in a N-S direction, the Bayanobo Group and iron ore beds within it were deposited. This is the base deposit formed during the first metallogenic stage. Later, at the end of the Paleozoic to the beginning of the Mesozoic, the neighboring Mongolian geosyncline closed, and its intense folding movement and the accompanying magmatism spread to this area. In addition, the deep fracture on the boundary was reactivated. As a consequence, the Bayanobo Group and other sediments of platform type were tightly folded and intruded by granite. At the beginning of the Mesozoic, the area entered into a diwa stage marked by emplacement of granite. The deposits of rich rare earths and niobium with probably a part of iron ore were formed chiefly during this stage. The Sinian iron beds, i.e., the base deposit, were also subjected during this stage to metasomatism and hydrothermal alteration in varying degrees in different parts, or were intersected by subsequent magnetite. According to the samples analysed, the absolute age of the granite is mostly 340-220 m.y., but that of the aegirine-type iron ore at level 1,722 m in the Chief Mine was determined as 207 m.y., i.e., Indosinian. Furthermore, in respect to the petrochemical characteristics of the granite at this mine, its silica content commonly exceeds 73%, and in some cases is higher than 80%; alkalinity is slightly increased with more than 8% K_2O+Na_2O; but iron and magnesium contents are low. These features are similar to those peculiar to diwa-type granite, but different from those characteristic of Hercynian granite in the Mongolian paleogeosyncline. Petrochemically the latter may be attributed to geosynclinal type (low SiO_2 and alkali contents, but high iron and magnesium contents). Therefore, the age of the granite in this area is worthy to be considered further.

The iron deposit in the Shihuiding Formation at the Shilu Mine is a polygenetic compound ore deposit formed also under such a tectonic condition. Originally it was iron ore beds formed during a platform stage in the Paleozoic. Later it was subjected to contact metasomatism, reworking and ore superimposition by the Yanshanian (Mesozoic) diwa-type granodiorite and, finally, changed into a partially enriched deposit as we see it today.

(3) A platform ore deposit changed and richened by diwa tectonism

Polygenetic compound ore deposit of this type originally was a platform-type sedimentary deposit, changed and richened later, during a diwa stage, by the action of heated water, magmatic processes and structural deformations.

The Huangmei iron deposit in Hubei Province is situated in the Central China diwa region. In the Middle Carboniferous-Early Permain, this area was in the maximum-stability period of a platform stage. During Huanglong and Qixia ages blue-gray siderite ore was deposited. This was the original deposit. In the middle Mesozoic, the area entered into a diwa stage. This time tectonic movements intensified, heat flow increased, and underground hot water brisked up. They reworked the original deposit and turned it into a yellowish siderite deposit with coarse ooides and raised the tenor as we see it today. With the proceeding of faulting and folding, siderite, being reworked, migrated and filled in cracks and other favourable structural positions and obtained the features of a hydrothermal endogenic ore.

(4) A diwa ore deposit changed and richened by diwa tectonism

The process of diwa development may be divided into three periods, namely, the initial-mobility period, maximum-mobility period and residual-mobility period. The most intense magmatism, tectonism and metamorphism appear in the maximum-mobility period. Sedimentary or volcano-sedimentary ore deposits formed in the initial-mobility period may be changed and sometimes richened later, particularly in the maximum-mobility period.

The above-mentioned Dangqu siderite deposit of Jurassic age in Xizang was formed in the initial-mobility period of diwa development. Later it was changed by diwa hydrothermal process and turned into a polygenetic compound ore deposit.

(5) A pre-diwa ore deposit reconstructed and enriched by diwa tectonism

Polygenetic compound ore deposit under such a tectonic condition is formed by the action of hypogene heated water generated during a diwa stage. The heated water mobilizes ore substances dispersed in structural layers representing various pre-diwa stages and enriches them to form an economically valuable deposit.

The above-mentioned so-called telethermal deposits of lead, zinc, antimony, mercury and uranium in Hunan, Guangxi and Guizhou Provinces are of this type.

4. Regularities in distribution of polygenetic compound ore deposits and direction for their prospecting

(1) The diwa region is a metallotectonic one in which polygenetic compound ore deposits spread most widely

Polygenetic compound ore deposits may occur in any tectonic regions, but develop most abundantly in diwa regions. The reasons are as follows.

1) The diwa region is the latest tectonic element known presently. Therefore, it has experienced the most complicated process of crustal development and the most stages of development and consequently the most metallogenic stages. In addition to this, the diwa region has inherited the most varied and richest useful minerals, all of which can serve as base deposits.

2) The diwa region is a mobile one, which is characterized by intense tectonism, magmatism and metamorphism. Moreover, high terrestrial heat flow in it may result in an increase in the temperature of heated water. These are the favourable factors for later mineralizations. They can either form a new deposit superimposed on an older one, or rework, reform and enrich a pre-existing deposit or concentrate ore-forming elements dispersed in older sedimentary rocks.

The above-mentioned two reasons are favourable conditions for the formation of polygenetic compound ore deposits.

(2) Stratigraphic control is the most characteristic feature

Polygenetic compound ore deposits are mainly based on strata-bound ore i.e., sedimentary or volcanogeno-sedimentary, metasedimentary or volcanogeno-metasedimentary) deposits or original ore deposits (or source beds) confined to a definite stratigraphic position. Therefore, after superimposition, reworking and reconstruction by later mineralizations, polygenetic ore deposits as we see them today still preserve the features of strata-bound deposits.

Moreover, subsequent superimposed deposits or richening and enrichment caused by later mineralizations also display the features of stratigraphic control. For example, contact-metasomatic superimposed deposits generated by later magmatic processes often possess the features of selective replacement, thus being confined to a definite stratigraphic position. The Daye iron deposit of skarn type in Hubei Province is a famous example. Here the Jialingjiang Limestone of early Triassic age served as a favourable country rock. Magnetite formed by ore superimposition and reworking of the country rock is just found in this stratigraphic position. Mineral deposits generated as a result of enrichment by hypogene hot water are often seen in highly porous and permeable rocks, especially in those between slightly permeable rocks.

(3) Magmatic control sometimes is a necessary factor

For the superimposed-richened or reworked-richened ore deposits formed by later magmatic intrusion (or eruption), besides stratigraphic control (if any) of the original ore deposits, it is necessary to consider a magmatic control at the same time. As the subsequent mineralizations being superimposed on or reworking the original ore deposits result from magmatism, they are inevitably controlled by the petrochemical composition, position, morphology, trend, regularity in distribution and contact zone of the related igneous body.

We cite the Shilu iron deposit as example. Here the base deposit, i.e., a geosynclinal-type metasedimentary iron deposit, is controlled by a WNW-ESE trending geosynclinal trough. Furthermore, the superimposed orebodies of skarn type (mostly rich ores) are closely associated spatially with the Mesozoic diwa-type granodiorite bodies intruded into the Shilu Group. Therefore, when we search for Shilu-type iron deposits beyond the Shilu ore district, we must pay attention to the regularities governing the distribution of both the Shilu Group and the intrusions. The zone where the granodiorite intrusions came into contact with the corresponding geosynclinal sedimentary formation, i.e., the zone where two ore-forming factors of two tectonic stages of development are linked together, may be the promising area for the search for Shilu-type iron deposits.

(4) Structural control also palys an important role

Structural deformations often participate in the formation of polygenetic compound ore deposits. Therefore, when we search for this type of mineral deposits associated with structures, it is also reasonable to study regularities in structural control. This is true, of course, for those polygenetic compound ore deposits whose original deposits or source beds are directly or indirectly controlled or influenced by structures (e.g., volcanogeno-sedimentary and sedimentary ore deposits in various basins) or whose later reworking and reconstruction are directly caused by tectonic movements. Even if for those polygenetic compound ore deposits whose late superimposition, reworking and reconstruction are caused by magmatism and the action of underground water, it is also necessary to pay attention to structural control, because magmatism and underground water are often controlled by structures too. In this case, structural control is an indirect factor for the formation of these ore deposits. For example, the later magmatism resulting in the formation of the Shilu iron deposit is controlled by block-faulting developed here during the diwa stage. Intrusions are usually localized along these fractures, especially at the intersections of two sets of fractures. NNE-trending fractures are plentiful in this area, but WNW-trending fractures also exist at the same time. Study of the orientation, attitude and intersection of these fractures is conducive to the search for Shilu-type iron. For example, the contact of a WNW-trending geosynclinal trough with diwa granodiorite intrusions will be a promising place.

5. Theoretical and practical significance of studying polygenetic compound ore deposits

(1) Theoretical significance

1) In classifying the genetic types of mineral deposits. According to traditional genetic concepts, mineral deposits are divided only into endogenic and exogenic.

After a thorough study, however, many ore deposits are found to be similar both to endogenic and exogenic ones with the result of difficulty in making a clear distinction between them. Frequently they can not be ascribed to endogenic or exogenic category. The reason is that these ore deposits are by no means monogenetic. Therefore, the presentation and expounding of polygenetic compound ore deposits can help us to solve such a difficult problem.

2) In defining the direction for prospecting. Previous theories concerning defining the direction for prospecting were based on concepts of monogenetic ore deposits. For example, in regard to endogenic ore deposits, one often devotes much attention to chemical composition of magmatic rocks, structure of intrusions, regularities in their distribution and, finally, country rock conditions. In regard to exogenic ore deposits, one frequently attaches primary importance to ore-bearing stratigraphic horizons and features of sedimentary formations. These theories in most cases have encountered difficulties in such areas as China with complicated tectonic features and genesis of mineral deposits. Study of polygenetic compound ore deposits enables us to open up a new domain in the theoretical study of the direction for prospecting. According to a preliminary analysis, at least the following points need to be paid attention to.

a. Necessity of a comprehensive study and analysis.

Based on the complex nature of the factors controlling the formation of polygenetic compound ore deposits, in studying the regularities in distribution of these deposits and the direction for their prospecting, we shouldn't over-emphasize one regularity in ore control at the expense of the others. In other words, we shouldn't study them according to the old prospecting concepts of monogenetic ore deposits and empirical regularities, but ought to link stratigraphic control, magmatic control and structural control together, comprehensively study and consider them. The above-mentioned comprehensive control of the Shilu iron deposit is enough to illustrate this.

b. Making a concrete analysis of concrete conditions.

Polygenetic compound ore deposits are varied. As mentioned above, there are many models and types of metallogenesis as well as various tectonic conditions of ore formation. Therefore, in searching for ores we must deal with different mineral deposits in different ways in accordance with the history of metallogenic development, tectonic setting and conditions of formation of concrete mineral deposits, and explore the direction for prospecting and ore indications corresponding to the reality.

(2) Practical significance.

Study of polygenetic compound ore deposits is of great economic importance. The principal reasons are as follows.

1) Being increased in the tenor of ore, these ore deposits are often of larger economic value than their original deposits. As for source beds, they may become economically valuable from the concentration of originally dispersed ore-forming elements. Therefore, such ore deposits can be regarded as important targets for the search for rich ore deposits. Study of the process of their formation, peculiarities of their occurrence and ore indications will contribute to the search for ores with greater, faster, better and more economical results.

2) They are mostly formed in such a tectonic region as diwa region, with such complex geologic history as in China, and abundantly occur in it. Diwa regions spread widely in the world. Therefore, polygenetic compound ore deposits are plentiful. Diwa regions in China occupy more than two thirds of the whole country's area. Clearly, study of the metallogeny and distribution of these ore deposits will contribute to accelerating the prospecting for and prognosticating of mineral resources.

By the above-mentioned reasons, in the last few years polygenetic compound ore deposits have attracted the high attention of many investigators in the fields of economic geology, geochemistry and particularly metallogeny both at home and abroad. Much attention should be devoted to metallogenic investigations.

Metallogeny, as is defined here, deals with the process of formation of mineral deposits and the regularities in their distribution from the tectonic point of view. It combines economic geology, a relatively narrow discipline, with tectonics to explore metallogenic theories. The above-mentioned polygenetic compound ore deposits, genesis of which is long debated and the direction for their prospecting is difficult to define, can be understood fully, if we use the metallogenic methods of investigation, carry out a comprehensive study on the basis of the new tectonic theory, probe them from the viewpoint of their development and connection, and, finally, make a concrete analysis of concrete conditions.

CHAPTER 6
RELATIONS OF TECTONICS TO EARTHQUAKES IN CHINA [1]

The earthquake is a form of tectonic movement and also one of the important expressions of neotectonic and especially recent movement. The preparation, occurrence, distribution and migration of earthquakes are all intimately related to tectonic movement and its attribute. The main part of the crust in China at the present stage belongs to mobile regions (including geosynclinal and diwa ones) and, therefore, earthquakes occur frequently. Moreover, different areas are located in different tectonic positions and, therefore, the distribution of earthquakes is uneven and their intensity and frequency vary from place to place (Fig. 6.1). For an understanding of the laws governing the activity and distribution of earthquakes, we discuss below the relations of tectonics to earthquakes in China from one angle.

I. Crustal development and seismicity

In the light of the progression theory[13], crustal structure develops with transformation and alternation between mobile and stable regions, with many stages and an unevenness character. Being an expression of tectonic movement, earthquakes must be conditioned by the general laws governing the development of the crust.

1. Transformation between mobile and stable regions and alternation between high and low seismicity

Firstly, according to the law of progression with transformation between mobile and stable regions, the earth's crust in the process of its evolution has experienced several tectonic stages of different character, among which geosynclinal, platform and diwa stages are intimately related to modern earthquakes (Fig. 6.2.). As mentioned above, the geosynclinal stage is one having a high crustal mobility and exceedingly strong tectonic movement. During this stage, regional horizontal stress mostly occupies a dominant position and has resulted in strong orogenic movement with active folding and faulting. As known from the present-day geosynclinal regions, it is a stage with many large earthquakes (see below). Ancient earthquake relics in geologic history have also proved this idea.

After energy is released in various ways (including seismological), the geosynclinal stage transforms itself into its opposite, the platform stage, a stage of tectonic quiescence marked in the greater part of the time by slow oscillatory movement affecting a large area, and by inactive folding and faulting. Therefore, seismic activity is weak, as can be seen from the present-day platform regions (see below). Afterwards, energy reaccumulates, and this relatively stable platform stage also transforms itself into its opposite, the diwa stage, a stage with a new,

---

1. Wei Bailin.

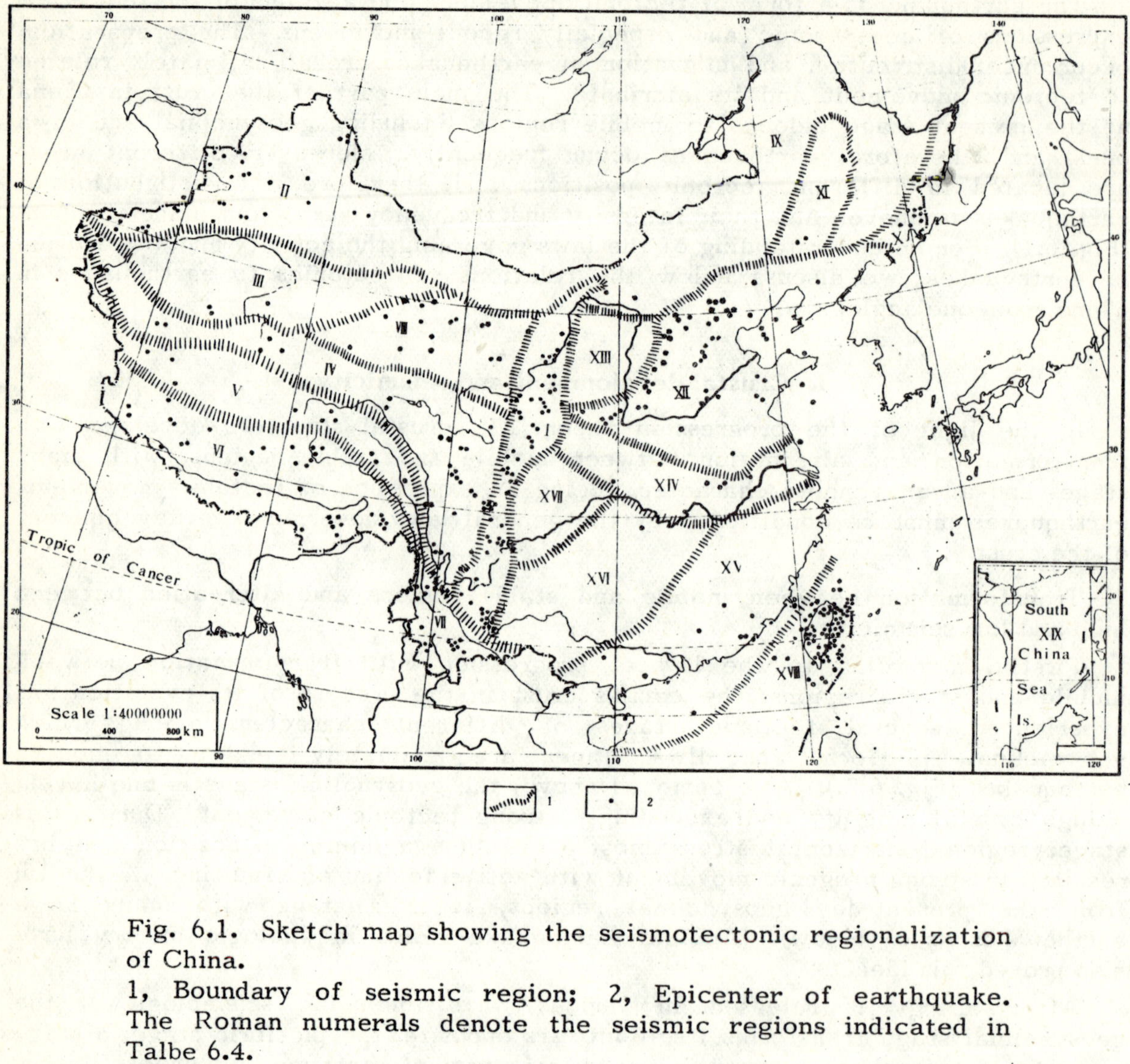

Fig. 6.1. Sketch map showing the seismotectonic regionalization of China.
1, Boundary of seismic region; 2, Epicenter of earthquake. The Roman numerals denote the seismic regions indicated in Talbe 6.4.

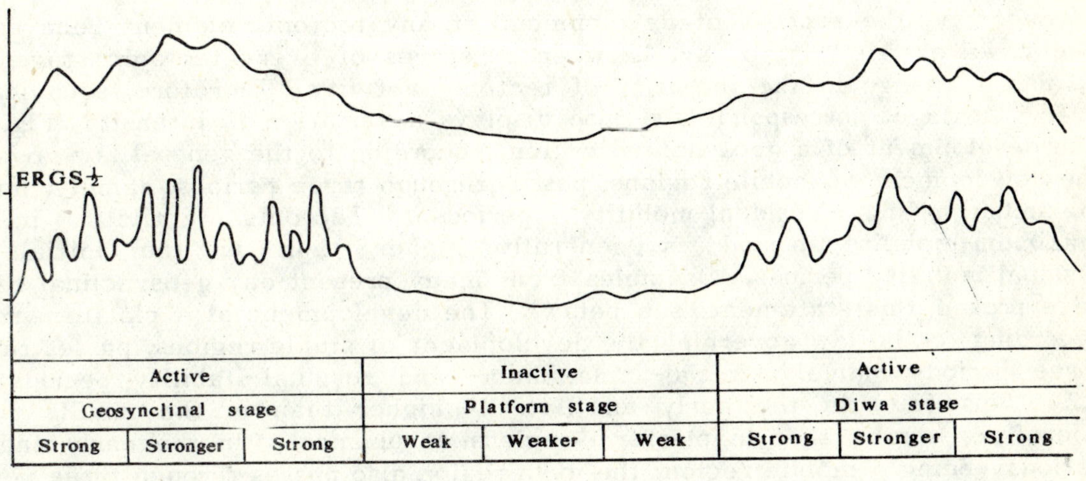

Fig. 6.2. Schematic diagram showing the process of transformation between tectonically and seismically active and quiet periods.
W denotes the total energy of tectonic movement. J indicates seismic strain energy.

exceedingly high crustal mobility. During this stage, horizontal movements again occupy a dominant position in the greater part of the time, orogenic processes occur very strongly, the regional stress field has undergone adjustment, and folding and faulting take place actively, which not only cause the reactivation of old but also produce new fractures. Therefore, this is also a stage of intense seismicity. Both the present stage of diwa regions (see below) and the ancient earthquake evidence in geologic history have proved this argument.

It can be thus concluded that along with transformation between tectonic elements of different character, i.e., alternation of mobile regions with stable ones, seismicity also shows a variation in intensity, i.e., intense seismicity alternates with weak one (Fig. 6.2).

Secondly, the process of development of any tectonic element behaves also like a wave. That is to say, between substages of every tectonic stage there is also a change in the intensity of tectonic activity. Therefore, seismicity in every substage, correspondingly, also displays a variation in intensity (Fig. 6.2). The development of a geosynclinal region, according to the general law governing the development of mobile regions, passes through three periods, namely, initial-, maximum- and residual-mobility periods. Tectonic activity in the maximum-mobility period is generally higher than in the initial- and residual-mobility periods. Examples from many present-day geosynclinal regions have proved this statement (see below). The development of a platform region, according to the law governing the development of stable regions, passes through three periods, namely, initial-, maximum- and residual-stability periods (Fig. 6.2). Seismicity in the early and late is higher than in the middle period. Therefore, earthquakes in platform regions occur mostly in the early and late periods. Being a mobile region, the diwa region also passes through three periods, namely, initial-, maximum- and residual-mobility periods (Fig. 6.2). Seismicity attains the greatest intensity in the maximum-mobility period.

Finally, in every substage of crustal development tectonic process also shows the above-mentioned regularity, giving rise to further minor stages of varying intensity. Consequently, seismicity in these minor stages also differ in intensity, i.e., an active period alternates with a quiet one. Beginning from 1011, for example, four active periods separated by three quiet periods have been distinguished in the North China diwa region (Table 6.1).

Even in an earthquake sequence an active moment also alternates with a calm one. This is a still minor variation in the intensity of seismicity, as shown by the earthquake sequence map of Xingtai (Fig. 6.3).

2. Transformation between mobile and stable regions and migration of earthquakes

All regions in the crust advance forward according to the law of progression with transformation between mobile and stable regions, but the transformation varies in time from region to region. That is to say, the development of various parts of the earth's crust is uneven. Therefore, seismicity closely related to various stages of crustal development displays a variation in intensity not only in time but also in space (in plane). The latter case is generally expressed in the migration of earthquakes. On a regional scale, though earthquakes generally occur under the action of an united tectonic stress, strain-energy related to earthquakes frequently accumulates more in some localities, but less in others,

Table 6.1. Alternation between seismically active and quiet periods in the North China diwa region

Active and quiet period	Number of earthquakes			
	M=8	M=7.9 - 7	M=6.9 - 6	M=5.9 - 5
The 1st active period (1011 - 1076)		1	3	4
The 1st quiet period (1077 - 1289)			1	
The 2nd active period (1290 - 1368)	1		6	8
The 2nd quiet period (1369 - 1483)				2
The 3rd active period (1484 - 1730)	4	3	19	60
The 3rd quiet period (1731 - 1811)				11
The 4th active period (1812 - present)		6	18	70

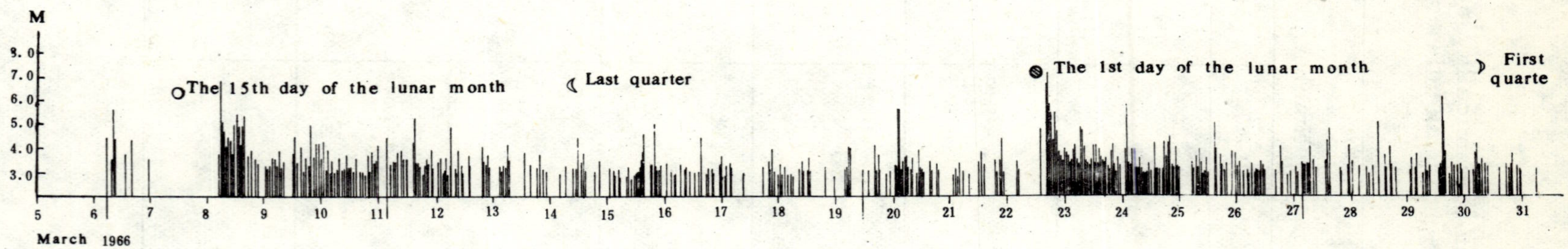

Fig. 6.3. Diagram showing the concentration and sparseness of seismic events in the Xingtai earthquake sequence.

because various parts of the earth's crust are in different stages of tectonic development and in different tectonic conditions and possess different crustal structure and different physical properties of medium. Thus "active" and relatively "quiet" regions can be formed in the earth's crust. When regional tectonic stresses further strengthen their action, strain-energy accumulated in an "active" region may be released by earthquakes, and the region appears to be seismically active, while a "quiet" region seems to be seismically calm due to less strain-energy accumulation. When the tectonic stress continues to act, energy in the original "active" region progressively decreases due to its release, strain-energy needs to be accumulated again, seismicity reduces, and the region gradually changes into a new "quiet" one. However, the original "quiet" region in turn transforms itself into a new "active" one due to strain accumulation which increases so rapidly as to break through the elastic limit of a rock in the locality, resulting in the failure of the rock and the occurrence of an earthquake. The transformation between "active" and "quiet" regions generally takes place many times, giving rise to their migration in space. In this case the migration of "active" regions is the foundation of the migration of earthquakes. There are three main transformation-migration types.

(1) Directional transformation-migration type. A seismically "active" region migrates and transforms in the trend of a structural line, as seen in the Changde-Guiding seismotectonic zone (Fig. 6.4).

(2) Reciprocating transformation-migration type. A seismically "active" region migrates and transforms back and forth in opposite directions within a seismotectonic region or zone, as seen in the N-S diwa intense seismic region (Fig. 6.5).

(3) Jumping transformation-migration type. There are several parallel seismotectonic zones or lines within a seismotectonic region or zone. A seismically "active" region jumps from one zone or line into the other. For example, earthquakes in the central Hebei intense seismotectonic zone migrate along NE-trending fractures having an en echelon arrangement (Fig. 6.6).

II. Control of tectonics on earthquakes

1. Seismicity conditioned by tectonic character

As seen from Fig. 6.1, great earthquakes in China occur mostly in present-day mobile regions and seldom in stable ones. This suggests that the magnitude and frequency of earthquakes depend on tectonic character. The relationship may be summarized as follows.

(1) Regions of different tectonic character show different seismicity

Regions of different tectonic character have different intensity and type of crustal movement, different material composition, different construction and different structure. They differ from each other also in the accumulation of energy and strain produced under the action of regional tectonic forces. As a result, they necessarily differ from each other in the magnitude, frequency and characteristic features of earthquakes related to the above-mentioned factors.

The geosynclinal region is a mobile region. Crustal movement in it is intense, tectonic stress is strong and varies greatly, and strain-energy in rocks can

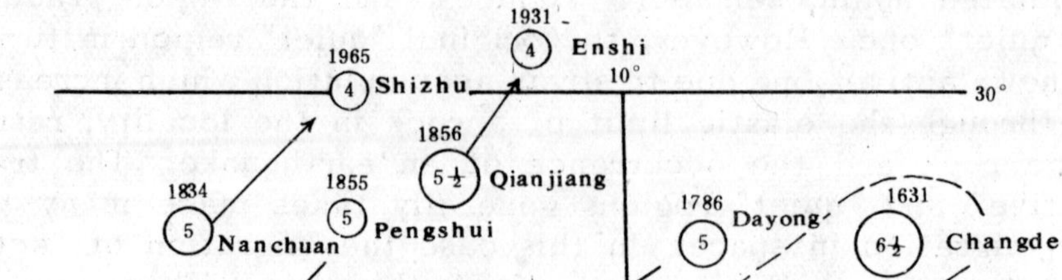

Fig. 6.4. Directional migration of earthquakes in the Changde–Guiding seismotectonic zone.

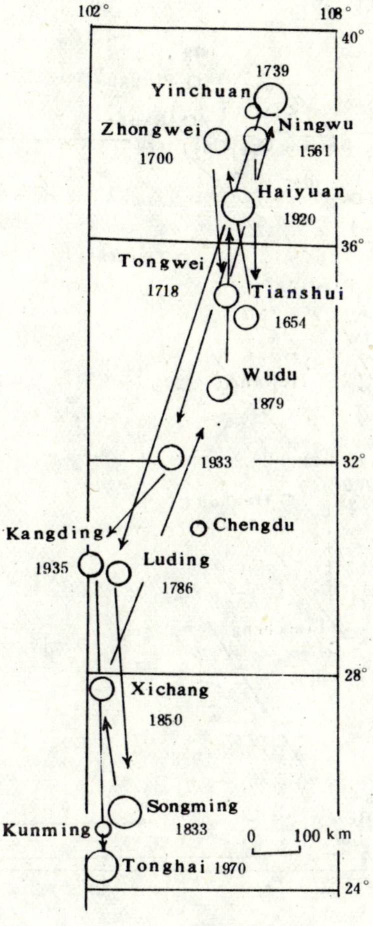

Fig. 6.5. Reciprocating migration of M > 7 1/2 earthquakes in the N-S diwa intense seismic zone.

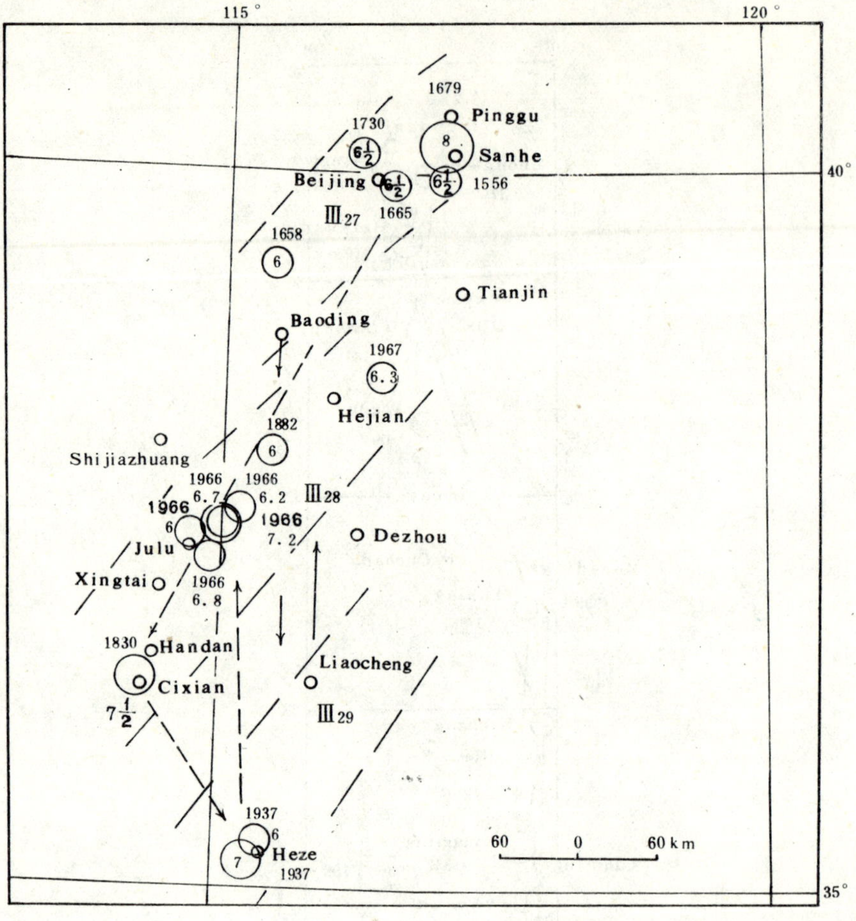

Fig. 6.6. Seismotectonic lines and jumping migration of M $\geq$ 7 earthquakes in the Central Hebei intense seismotectonic zone. 1658-1730: seismically active region III_{27}, quiet regions III_{28} and III_{29}. 1830-1882: seismically active region III_{28}, quiet regions III_{29} and III_{27}. 1937: seismically active region III_{29}, quiet regions III_{27} and III_{28}. 1966: seismically active region III_{28}, quiet regions III_{29} and III_{27}. Arrow indicates the direction of migration of seismically active region, while dotted line arrow denotes the direction of migration of M $\geq$ 7 earthquakes.

be accumulated and released rapidly. Therefore, earthquakes have a great magnitude and a high frequency. Regions presently having such a tectonic character generally show a high seismic activity, especially when they are in the maximum-mobility period of development. We may take the Taiwan geosynclinal region as example, which is presently still in its maximum-mobility period. In the period of 68 years (from 1904 to 1972), earthquakes of M=8 occurred twice; those of $M \geqslant 6$, 206 times.

The platform region is a stable region. It has a well-consolidated folded basement. Crustal movement in it occurs slowly, dominated by downwarping and upwarping affecting a large area. It reacts to regional tectonic forces slowly, and the accumulation and release of strain-energy take place also slowly. Therefore, earthquakes generally have a small magnitude and a low frequency. Example is the Yishan (Ordos) platform region, where the largest earthquake has a magnitude of 5 1/2. At the edge of a platform region adjacent to a mobile one earthquakes can attain magnitude 6, as seen at the edge of the Songliao platform region.

The diwa region is another mobile region different from the geosynclinal one. Crustal movement in it is intense, and block-faulting is common. Therefore, under the action of strong and varied regional tectonic forces, the accumulation and release of strain-energy in rocks take place rapidly. Earthquakes appear to have a great magnitude and high frequency. This feature is particularly prominent in diwa regions being presently in the maximum-mobility period of development. For example, on the two sides of the Tianshan diwa region, as calculated only from Chinese territory, earthquakes of $M \geqslant 8$ have occurred twice since 1902.

The differences in the seismic characteristics of tectonic regions of different character are expressed not only in the above-mentioned seismicity level, but also in other aspects (Table 6.2).

(2) Regions having the same tectonic character but being in different periods of development show different seismicity

As mentioned above, any tectonic element experiences a process of generation, development and dying out. For example, both geosynclinal and diwa regions have initial-, maximum- and residual- mobility periods; platform region, initial-, maximum- and residual-stability periods. In the process of their wave-like development, every period differs from other in intensity and features of tectonic activity and, as a result, in seismicity genetically related to them. Thus two regions, having the same tectonic character but being now in different period of development, may be different in their seismicity. For example, both the Yarlung Zangbo River-Himalayan region and Kunlun region are geosynclinal, but the former at present is in the maximum-mobility period and, therefore, seismically very active, whereas the latter has entered into the residual-mobility period and, therefore, seismically relatively quiet. Such a difference in seismicity for different periods of development also exists in diwa and platform regions.

(3) Regions having the same tectonic character and being in the same period of development but having different features of historical development show different seismicity

By the difference in features of historical development we mean that some

Table 6.2 Seismic characteristics of geosynclinal, platform and diwa regions in China

Seismic characteristics		Mobile region		Stable region
		Geosynclinal	Diwa	Platform
Seismicity level	Magnitude	Large(M max 8)	Middle large (M max 8)	Small (M max 8)
	Frequency	High	Middle high	Lower
	Recurrence rate	High	Middle high	Lower
Depth to focus		30-70 km, the greatest 270 km	5-30 km for Cathysian age, 30-60 km for Central Asia age	20-30 km
Distribution of epicenters		Linear of belt-shaped	Belt-shaped or en echelon	Random
Migration of epicenters		Linear reciprocating migration along structural line	Jumping and directional migration as well as linear directional or reciprocating migration	Zonation indistinct
Type of earthquake sequence		Foreshock-main shock-aftershock or earthquake swarm type	Foreshock-main shock-aftershook and main shock-aftershock as well as earthquake swarm type	Isolation type
Field of influence by earthquakes		Isoseismal lines generally form an ellipse with a long axis	Isoseismal lines generally form an ellipse with a short axis	Isoseismal lines generally form a concentric circle

regions display significant magmatic activity but others corresponding to them in tectonic character and in development period do not, or some regions exhibit specific structural styles but others corresponding to them in tectonic character and in development period do not. Such a difference in features of historical development creates different seismogenic settings for regions having the same tectonic character and being in the same period of development. Therefore, they must differ from each other in seismicity. For example, the North China, SE, NE, and Yunnan-Guizhou diwa regions, though all in the residual-mobility period of diwa development, stand in contrast with each other in seismicity: the first is seismically active, the second less active, but the third, fourth and the western part of the SE diwa region show low seismicity. The reasons for such a difference are as follows: 1) The latter regions are structurally unique, which have not experienced prominent differential movement and lack such large active grabens as the Fenhe-Weihe graben, such large-displacement active fractures as the cis-Taihang fracture zone, and such large-scale and deeply cutting active structures as the Yingkou-Tancheng deep fracture. 2) In northeast and south China, magmatic intrusion and volcanic extrusion took place violently during Mesozoic-Cenozoic times and energy was released mostly as heat energy. Moreover, the intruding magmas have cemented the original major deep-seated fractures, and the Earth's stress is difficult to find a suitable place to accumulate.

2. Distribution of earthquakes controlled by active fractures

In China, not only tectonic regions of different character show different seismicity, but also the distribution of earthquakes in tectonic regions of whatever character is uneven. Most earthquakes occurred in active fracture zones or near them. The direction of the long axes of new fractures, meizoseismal area, and isoseismal lines created by an earthquake coincides mostly with the strike of the corresponding master fracture. The arrangement and migration of epicenters are generally intimately related to a regional major deep-seated fracture. Focal mechanism solutions also indicate that one of two nodal planes is parallel to the strike of the corresponding fracture in the study area. On these grounds we can conclude that the distribution of earthquakes is controlled by active fractures. The control is also reflected in the following two aspects.

(1) Along an active fracture earthquakes are aligned in an earthquake zone, in which the magnitude of earthquakes bears a close relation to the intensity of recent movement along the fracture. Modern active fractures are the sites of the greatest change in stress and strain in locality. Precise leveling survey indicates that on the two sides of an active fracture there generally appear linear anomalies of crustal deformation, which reflect the intensive accumulation of strain-energy at the corresponding places. Earthquakes closely related to strain-energy must occur along active fractures. The degree of accumulation of strain-energy depends chiefly on the intensity of movement along a fracture. For example, the Tianshan and Himalayan Mountains and the Fenhe-Weihe graben are all the zones of intense vertical differential movement since the Cenozoic. Therefore, great earthquakes have frequently occurred and have a linear distribution, forming the WNW-trending Tianshan, the WNW-trending Himalayan and the NE-trending Fenhe-Weihe seismotectonic zone respectively. The E-W-trending Bayanobo-Kaiyuan fracture zone, though large in size, is less active since the Cenozoic and, as a result, shows low seismicity. In west Hunan, Guizhou

and Guangxi, in spite of the existence of large fracture zones, great earthquakes have not occurred because of the absence of intense neotectonic movement.

(2) The size of an active fracture zone determines the degree of concentration of earthquakes in the corresponding earthquake zone. A large active fracture zone consists generally of many deep-seated, major and common fractures. The mode of its activity is very complex. Various fractures in it differ from each other in the intensity and time of their activity and in the direction of displacement. Consequently, the accumulation and release of strain-energy in the zone appear to be a very complex process. On the contrary, a small fracture zone is simple, both in its construction and mode of its activity and in the process of accumulation and release of strain-energy. Thus it can be seen that the size of a modern active fracture zone influences not only the magnitude of earthquakes in the zone, but also significantly the degree of concentration of earthquakes.

1) Global major deep-seated fracture zones are belts of most densely distributed earthquakes. For example, the circum-Pacific geosynclinal and diwa belt, being at present just in the maximum-mobility period of development, and the Alpine-Himalayan geosynclinal and diwa belt, having not terminated the maximum-mobility period yet at present, are two global major deep-seated fracture zones and at the same time the conjunction of several gigantic crustal bodies. Almost all the world's earthquakes are concentrated in these two belts. Earthquakes occurring in the former belt account for 80% and those in the latter for 15% of the total number of the world's earthquakes.

2) Nation-wide major deep-seated fracture zones are belts of densely distributed earthquakes. The N-S diwa region in China extending from north to south in a long belt lies just at the Yinchuan-Kunming major deep-seated fracture zone (some fractures being concealed basement fractures). This major deep-seated fracture zone has a long history of development, and serves as a tectonic conjunction zone between the eastern and western crustal bodies of China. It shows high and dense seismicity. Earthquakes of magnitude 6 occurring in the zone account for one-sixth of the total number of the earthquakes with the same magnitude in China.

3) Regional major deep-seated fracture zones are belts of relatively densely distributed earthquakes. By the regional major deep-seated fracture we mean what separates two first-order tectonic regions. For example, the Fenhe-Weihe graben lying between the North China diwa region and the Yishan (Ordos) platform region has experienced three earthquakes of magnitude 8; the E-W tectonic zone lying between the Tianshan and South Xinjiang diwa regions has been struck by 22 earthquakes of magnitude $\geqslant 6$.

4) Local major deep-seated fracture zones are belts of less densely distributed earthquakes. Fracture zones of this kind serve generally as boundaries between second-order or minor tectonic regions. They are smaller than the above-mentioned three kinds, and earthquakes in them have a less dense distribution. Example is the cis-Taihang major fracture.

3. Great earthquakes localized only in some definite structural positions

In any seismotectonic zone great earthquakes do not occur everywhere but are localized in some definite structural positions. The results of studies by Chinese seismologists reveal that earthquakes are liable to occur in the following structural positions.

(1) Fractures in a seismotectonic zone being in an active period at present. Examples are the Yingkou-Tancheng and Yinchuan-Kunming major deep-seated fractures.

(2) Right-lateral slip faults in the eastern crustal body where NE tectonic systems occupy a dominant position and left-lateral slip faults in the western crustal body where NW tectonic systems predominate. Faults in the Xingtai area may be cited as examples for the former and the Xianshui River fault in Sichuan Province for the latter.

(3) Areas of the most intense differential movement along a fracture zone on the most deeply subsided side of an inclined basin. For example, in the area of the most intense differential movement along a fracture on the southern, deeply subsided side of the Weihe River fault depression, North China diwa region, an earthquake of magnitude 8 occurred in 1856 (in Huaxian County).

(4) Two sides of an uplift created by transverse fractures in a fault depression. For example, the Zhaoxian, Shanxi Province, earthquake of 1303 with magnitude 8 occurred in a step-like fracture zone on the south side of a transverse uplift created by the Hanhouling fracture and others in the Fenhe-Weihe graben.

(5) The tip of the acute angle of an active tectonic basin. For example, the Pinglu earthquake of 1815 with magnitude 6 occurred at the tip of the northeast acute angle of the Lingbao diwa basin, which is situated at the intersection of two fractures.

(6) Mesozoic-Cenozoic fault depressions developed, along an active fracture zone, like a string of beads or in an en echelon manner. For example, in depressions developed like a string of beads in Midu-Erhai-Eryuan-Jianchuan, Yunnan Province, several earthquakes of magnitude > 6 occurred, and among them the Dali earthquake in the Erhai depression reached a magnitude of 7.

(7) The sharply contrasting zone between an uplift and a depression on the two sides of an active fracture, i.e., the zone of high velocity gradient of neotectonic movement. For example, the Manas, Xinjiang, earthquake of 1906 with magnitude 8 occurred in the structurally and topographically sharply contrasting zone along an active fracture between the Tianshan uplift and the Junggar depression.

(8) The gap between active fractures lying on the same extension. For example, the gap between the Xinyang and Shangxian-Nanyang active fractures was struck by the Nanyang earthquake of magnitude 6.5 in A.D. 46.

(9) The intersection of two or more sets of active fractures. For example, the Mabang, Hebei Province, earthquake of 1679 with magnitude 8 occurred at the intersection of the Daxing-Tongxian and Xiadian-Pinggu fractures; the Yangjiang, Guangdong Province, earthquake of 1969 with magnitude 6.2 also occurred at the intersection of two sets of fractures: NE- and NEE-trending.

(10) The convergent point of several, or sets of, fractures. For example, earthquakes in Jianchuan and Eryuan, Yunnan Province, occurred in such a structural position.

(11) The intersection of a deep structure with a surface one. For example, earthquakes in Tianshui, Gansu Province, occurred at the intersection of a WNW-trending surface structure with a N-S-trending concealed one.[1]

(12) The sharply bending (along strike or dip) position of an active fracture. For example, the Heyuan, Guangdong Province, earthquake[2] of 1962 with magnitude 6.2 occurred in an area where the strike of the southern segment of the Heyuan-Shaowa major fracture takes an S-shape bend. The Zayu, Xizang, earthquake of magnitude 8 1/2 also occurred in an area where the strike of the structural line takes a sharp bend from WNW to almost N-S.

(13) The area of intense horizontal movement along an active fracture. The Luhuo, Sichuan Province, earthquake of 1973 with magnitude 7.9 may be cited as example. When the earthquake occurred, the Xianshui River fracture was strongly reactivated. Cracks formed along the fracture have a horizontal displacement of 2.5 m. The results of focal-mechanism studies also suggest that the horizontal component of the seismic slip is the most pronounced.

III. A preliminary view on the relationships between seismicity and tectonics of China

As mentioned above, since the beginning of the Mesozoic, China has been divided, taking the N-S diwà region (Yinchuan-Kunming axis) as a line of demarcation, into two evolution systems, the eastern and the western, forming two crustal bodies. The tectonic characteristics and development history of these two crustal bodies as well as the influence of the two global crustal bodies, the Pacific and the Mediterranean, have made the different parts of the eastern and the western crustal bodies of China distinct from each other in seismicity (Table 6.3).

The N-S diwa region lying between the eastern and the western bodies is a N-S-trending major deep and concealed fracture zone and the seat of the gigantic N-S tectonic system. Fractures of NE tectonic systems in east China and those of NW tectonic systems in west China meet with the N-S fracture zone to form a feather-like symmetry. Therefore, the N-S fracture zone, especially its intersections with fractures of various tectonic systems, shows high seismicity and is one of the intense earthquake zones within the mainland of China. It has an average crustal thickness of 50 km. The greater part of the zone is in the residual-mobility period of diwa development. Focal depths are small (5-20 km) in the south but large (12-25 km) in the north. Earthquakes migrated from north to south and vice versa (Fig. 6.5).

The western crustal body of China has a crustal thickness of 50-70 km. It is occupied entirely by mobile regions. Among them, geosynclinal regions have all entered into the fold belt period, while diwa regions, for the most part, have just entered into the maximum-mobility period. Moreover, the Yarlung Zangbo River-Himalayan geosynclinal fold region and the Southern Himalayan diwa region are part of the global E-W seismotectonic belt. Therefore, this crustal body

---

1. Arguments for this N-S-trending concealed deep-seated fracture will be described in the text below.
2. The water storage of the Xinfeng River reservoir played a trigger role for the earthquake.

Table 6.3. Characteristics of seismicity of China

Seismic characteristics	The western crustal body		The Conjunction zone	The eastern crustal body	
	Diwa regions of Central Asia age in the majority and geosynclinal regions in the minority	Yarlung Zangbo River-Himalayan geosynclinal region, S. Himalayan diwa region	N-S diwa region	Diwa regions of Cathysian age in the majority and platform regions in the minority	Taiwan geosynclinal region
Level Intensity Frequency	Middle high Large Middle high	High Large High	High Large high	Low Large Low	Highest Large Highest
Distribution of earthquakes	In a belt trending WNW	In a belt trending E-W and NW	In a belt trending N-S	In an en echelon manner trending NE	In a belt trending NNE
Migration of earthquakes	Linear reciprocating migration in a seismic zone along a NW tectonic system	Linear reciprocating migration in a seismic zone along a NW tectonic system	Reciprocating migration along a N-S structural line	Jumping migration along a NE tectonic system, and reciprocating and directional migration along a structural line	Clockwise migration along the circum-Pacific island arc belt
Depth to focus	Mostly 10-45 km	Mostly 20-70 km	12-25 km in the north and 5-20 km in the south	20-30 km in northeast, 15-20 km in north, 12-14 km in central and 5-8 km in south China	40-100 km

Direction of the principal compressive stress	Mostly NE-SW	Nearly N-S in the majority	NE-SW and N-S in the north, E-W in the center and NW-SE in the south	Mostly ENE-WSW	Nearly E-W in the majority
Strike of the master fault and direction of its slip	WNW; mostly sinistral, but dextral when seismic slip is diagonal to the master fault	WNW; mostly sinistral	NNE and NNW; mostly dextral in the north but sinistral in the south	NNE and NE; mostly dextral	NNE; dextral in the west but sinistral in the east
Seismic moment	Log Mo=0.97M+19.6		Log Mo=1.09M +18.6	Log Mo=0.71M +0.21	Log Mo=0.94M +19.8

Note: Mo is seismic moment (in dyne · centimeter) and M is surface wave magnitude.

shows high seismicity. Since 1889 to the present earthquakes of magnitude ⩾ 8 have occurred seven times in west China. Most earthquakes have focal depths of 10-45 km, except in the Karakorum region, where some earthquakes occurred at greater depths. The earthquakes are distributed along several major deep-seated fracture zones, NW- and WNW-trending, in Altay and Urumqi, at the southern foot of the Tianshan Mountains, at the northern edge of the Kunlun Mountains, at the northern and southern edges of the Altun Mountains, at the northern edge of Qilian Mountain, at the northern edge of the western Qinling Mountains, in the Bayanhar Mountains and Gandise Range, and in the Yarlung Zangbo River. These major deep-seated fractures are mostly boundaries of the first and second order tectonic elements. Geologic development of these tectonic elements must have affected the seismic activity of the major deep-seated fractures. Earthquakes migrated along the strike of the fractures in a linear pattern.

The Yarlung Zangbo River-Himalayan geosynclinal region lying on the southern border of the western crustal body is a junction zone between two global crustal bodies and is also a Cenozoic geosynclinal belt. Neotectonic movements are significant and Cenozoic strata have been subjected to intense folding, metamorphism and faulting. This region is the most intense earthquake zone in the western crustal body, and earthquakes are distributed mainly in its northern part in the vicinity of the Yarlung Zangbo River deep fracture zone. The zone makes an arcuate turn to southeast to meet the Tengchong area, west Yunnan Province, and extends further southwards into Burmese territory. Great earthquakes frequently occur in this region. Focal depths generally range from 20 to 70 km, but few earthquakes occur at a depth of 150 km.

The eastern crustal body of China has a crustal thickness of 30-40 km. In Sichuan and Yishan (Ordos) it is less than 30 km, but in some regions it can reach 45 km. This crustal body is occupied mainly by diwa regions of Cathysian age and has left only three small platform regions, the Songliao, Yishan (Ordos), and Sichuan, and a geosynclinal region, the Taiwan. Earthquakes occurring in this part of the mainland of China, the main body of which consists of diwa regions of Cathysian age, have a great magnitude but a low frequency. The distribution of earthquakes is uneven. Few great deep-focus earthquakes occur in the Hunchun area, northeast China, under the influence of the circum-Pacific seismic belt, but most earthquakes are concentrated in the North China diwa region and the easthern part of the SE diwa region, localized in several NE-trending major deep-seated fractures. These are: the Changle-Zhao'an, the Yingkou-Tancheng, the cis-Taihang, the Fenhe-Weihe, etc. The intersections of these fractures with fractures of other trends show high seismicity. Focal depth increases from south to north: 5-8 km in southeast, 12-14 km in central, 15-20 km in north, and 20-30 km in northeast China. It decreases from east to west, i.e., from coast to inland: 30 km along the coast and 10-20 km inland. Earthquakes in every seismotectonic zone migrated according to the directional or reciprocating migration type. Earthquakes in various seismotectonic zones within a seismotectonic region often jump in an en echelon manner from one zone to another. Taking east China as a whole, earthquakes of magnitude ⩾ 8 cut obliquely across the Cathysian tectonic system and migrated from east to west: the first active period showed high seismicity, and seismic energy was released through earthquakes of magnitude ⩾ 8. As calculated from the sources of 1604, seismic activity was concentrated chiefly in the Quanzhou-Zhao'an zone in that year, but migrated to the Yingkou-Tancheng and Shijiazhuang-Sanhe zones successively in 1665 and 1679; finally it moved to the Fenhe-Weihe zone in 1695.

The second active period showed relatively low seismicity, and seismic energy was released generally through earthquakes of magnitudes 6-7. This period began with the Nan'ao earthquake of magnitude 7 1/4 in the Quanzhou-Zhao'an zone in 1928. In 1910-1932, it migrated to the Yingkou-Tancheng zone, where earthquakes of magnitude 6-6 3/4 occurred six times. In 1937, the earthquakes migrated westwards to central Hebei, where in that year an earthquake of magnitude 7 and another earthquake of magnitude 6 3/4 occurred in Heze. In 1966-1975, earthquakes were concentrated in central Hebei and the Yingkou-Tancheng zone, where five earthquakes of magnitudes 7.2, 6.3, 7.4, 7.3 and 7.9 occurred in Xingtai (1966), Hejian (1967), Baihai (1969), Haicheng (1975), and Tangshan (1976), respectively. During that period, in the process of the general westward migration of earthquakes, three earthquakes migrated in an opposite, eastward direction.

The Taiwan geosynclinal region lying at the southeastern edge of the eastern crustal body is part of the modern circum-Pacific geosynclinal and diwa belt. The western part of the region consists of continental crust with a thickness of 30-70 km. The eastern edge adjacent to the West Pacific belt of trenches and island arcs (major deep fractures) is made of oceanic crust. Therefore, the region is situated in a transition zone between continental and oceanic crust and is at present in the maximum-mobility period of geosynclinal development. As a result, neotectonic movement in the region is particularly intense, and seismicity level reaches its climax. Earthquakes occur mostly at the depths of 40-100 km, with a maximum of 270 km, and belong to intermediate to deep-focus earthquakes. The earthquakes migrated clockwise along the Pacific belt of island arcs, jumping from one seismotectonic zone to another.

As known from Chapter 4, various tectonic systems and their distribution pattern in China have been created by two sets of horizontal shear forces oriented N-S and two sets of horizontal compressive forces oriented E-W and N-S. The magnitude and direction of these forces varied historically from age to age. Modern tectonic forces obtained by focal mechanism solutions are not in conformity with paleotectonic forces acting in geologic history. In the eastern crustal body, not sinistral force moment but NE-SW oriented horizontal compressive force occupies a dominant position at the present time. Therefore, seismic slip occurring along NNE-trending master faults is of dextral sense. Typical examples are some earthquakes in Xingtai, Hejian, and west Taiwan. The situation of the western crustal body is more complex. North of the Yarlung Zangbo River-Himalayan geosynclinal region, the axis of the principal compressive stress is oriented NE-SW. In the Yarlung Zangbo River-Himalayan geosynclinal region and the Southern Himalayan diwa region, at a depth of more than 70 km, the axis of the principal compressive stress is oriented nearly N-S. At a depth of less than 70 km, in the Himalayas and south of them, the axis of the principal compressive stress is perpendicular to the dominant structural trend, i.e., oriented nearly N-S, and seismic slip appears to have a bias for seismic strike-slip and thrust faulting. North of the Himalayas, the axis of the principal compressive stress strikes parallel to the dominant structural trend, i.e., oriented nearly E-W, and seismic slip has a preference for strike-slip and normal-type faulting.

Obviously, this state of tectonic force in the western crustal body indicates that dextral force moment still occupies a dominant position. Stress state stands differently from each other in various segments of the N-S diwa region: the axis of the principal compressive stress in the northern part shows a NE-SW or N-S direction; in the central part, a E-W direction; and in the southern part, a NW-SE direction. Such a stress distribution is related to an NE-SW compression in the eastern and a dextral motion in the western crustal body.

IV. Seismotectonic regionalization of China

Based on the above-mentioned points it can be seen that the preparation and occurrence of earthquakes in a region are closely related to its tectonic character, and the distribution and migration of earthquakes are generally controlled by specific structural positions. Therefore, when studying the laws governing the time and place of occurrence of an earthquake, its magnitude and relation to structure, one must deal with the problem of seismotectonic regionalization.

There are various ways of seismotectonic regionalization. Our seismotectonic division is based mainly on: (1) tectonic character of the various discussed regions at the present stage, present period of their development and magnitude of neotectonic movement; (2) regional tectonic features, relation of surface structures to deep ones, and depth, length and attitude of modern active fractures; (3) relation of time, place and intensity of seismic events to structure; and (4) variation of regional stress field through time and its features at the present stage.

The following is our seismotectonic regionalization scheme of China made on the basis of the results of seismotectonic studies according to the theory of progression with transformation between mobile and stable regions.

1. Principles of seismotectonic regionalization

(1) Classification and nomenclature of earthquakes on the basis of magnitude

As a rule, earthquakes can be classified on the basis of magnitude as follows: intense earthquake $(M) \geq 8$; $(M) 7 \leq$ strong earthquake $(M) < 8$; $(M) 6 \leq$ moderate to strong earthquake $(M) < 7$; $(M) 5 \leq$ moderate earthquake $(M) < 6$; weak earthquake $(M) < 5$.

(2) Classification and nomenclature of seismic areas

Seismic areas fall into three categories according to their seismicity, tectonic features and areal extent.

1) 1st-order seismic region: seismotectonic or seismic region.

This refers to a region homogeneous in tectonic character or type, in development period, in tectonic features and in seismicity, whereas the different parts of it may still differ from each other in intensity of neotectonic movement and in seismicity level. Its name consists of three parts: place, tectonic character, and largest magnitude, e.g., N-S diwa intense seismic region, Kunlun geosynclinal strong seismic region, and Yishan (Ordos) platform weak seismic region.

2) 2nd-order seismic region; seismotectonic zone.[1]

By this we mean a zone of earthquake swarms within a seismic region arranged zonally along a definite regional structural line. Various seismotectonic zones are distinguishable from each other in features of structural lines, in intensity of neotectonic movement and in seismicity level. The name of the seismotectonic zone has two significances: place and largest magnitude, e.g., Fenhe-Weihe intense seismotectonic zone and Taihang Mountains moderate seismotectonic zone (within the North China diwa intense seismic region).

3) 3rd-order seismic region: seismotectonic line.[2]

This refers to a seismogenic structure within a seismotectonic zone. It is composed of one or a set of active fracture(s). Its name also has two significances: place and largest magnitude, e.g., Yingkou-Tancheng intense seismotectonic line (within the Liaoning-Shandong intense seismotectonic zone).

(3) Principles of attributing epicenters

1) Epicenter lying in a certain 1st-order tectonic region is attributed to a seismic tectonic region.

2) Epicenter lying at the boundary between two tectonic regions of different tectonic character (mobile and stable regions) and in the vicinity is attributed to a mobile region.

3) Epicenter lying at the boundary between two tectonic regions of the same tectonic type (geosynclinal and diwa regions) or character (both are geosynclinal regions or both are diwa regions) and in the vicinity is attributed, according to the time, place, magnitude and focal mechanism of the earthquake, to that tectonic region to which the structural line is parallel and to which the intensity of neotectonic movement corresponds.

(4) Seismotectonic regionalization of China

Based on the above-mentioned principles of seismotectonic regionalization, we divide the crust in Chinese territory into 19 seismotectonic regions and 42 zones (Table 6.4, Fig. 6.1) with brief description.

The N-S diwa intense seismic region (I)

It is principally in agreement with the N-S diwa region in geographical limits and is a narrow and long belt composed of two sets of NNE- and NNW-trending major deep-seated fractures which intersect each other in a N-S direction as a whole and have an X-shape. At the present stage, it is in the residual-mobility period of diwa development. However, it is still active under the influence of the diwa tectonic movements of the western crustal body of China. Neotectonic movement is expressed as strong vertical differential movement. The region shows intense and high-frequency seismicity. There occurred three earthquakes of magnitude 8, and earthquakes of magnitude 6 and above have been reported

1. Not plotted on the map because of the small scale.
2. Not plotted on the map because of the small scale. It is described in the book occasionally.

Table 6.4. Seismotectonic regionalization of China

Seismic region	Seismotectonic zone	Seismotectonic line
I. N-S	Kang-Dian	Xiaojiang, Kangding-Yipinglang, Luding-Yimen, Eshan-Tonghai, Shiping-Jianshui, Mabian
	Songpan-Jingtai	Songpan-Jingtai, Wenxian-Haiyuan, Longmen Mountain
	Liupan-Helan	Yinchuan, Liupan Mountain
II. North Xinjiang	Altay	Fuyun
	Junggar	
	Northern Tianshan	Wenquan-Urumqi, Tarqiyiling
	Southern Tianshan	Wuqia-Kalpin, Baicheng-Korla
	Alxa	
III. South Xinjiang	Tarim	
	Altun	cis-Altun
IV. Kunlun-Bayanhar	Kunlun	Gonggel Mountain
	Burhan Budai Mountain	
	Bayanhar	Luhuo-Xianshui River
	Hoh Xil	Zhongdian-Xiaguan
V. North Xizang		

VI. Gandise-Himalayan	Gandise	Nyainqentanglha
	Himalayan	Yarlung Zangbo River-Zayu
VII. West Yunnan	Wuliang Mountain	
	Nushan Mountain	
VIII. Qinghai-Gansu	Corridor	Corridor-Nanshan, Shandan
	Qilian Mountain	Nanshan
	Qaidam	
IX. NE	Hohhot, Hinggan Mountains, Jilin	
X. Hunchun		
XI. Songliao		
XII. North China	Liaoning-Shandong	Yingkou-Tencheng
	Hebei-Shandong	
	Central Hebei	Pinggu-Shijiazhuang, Cixian-Hejian, Liaocheng
	Taihang Mountain	
	Fenhe-Weihe	Weihe, Fenhe, Yuanping-Daixian, Datong
	Xuchang-Huainan	
XIII. Yishan(Ordos)		

XIV. Central China	Qinling-Dabie, Nanjiang, Daba-Jianghan	
XV. SE	Quanzhou-Hainan	Changle-Zhao'an
	Fujian-Guangdong	Heyuan-Shaowu, Yangjiang
	Yunkai Mountain	Lingshan
	Jiangxi	
XVI. Hunan-Guangxi-Guizhou	Yueyang-Wuzhou, Changde-Guiding, Yunnan-Guangxi	
XVII. Sichuan		
XVIII. Taiwan	East and West Taiwan	
XIX. South China Sea		

to be more than 100 times. Earthquakes group into a belt striking in a N-S direction. Intense earthquakes occurred at the turns of the north and south ends. Strong earthquakes occurred at the intersections of several sets of fractures and at their turns (e.g., in Xiji, Haiyuan, Guyuan, Tianshui, Kangding, Luding and Tonghai). Earthquakes migrated in opposite directions along the N-S structural trend (Fig. 6.5).

As shown in Fig. 6.5, within the N-S intense seismic region, earthquakes of magnitude $\geq$ 7 1/2 migrated to and fro three times in a N-S direction in the period of 410 years (from 1561 to 1971). At the first time an earthquake migrated from north to south in a time span of 94 years (1739-1833). At the second time it migrated from south to north in a time span of 87 years (1833-1920). At the third time it migrated again from north to south in a time span of 50 years (1920-1970). At present it is probably at the fourth time, migrating from south to north[34, 89, 92].

Three seismotectonic zones can be distinguished in the region.

The Kang-Dian intense seismotectonic zone. Situated south of the Luding-Ya'an line, it consists of a series of N-S-trending active fractures cut by NW-trending fractures at the north and south ends. Along the principal fracture zone there are Cenozoic diwa basins arranged like a string of beads; some of them remain to be lakes at present. Vertical differential movement is apparent in various fracture zones. For example, Cenozoic sediments in the Xichang basin attain a thickness of more than 200 m, but the Lower Pleistocene Xigeda Bed has been raised to an elevation of over 1,000 m in the Daqing-Liangzi area and, consequently, the amplitude of the vertical movement reaches more than 1,000 m. The zone shows high seismicity. Epicenters are concentrated towards the structural intersections at the north and south ends. An earthquake of magnitude 8, another of 7.7 and forty-seven of > 6 have occurred. The zone is divisible into one intense (Xiaojiang), three strong (Kangding-Yipinglang, Luding-Yimen and Eshan-Tonghai), and two moderate to strong (Shiping-Jianshui and Mabian) seismotectonic lines. It is now at the beginning of a seismically quiet period in the third strain accumulation and release cycle. The probability of occurrence of an earthquake of magnitude > 7 is quite small.

The Songpan-Jingtai strong seismotectonic zone. Situated north of the Luding-Ya'an line and west of the Liupan-Helan Mountain line, it is a N-S-trending concealed tectonic zone (composed of the Songpan-Jingtai and Wenxian-Haiyuan concealed fractures). Though it shows an E-W direction on the surface and exhibits signs of a N-S trend only at Minshan Mountain, telluric current and magnetic data indicate that in the Tianshui area a N-S-trending structure exists at a depth of more than 7 km. In addition, the following phenomena can also be used to support the above viewpoint: (1) West of the zone the land surface has been raised up to 3,500-4,000 m above sea level, but east of the zone it suddenly descends down to 2,500-3,000 m, the zone being a topographic demarcation line. (2) Cenozoic diwa basins have an en echelon arrangement in a N-S direction. (3) Rivers from Minxian to Lanzhou run in a N-S direction, e.g., the Taohe River, the Zuli River, the Bailong River (near Wudu), the Hulu River and the Jialing River. (4) Epicenters have a N-S alignment along the zone. (5) The long axes of the meizoseismal areas of moderate to strong earthquakes are oriented N-S. (6) A set of nodal planes of the 1954 east-of-Minqin earthquake with magnitude

7 strikes NW 4°.

Seismic activity of this zone is concentrated towards Tianshui and Wudu, where a N-S-trending concealed structure intersects an E-W-trending surface structure, and towards the south and north ends, e.g., Minqin and Songpan. Earthquakes migrated in the following way. In 1125-1352 earthquakes of magnitude 7 were concentrated towards the Lanzhou-Huining area. In 1654-1878 they migrated southwards to Tianshui and Wudu, and after 1933 jumped to the south and north ends, e.g., the 1933 Songpan earthquake of magnitude 7 1/2 at the south end and the 1954 east-of-Minqin earthquake of magnitude 7 at the north end. The zone may be divided into the Songpan-Jingtai and Wenxian-Haiyuan strong and the Longmen Mountain moderate seismotectonic lines. It is now in the middle part of a seismically quiet period in the third strain accumulation and release cycle. We suspected that the zone was experiencing a dilatancy stage precursory to an M > 7 earthquake. Indeed, an earthquake of magnitude 7.2 occurred between Songpan and Pingwu on July 16, 1976.

The Liupan-Helan intense seismotectonic zone. It is composed of a series of parallel N-S-trending fractures, the Helan Mountain fracture, the Zhuozi Mountain fracture and the Yinchuan graben. Along the Helan Mountain fracture ultrabasic rocks have an intermittent distribution. The Yinchuan diwa is a Mesozoic-Cenozoic fault depression, in which Quaternary sediments have a thickness of 1,600-1,800 m. In the Liupan Mountain area, a set of NW-trending fractures intersects it. Therefore, earthquakes occur concentratively. The zone shows considerably high seismicity, and two earthquakes of magnitudes 8-8 1/2, four of 7-7 1/2, and eleven of 6-6 1/2 have occurred. In 1561-1852, seismicity was concentrated chiefly in Pingluo and Zhongwei, but after 1920 it migrated to Xiji, Maiyuan and Guyuan. In recent years the zone is in the early to middle part of a seismically quiet period in the third strain accumulation and release cycle, and M > 7 earthquakes are unlikely to occur. Within the zone two intense seismotectonic lines, the Yinchuan and the Liupan Mountain, can be singled out.

The North Xinjiang diwa intense seismic region (II)

It is identical on the whole with the Altay, Junggar, and Tianshan diwa regions in geographical limits. Structural lines in it trend NW and WNW, but NE at Jieshan Mountain, Junggar Basin. At present it is in the maximum-mobility period of diwa development. The region shows severe neotectonic movement, sharp tectono-topographic contrast and high seismicity [92].

The region can be divided into four seismotectonic zones.

The Altay intense seismotectonic zone. It corresponds principally to the Altay Mountains (extending eastwards out of Chinese territory). Structural lines trend approximately NW. The zone shows high seismicity. From 1931 to the present five earthquakes of magnitude > 6 and one of 8 have occurred. There is an intense seismotectonic line, the Fuyun, in the zone, where an earthquake of magnitude 8 has occurred. At present the zone is in the middle part of a seismically quiet period in the third strain accumulation and release cycle, and the probability of occurrence of an earthquake of magnitude > 7 is small.

The Junggar moderate to strong seismotectonic zone. It is a triangle-like basin bordered by fractures. Neotectonic movement is expressed as large-scale downwarping responsible for the accumulation of thick Quaternary sediments. The zone shows low seismicity. From 1932 to the present only two earthquakes of magnitude > 6 have occurred.

The Northern Tianshan intense seismotectonic zone. It is separated from the Junggar moderate to strong seismotectonic zone by the Urumqi deep-seated fracture in the north and extends southwards to Narat Mountain. Neotectonic movement is characterized by rapid uplifting. On the ground of the planation surface formed at the end of the Paleogene, the zone began to rise in Pliocene time with a speed of 100-130 m/m.y., which has increased up to 1,000 m/m.y. in the Quaternary. Fractures are inherited from old structural lines and developed in a WNW direction, producing a series of Cenozoic basins. Ten earthquakes of magnitude 6 and one of > 8 have occurred. At present the zone is at the end of a seismically quiet period in the second strain accumulation and release cycle and will enter into an active period. In recent years the probability of occurrence of an earthquake with magnitude > 7 is large. Within the zone the Tarqiyiling strong and the Wenquan-Urumqi intense seismotectonic lines have been singled out.

The Southern Tianshan intense seismotectonic zone. Situated south of Narat Mountain, with a fracture in between it adjoins the South Xinjiang diwa region in the south. Neotectonic movement is similar to that in the Northern Tianshan intense seismotectonic zone. Structural lines trend WNW, but turn into ENE in Kalpin and westwards almost E-W. They meet with several sets of structural lines near Wuqia, where seismicity appears to be of great intensity and high frequency. One earthquake of magnitude 8 and fifteen of > 6 (calculated only in Chinese territory) have occurred. At present it is in the middle part of a seismically quiet period in the third strain accumulation and release cycle, and the probability of occurrence of an earthquake of magnitude > 7 is small. Within the zone the Wuqia-Kalpin intense and the Baicheng-Korla strong seismotectonic lines have been singled out.

The Alxa moderate seismotectonic zone. It occupies the Alxa area. Structural lines trend approximately ENE. The zone shows low seismicity. The greatest known earthquake only has a magnitude of 5.8.

The South Xinjiang diwa strong seismic region (III)

It is identical on the whole with the South Xinjiang diwa region in geographical limits and at present is in the maximum-mobility period of diwa development. Neotectonic movement is characterized by block-faulting and differential downwarping and upwarping. Mesozoic-Cenozoic sediments in the central part of the Tarim Basin attain a thickness of more than 10,000 m. The greatest earthquake has a magnitude of 7 1/4. Seismicity has a low frequency, and only four earthquakes of magnitude > 6 have occurred since 1922 [92].

The region can be divided into two seismotectonic zones.

The Tarim moderate to strong seismotectonic zone. It occupies the Tarim Basin. Neotectonic movement is characterized by unbalanced subsidence, which has given rise to thick Quaternary sediments. A set of X-shaped fractures, ENE- and WNW-trending, has been formed in the basin. The zone shows low seismicity, and only an earthquake of magnitude 6 took place in 1927.

The Altun strong seismotectonic zone. It occupies the Altun Mountains and consists of a series of ENE-trending structures. Neotectonic movement is dominated by intense uplifting. One earthquake of magnitude 7 1/4 and two of 6-6 3/4 have occurred.

The Kunlun-Bayanhar geosynclinal strong seismic region (IV)

It includes the Kunlun and Bayanhar geosynclinal regions and at present is in the fold belt period of geosynclinal development. Structural lines trend NW at the two ends and almost E-W in the center, forming an overturned S-shape. Neotectonic movement is dominated by uplifting. The region shows moderate to high seismicity. The largest earthquake attains a magnitude of 7.9. Since 1481, sixty-one earthquakes of magnitude > 6 have occurred [92].

Within the region four seismotectonic zones can be singled out.

The Kunlun strong seismotectonic zone. It is separated from the South Xinjiang diwa strong seismic region by the southern Tarim border fracture on the north and from the North Xizang diwa moderate seismic region by major fractures on the south; eastwards, it terminates at Alke Mountain. Neotectonic movement appears as intense uplifting. Basic, intermediate and alkaline rocks were intruded along fractures in the western Kunlun Mountains during Quaternary time. Since 1910, fourteen earthquakes of magnitude > 6 have occurred along E-W-trending structures. The greatest magnitude is 7 1/2.

The Burhan Budai Mountain strong seismotectonic zone. It is separated from the Bayanhar strong seismotectonic zone and the Qinghai-Gansu diwa intense seismic region both by major fractures on the south and north. There has been developed a set of almost E-W-trending active fractures, along which basic rocks, Cenozoic basins and lakes are aligned in a line. Earthquakes of magnitudes 7 1/2 and 7 occurred east of Toksun lake in 1937 and 1963 respectively.

The Bayanhar strong seismotectonic zone. Its northern limit lies south of the Kunlun and Burhan Budai Mountain strong seismotectonic zones, and its southern limit is defined by the Yushu and Litang fractures and adjoins the Hoh Xil strong seismotectonic zone. Structural lines trend approximately NW. The zone shows high seismicity. Four earthquakes of magnitudes 7.5-7.9 and twenty of > 6 have occurred. There is a conspicuous strong seismotectonic line, the Luhuo-Xianshui River line, which extends for more than 300 km and appears to be a northwesterly directed sinistral strike-slip fault. The Luhuo earthquake of 1973 with a magnitude 7.9 occurred just at this fault, along which a 80 km-long crack was generated. At present the zone is in the early part of a seismically quiet period in the third strain accumulation and release cycle, and the probability of occurrence of an earthquake of magnitude ⩾ 7 is small.

The Hoh Xil strong seismotectonic zone. Separated from the North Xizang diwa moderate to strong seismic region by fractures, it stretches along the Hoh Xil Range and extends southwards into Luli Mountain. Structural lines trend NW in the Hoh Xil Range but NNW in Luli Mountain. Neotectonic movement is very intense. Since the beginning of the Pliocene the planation surface has been raised up to 3,400 m. Seismicity in the southeastern part is higher than in the northwestern. The former has been struck by two earthquakes, while the latter by one only. Since 1481, twenty-three earthquakes of magnitude ⩾ 6 and three ≥ 7 have occurred. Many earthquakes occurred at the Zhongdian-Xiaguan strong seismotectonic line. At present the zone is in the early part of a seismically quiet period in the fourth strain accumulation and release cycle, and the probability of occurrence of an earthquake of magnitude ≥ 7 is small.

The North Xizang diwa moderate to strong seismic region (V)

Separated from the Kunlun-Bayanhar geosynclinal strong seismic region by a major fracture on the north and from the Gandise-Himalayan geosynclinal intense seismic region on the south, it occupies the great part of the North Xizang diwa region. Structural lines trend approximately NW to WNW. Since 1914, eight earthquakes of magnitudes 6-6 1/2 have occurred [92].

The Gandise-Himalayan geosynclinal and diwa intense seismic region (VI)

It includes the Gandise and Yarlung Zangbo River-Himalayan geosynclinal regions, Southern Himalayan diwa region and part of the North Xizang diwa region, being a part of the global Alpine-Himalayan tectonic belt and also a zone of variable crustal thickness, intense neotectonic movement and high seismicity.

Within the zone two seismotectonic zones can be singled out.

The Gandise intense seismotectonic zone. Situated north of the Yarlung Zangbo River deep-seated fracture, it is a geosynclinal fold belt inversed in the Yanshanian orogeny. Structural lines trend almost E-W. On the foot of the Nyainqentanglha Range there is a major fracture extending for tens of kilometers. Along the fracture a series of hot springs outcrop in a line. The zone shows relatively high seismicity, and the earthquake of 1951 with magnitude 8 occurred just at the turn of the fracture. Since 1921, twenty-three earthquakes of magnitude > 6 have occurred in the zone. At present the zone is in the late part of a seismically quiet period in the second strain accumulation and release cycle, and will probably enter into an active period in the coming years. The probability of occurrence of a M ≥ 7 earthquake is large.

The Himalayan intense seismotectonic zone. It occupies the Himalayas south of the Yarlung Zangbo River deep-seated fracture. Neotectonic movement is characterized by intense uplifting, which has caused the zone to rise since Eocene time from sea up to a height of 8,848.13 m. The zone shows high seismicity and has experienced an earthquake of magnitude 8 1/2. Since 1911, thirty-three earthquakes of magnitude ≥ 6 have occurred (only in Chinese territory). Within the zone there exists the conspicuous Yarlung Zangbo River-Zayu intense seismotectonic line. At present the zone is in the middle to late part of a seismically quiet period in the second strain accumulation and release cycle, and will probably enter into an active period. Therefore, it is in danger of occurrence of an M > 7 earthquake. Indeed, on May 29, 1967 two earthquakes of magnitudes 7.5 and 7.6 occurred in Longling and Luxi, Yunnan Province.

The West Yunnan diwa strong seismic region (VII)

It coincides principally with the West Yunnan diwa region in geographical limits and at present is in the maximum-mobility period of diwa development. Structural lines trend almost N-S. Neotectonic movement is expressed as vertical differential movement. The Bilu Mountain planation surface between the Lancang

River and the Nujiang River has been uplifted with a height of 2,900 m. Since 1923, one earthquake of magnitude 7 and eight of 6-6 3/4 have occurred [92].

Within the region two seismotectonic zones can be singled out.

The Wuliang Mountain moderate to strong seismotectonic zone. Situated east of the Lancang River, it consists of a series of NNW-trending fractures. Since 1884, six earthquakes of magnitudes 6-6 3/4 have occurred along the fractures and among them four occurred in the Pu'er area.

The Nushan Mountain strong seismotectonic zone. It is situated west of the Lancang River fracture zone and shows more intense neotectonic movement than in the Wuliang Mountain moderate to strong seismotectonic zone. Along the Lancang River fracture there exist ultrabasic rocks. Structural lines trend N-S. Since 1876, three earthquakes of magnitude 7 and four of 6-6 3/4 have occurred.

The Qinghai-Gansu diwa intense seismic region (VIII)

It is bounded to the north by the Alxa Right Banner, to the south by the Qaidam Basin, to the west by the Altun major fracture, and to the east by the Jingtai-Songpan line. The greater part of it has already entered into the maximum-mobility period of diwa development, and only a part of it, the western Qinling Mountains, is still in its geosynclinal fold belt period. The reigon is experiencing intense neotectonic movement, which is expressed in the following: (1) Inherited from ancient structural lines, fractures have been reactivated and cut across the Quaternary. Along the fractures occur a series of diwa basins and lakes, e.g., the Jinquan diwa and the Minle diwa. (2) Since the end of the Tertiary, Qilian Mountain has been risen, while its two sides, Hexi Corridor and Qaidam Basin, deeply subsided, e.g., Mesozoic-Cenozoic strata have a thickness of 6,000 m in the Jinquan diwa. The region shows high seismicity. Since 1900, one earthquake of magnitude 8, two of 7 1/4-7 1/2, and ten of 6-6 4/5 have occurred in the region [92].

Within the region three seismotectonic zones can be singled out.

The Corridor intense seismotectonic zone. It is situated north of the cis-Qilian fracture. Neotectonic movement is characterized by intense subsidence. There are a series of diwa basins distributed along the cis-Qilian fracture, which has cut across the Quaternary. The zone shows high seismicity. It has been reported that one earthquake of magnitude 8, two of 7 1/4-7 1/2, and six of 6-6 3/4 have occurred in the zone. The Shandan strong and the Corridor-Nanshan intense seismotectonic lines have been singled out. At the intersection of the latter with a NNW-trending fracture the famous Gulang earthquake of magnitude 8 occurred in 1927. At present the zone is in the middle to late part of a seismically quiet period in the second strain accumulation and release cycle, and will probably enter into an active period in the coming years. Therefore, an $M \geqslant 7$ earthquake will be possible to occur.

The Qilian Mountain strong seismotectonic zone. It is situated south of the cis-Qilian fracture and north of the Qaidam Basin. The zone as a whole is experiencing uplifting. Structural lines are inherited from the NW direction and still active in Cenozoic time. Since 1917, one earthquake of magnitude 7 1/2 and eight of 6-6 3/4 have occurred. Within the zone there exists the prominent

Nanshan strong seismotectonic line.

The Qaidam moderate to strong seismotectonic zone. It is bounded by fractures. Neotectonic movement is dominated by intense subsidence. The Tertiary has a thickness of more than 3,000 m. Strike-slip faults are well-developed. The zone shows lower seismicity than the above-mentioned two zones. Since 1917, two earthquakes of magnitudes 6-6 4/5 have occurred at the intersections of two sets of fractures.

The NE diwa moderate to strong seismic region (IX)

It occupies a vast area of northeast China lying north of the Hetao area and the Bayanobo-Kaiyuan fracture and west of the Mishan-Dunhua fracture with the exception of the Songliao platform region, and corresponds approximately to, but somewhat less than, the NE diwa region in geographical limits. Neotectonic movements manifest themselves as low-amplitude upwarping and downwarping and basaltic eruptions (known three times). Locally Tertiary rocks have been subjected to tilting and faulting. The region shows low seismicity. No earthquake of magnitude > 6 took place, except at the boundary between the Yinshan area and the Songliao platform region, where an earthquake of magnitude 6 occurred [92].

Within the region three seismotectonic zones can be singled out.

The Hohhot moderate to strong seismotectonic zone. It lies south of the Chifeng fracture. The southern limit is defined by the southern boundary of the Hetao graben and the Zhangbei fracture. Westwards, Alxa delimits the zone. Structural lines trend chiefly E-W. In this zone the top of the North Xinjiang-Hinggan arcuate structure meets with an E-W tectonic system. Neotectonic movement is intense, accompanied by basaltic eruptions. Large-amplitude subsidence, with a depth of 4,000-5,000m in the south and 6,000-7,000 to 10,000 m in the north, took place along the Hetao graben-type diwa basin, where three earthquakes of magnitudes 6-6 1/4 have occurred since 1929.

The Hinggan Mountains moderate seismotectonic zone. It adjoins the Songliao platform region with the Baicheng fracture in between and terminates to the northeast at the Sunwu-Dedu fracture. The zone shows weak neotectonic movement and low seismicity. The largest earthquake does not exceed a magnitude of 5 1/2.

The Jilin moderate seismotectonic zone. It is separated from the Hunchun strong seismic region by the Mishan-Dunhua fracture on the east. The western limit extends from Kaiyuan and the Yingkou-Tancheng fracture (its northern extension) through Wuchang, Suihua and Bei'an to the Sunwu-Dedu fracture. Structural lines trend principally NE. Along fractures late Tertiary to Quaternary basalts have poured out. The zone shows weak differential movement and low seismicity. Only at the west edge adjacent to the Songliao platform region have two earthquakes of magnitude 6 occurred. The northern extension of the Yingkou-Tancheng fracture has been struck by many weak earthquakes, the greatest magnitude of which, however, does not exceed 5 3/4.

The Hunchun diwa strong seismic region (X)

It lies east of the Mishan-Dunhua fracture. On the south it adjoins the North China diwa region with the eastern extension of the Chifeng fracture in between. NE-trending fractures are tectonically active and along them Quaternary basaltic lavas have poured out. The region shows high seismicity. Earthquakes occur frequently. The largest magnitude reaches 7 1/4. Since 1905, twelve earthquakes of magnitude > 6 have occurred. Most earthquakes are concentrated towards Dongning and Hunchun. Hypocenters attain depths of 300-600 m. The zone is a part of the Pacific deep-focus seismic belt, which plunges with a very gently dipping tectonic plane into Chinese territory.

The Songliao platform moderate seismic region (XI)

It corresponds to the Songliao platform region in geographical limits bounded by fractures. At present it is in the residual-stability period of platform development. Neotectonic movement is dominated by downwarping and upwarping affecting a large area. The region shows weak differential movement and low seismicity. In 1958, an M = 5 1/5 earthquake occurred in Changchun. Only in the northwestern and southern parts adjacent to the NE diwa region have two M=6 earthquakes occurred.[92]

The North China diwa intense seismic region (XII)

It corresponds to the North China diwa region in geographical limits. Structural lines trend principally NNE or nearly NE. Neotectonic movement manifests itself chiefly as vertical differential movement in a NNE direction. Uplifts alternate with subsidences, both with different speed and amplitude. The region shows high seismicity, and five earthquakes of magnitudes 8-8 1/2 have occurred. Since 1011, the region has experienced three cycles of seismicity (Table 6.1) [48, 61, 92].

Within the region six seismotectonic zones can be singled out.

The Liaoning-Shandong intense seismotectonic zone. Situated east of the Yingkou-Tancheng fracture, it includes east Liaoning and east Shandong Provinces. Recent tectonic movements are expressed as differential uplifts and subsidences along a NNE-trending major fracture; from 1952 to 1971 the Liaodong Peninsula uplifted +70 to 100 mm, the Jiaodong Peninsula +20 mm, but the Penglai area subsided; Yexian and Anqiu, Shandong Province, uplifted +60 and +100 m, respectively, but Tancheng subsided -20 mm. The zone shows high seismicity, but the frequency of occurrence of earthquakes is low. It has been reported that one earthquake of magnitude 8, three of 7-7 2/5, and seven of 6-6 1/2 have occurred. In the zone there exists the Yingkou-Tancheng intense seismotectonic line composed of two NNE-trending deep-seated fractures. At present the zone has just entered into a seismically quiet period in the third strain accumulation and release cycle, and the probability of occurrence of an M > 7 earthquake is small.

The Hebei-Shandong strong seismotectonic zone. Situated west of the Yingkou-Tancheng and east of the Dezhou and Liaocheng fractures, it includes the Bohai diwa series and the West Shandong geodome series. The former is dominated by NE-trending structures and appears to be a subsiding area since Cenozoic time, while the latter by NW-trending structures and a geodome uplift for a long time. Since 1623, one earthquake of magnitude 7 1/2 and four of 6-6 1/4 have occurred.

The Central Hebei intense seismotectonic zone. Situated west of the Dezhou and Liaocheng fractures and east of the cis-Taihang fracture zone, it consists of a series of NNE-trending concealed fractures, depressions and uplifts having an en echelon arrangement. Leveling data in the period 1952-1971 indicate that the zone was subsided in a large area and that the amplitude of subsidence in the west (-30 mm) was larger than that in the east. The zone shows high seismicity. Earthquakes occur frequently. One earthquake of magnitude 8, four of 7-7 1/2, and twenty of 6-6 3/4 have been reported. The zone comprises three prominent seismotectonic lines.

The Pinggu-Shijiazhuang intense seismotectonic line. It is a NNE-trending active fracture zone and also a boundary between two second-order tectonic regions. In Tongxian and Sanhe NE- and NW-trending fractures meet with it. The Sanhe-Pinggu earthquake of 1679 with magnitude 8 occurred just at this extremely favourable position. The long axis of its meizoseismal area trends NE, so that it is closely related chiefly to the NE-trending fracture. Over a length of less than 180 km from Dingxing to Sanhe one earthquake of magnitude 8 and five of 6-6 1/2 have occurred since 1057.

The Cixian-Hejian strong seismotectonic line. It is composed of a set of NNE-trending fractures intersected by a NW-trending fracture in the vicinity of Cixian and Shenxian. The Cixian earthquake of 1830 with magnitude 7 1/2 occurred at the meeting of the Handan-Shijiazhuang fracture with a nearly E-W-trending fracture; the Xingtai earthquakes of 1966 with magnitudes 6 4/5 and 7 1/2 occurred at the intersection of a NE-trending fracture with a NW-trending one.

The Liaocheng strong seismotectonic line. It is composed of the Liaocheng fracture. The Heze earthquake of 1937 with magnitude 7 occurred just at the southwest end of the fracture. Three earthquakes of magnitude > 6 have occurred along the fracture.

From the three seismotectonic lines of the Central Hebei intense seismotectonic zone we can see the process of temporal and spatial transformation from a seismically active region into a quiet one and vice versa (Fig. 6.6). In the period 1658-1730, a seismically active region lay at the Pinggu-Shijiazhuang intense seismotectonic line; in the period 1830-1882, it migrated southwards to the Cixian-Hejian strong seismotectonic line; and in the period of 30 years from 1937 to 1967, it dropped on two lines, first on the Liaocheng and then on the Cixian-Hejian strong seismotectonic line. At present these two lines are in a period of seismic quiescence. The forthcoming seismic activity will probably migrate to the Pinggu-Shijiazhuang intense seismotectonic line, and an M > 7 earthquake will probably also migrate to that line. Indeed, on June 28, 1967, an earthquake of magnitude 7 4/5 occurred in Tangshan. After that the zone is in the early part of a seismically quiet period in the third strain accumulation and release cycle, and the probability of occurrence of an M > 7 earthquake in recent years is small.

The Taihang Mountain moderate to strong seismotectonic zone. Situated west of the cis-Taihang fracture and east of the Fenhe-Weihe graben, it is a geodome uplift composed of NNE- and NW-trending folds and faults. According to deformation data for the period 1952-1971, it rose +30 mm, while differential movement is not prominent. Seismicity is low and only one earthquake of magnitude 6 has been reported.

The Fenhe-Weihe intense seismotectonic zone. It is the boundary between the Yishan (Ordos) platform region and the North China diwa region, and consists of a series of NE-trending Cenozoic graben-type diwa basins arranged in an en echelon pattern. The basins subsided 1,066-5,000 m in Cenozoic time and 500-1,300 m in the Quaternary period. They are generally deep in the south and shallow in the north, with new fractures developed at their edges. The internal parts of these basins have been complicated by secondary transverse uplifts. Deformation data in the period 1952-1971 indicate that the largest amplitude of subsidence of the basins is 7.19 mm/year.

The process of transformation from a seismically active period into a quiet one and vice versa and the migration of earthquakes can be clearly identified in the zone. The first active period lasted 77 years (1291-1368), during which a seismically active region lay in the Fenhe graben and one earthquake of magnitude 8 and two of 6-6 1/2 occurred. In the second seismically active period, which lasted 67 years (1501-1568), the seismically active region migrated to the Weihe graben. During this period three earthquakes of magnitudes 8 7/10 and 6 3/4 occurred. In the third seismically active period, which lasted 57 years (1626-1683), the seismically active region migrated northwards to the Fenhe River, Lingqiu and Yuanping. During this period two earthquakes of magnitudes 7 and 7 1/4 occurred. In 1695 the seismically active region again migrated southwards to the Fenhe graben. The Pinglu earthquake of 1815 with a magnitude of 6 3/4 marks probably the beginning of the fourth seismically active period, and the seismically active region tends to migrate southwestwards (Fig. 6.7). The migration of $M > 7$ earthquakes shows the similar regularity. Within the zone the following seismotectonic lines can be singled out: the Weihe and Fenhe intense seismotectonic lines, the Yuanping-Daixian strong seismotectonic line and the Datong moderate to strong seismotectonic line. At present the zone is in the middle to late part of a seismically quiet period in the fourth strain accumulation and release cycle and will probably enter into a seismically active period in recent years. Therefore, an earthquake of magnitude > 6 is likely to occur.

The Xuchang-Huainan moderate to strong seismotectonic zone. Situated north of the Luonan-Xin'andian deep fracture, it is a Mesozoic-Cenozoic diwa basin. The zone trends in an E-W direction and shows low seismicity. Only two earthquakes of magnitudes 6 and 6 1/4 have been reported.

The Yishan (Ordos) platform weak seismic region (XIII)

Separated from surrounding diwa regions by fractures, it is at present in the residual-stability period of platform development. It shows weak neotectonic movement and low seismicity. Only at its edge there did an earthquake of magnitude 5 1/2 occur [92].

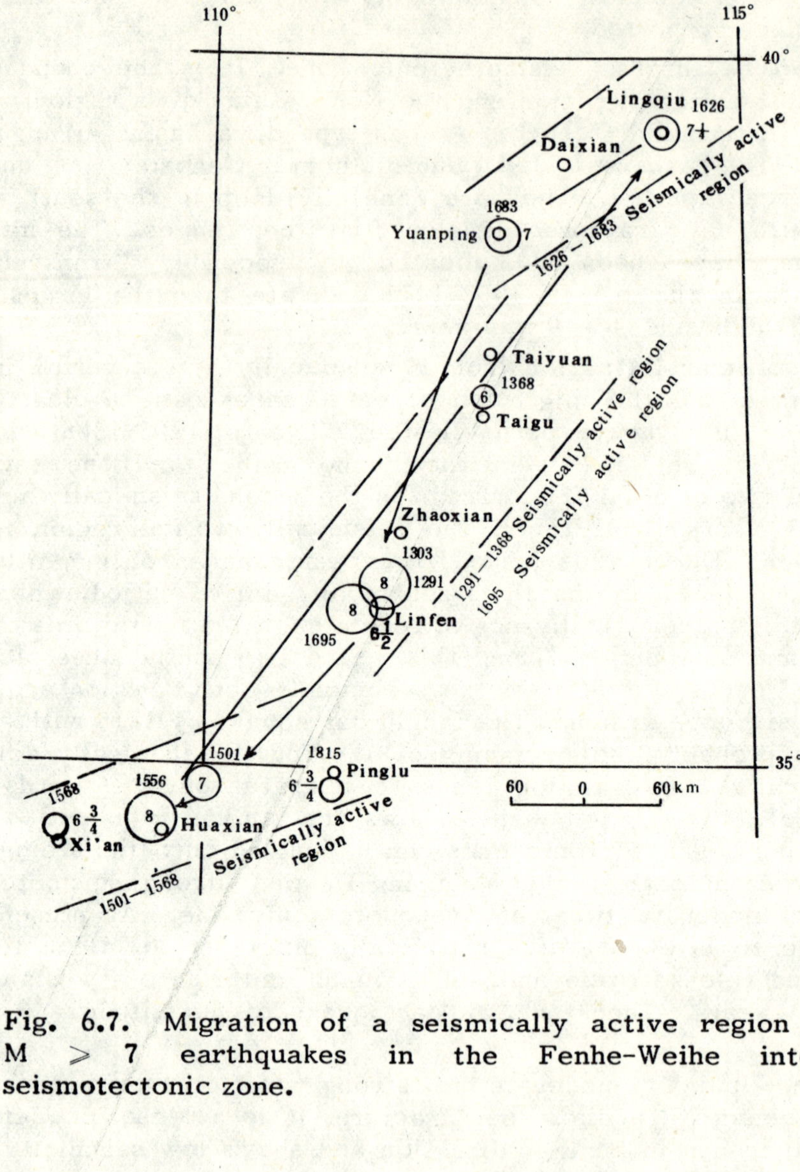

Fig. 6.7. Migration of a seismically active region and M ⩾ 7 earthquakes in the Fenhe-Weihe intense seismotectonic zone.

The Central China diwa moderate to strong seismic region (XIV)

It corresponds to the Central China diwa region in geographical limits. The principal structural line appears to be a southward-convex and nearly E-W oriented arcuate structure. The Qinling and Dabie Mountains have been rising for a long time, while the areas lying north and south of it subsided continuously. The region shows low seismicity with the largest magnitude of 6 3/4. As has been reported, nine earthquakes of magnitudes 6-6 3/4 have occurred in the region[92].

Within the region three seismotectonic zones can be singled out.

The Qinling-Dabie moderate to strong seismotectonic zone. Situatd south of the Luonan-Xin'andian deep fracture and north of the South Huaiyang deep fracture, it occupies principally the western wing and top of the southward-convex Huaiyang arc. The zone is characterized by a long uplift. In Cenozoic time a diwa basin was formed in Nanyang. Earthquakes are generally distributed at the edge of the basin. For example, both Nanyang and Huoshan have experienced M=6 earthquakes. According to historical records, five earthquakes of magnitudes 6-6 1/2 have occurred in the zone.

The Nanjing moderate to strong seismotectonic zone. Situated southeast of the Yingkou-Tancheng deep fracture and north of the Jiujiang-Wuxi fracture, it is a Mesozoic-Cenozoic diwa basin trending NE and extending into the Huanghai (Yellow) sea. Four earthquakes of magnitudes 6-6 3/4 have occurred in the zone. On may 5, 1974, an earthquake of magnitude 5 1/2 occurred in Liyang.

The Daba-Jianghan moderate seismotectonic zone. It includes Daba Mountain and the Jianghan Plain. The southern limit extends from Honghu through Yichang to Wanyuan and adjoins the SE diwa region. Structural lines trend approximately WNW and NW. The zone shows very low seismicity with the largest magnitude of not more than 5 1/2. In 1969 an earthquake of magnitude 4 7/10 occurred south of Baokang.

The SE diwa intense seismic region (XV)

It is part of the SE diwa region. Defined by fractures and Mesozoic-Cenozoic diwa basins having an en echelon arrangement, the western limit extends from Mufu and the Luoxiao Mountains, through Wuzhou, southwestwards to Pingxiang, and adjoins the Hunan-Guangxi-Guizhou moderate seismic region. At present the region is in the residual-mobility period of diwa development. The principal structural line belongs to the NE tectonic system. There are also NW and E-W structural lines intersecting it. Neotectonic movement is severe, and in Quaternary time there occurred volcanic extrusion, faulting, folding and terrace displacement. Since the Holocene the rate and amplitude of differential vertical movement have increased, sedimentation centers migrated and widespread hot springs occurred. The region shows high seismicity, and one earthquake of magnitude 8, three of 7-7 1/2 and nine of 6-6 1/2 have been reported. The seismicity tends to be weakened progressively from coast to island [20, 22, 49, 92, 96].

Within the region four seismotectonic zones can be singled out.

The Quanzhou-Hainan intense seismotectonic zone. Bounded by the Zhenghe-Haifeng fracture to the west, it extends southwestwards to the Leizhou

Peninsula. In this zone Mesozoic-Cenozoic volcanic activity occurred extensively, and Quaternary basaltic extrusion has a wide development. Structural lines trend principally NE; sometimes they are intersected by NW-trending fractures. The zone shows high seismicity. Since 1067, one earthquake of magnitude 8, three of 7-7 1/2, and four of 6-6 3/4 have occurred. The zone at present is at the end of a seismically quiet period in the second strain accumulation and release cycle and will probably enter into a seismically active period. In one of the following years an earthquake of magnitude > 6 is likely to occur. Within the zone there is the Changle-Zhao'an strong seismotectonic line, and the well-known Nan'ao earthquake occurred just along it.

The Fujian-Guangdong moderate to strong seismotectonic zone. The western limit of this zone extends from Shaowu through Heyuan and Zhaoqing to Hainan Island and with the Heyuan-Shaowu fracture and the Sihui-Wuchuan fracture in between it adjoins the Jiangxi moderate seismotectonic zone. Since 1806, three earthquakes of magnitudes 6-6 2/5 have occurred. There are the Heyuan-Shaowu and Yangjiang moderate to strong seismotectonic lines in the zone.

The Yunkai Mountain moderate to strong seismotectonic zone. It is situated west of the Sihui-Wuchuan fracture and east of the Guixian-Pingxiang fracture. There are a series of NE-trending fractures, Mesozoic granites and Cenozoic diwa basins. In 1936 an earthquake of magnitude 6 3/4 and in 1958 another earthquake of magnitude 5 3/4 occurred along the Lingshan moderate to strong seismotectonic line.

The Jiangxi moderate seismotectonic zone. It is separated from the Fujian-Guangdong moderate to strong seismotectonic zone by the Heyuan-Shaowu fracture on the east and from the Hunan-Guangxi-Guizhou moderate to strong seismic region by fractures arranged in an en echelon pattern on the west. NE-trending fractures and Mesozoic-Cenozoic diwa basins predominate and have an en echelon arrangement. The zone shows low seismicity with the largest magnitude not more than 5. M=5 earthquakes with low frequency occurred in Jing'an, Anfu, Shangrao and Jiujiang respectively.

The Hunan-Guangxi-Guizhou diwa moderate to strong seismic region (XVI)

This region, the eastern limit of which extends from the Mufu and Luoxiao Mountains to Wuzhou, is separated from the SE diwa intense seismic region by NNE-trending fractures, reaches the eastern edge of the Sichuan platform region on the west and with the Honghu-Yichang line as a demarcation line it adjoins the Central China diwa moderate to strong seismic region on the north. The region comprises the whole Yunnan-Guizhou diwa region and part of the SE diwa region. At present it is in the residual-mobility period of diwa development. Structural lines trend mostly NNE or NE and only in the south belong to the Guangxi arcuate structure. The region shows low seismicity and earthquakes occur infrequently. This is due to small differentiation of neotectonic movement. The largest earthquake reported has a magnitude of 6 1/2.

Within the zone three seismotectonic zones can be singled out.

The Yueyang-Wuzhou moderate to strong seismotectonic zone. The western limit extends from Huarong and Xinhua to Guilin. Structural lines trend NNE. Fractures and Cenozoic diwa basins have an en echelon arrangement. The zone shows low seismicity with the greatest magnitude not more than 5 1/2.

Earthquakes migrated roundly with a period of approximately 75 years. The latest is the Xiangxiang earthquake of 1936 with a magnitude of 5. The following period will probably migrate northwards (Fig. 6.8).

The Changde-Guiding moderate to strong seismotectonic zone. The western limit of this zone is defined by the Shizhu-Xishui line, while the southern limit extends from Yanjin and Anshun through Liuzhou to Wuzhou. Structural lines trend NE and consist of a series of NE-trending fractures and Cenozoic diwa basins (e.g., Dongting diwa, Taoyuan diwa and Yuan-Ma diwa). The zone shows low seismicity. The Changde earthquake of 1631 with magnitude 6 1/2 is the largest event. Seismic energy in the larger part of the zone has been released by moderate and weak earthquakes. Earthquakes regularly migrated in a definite direction from northeast to southwest (Fig. 6.4).

The Yunnan-Guangxi moderate seismotectonic zone. It lies south of the Changde-Guiding moderate to strong seismotectonic zone and west of the Yunkai Mountain moderate to strong seismotectonic zone. Structural lines trend principally WNW and E-W; NE-trending structures exist only in east Yunnan and east Guangxi. The zone shows low seismicity, and the largest magnitude does not exceed 5 1/2. Earthquakes migrated in a definite direction from southwest to northeast along NW- or WNW-trending structural lines.

The Sichuan platform moderate seismic reigon (XVII)

Situated in central Sichuan Province, it is surrounded by diwa regions. The region is a residual platform and now is in its residual-stability period. Neotectonic movement is characterized by downwarping and upwarping affecting a large area. Differential movement is not pronounced. The region shows low seismicity, and the largest magnitude does not exceed 5 1/2. Earthquakes are concentrated largely towards the southeastern edge of the region adjacent to the Yunnan-Guizhou diwa region, along the NE-trending boundary with the Yunnan-Guizhou diwa region and the Yibin fracture line. Seismicity is also observed at the northeast end of the Renshou fracture[92].

The Taiwan geosynclinal intense seismic region (XVIII)

It corresponds to the Taiwan geosynclinal region in geographical limits. It is a NNE-trending fold region of Himalayan age, a part of the circum-Pacific belt of island arcs and geosynclinal regions, and also a branch of the circum-Pacific seismic belt. Neotectonic movement is very severe. The Taiwan Mountains have been raised up to 3,950 m above sea level. The east coast is very steep and has dropped to a 3,000 m-deep sea basin with a contrast of 7,000 m. Therefore , the region shows high seismicity, and earthquakes occur frequently. Since 1900, two earthquakes of magnitude 8 and more than 140 earthquakes of magnitude > 6 have occurred. Delimited by a sinistral longitudinal valley fault, two seismotectonic zones can be singled out[92].

The East Taiwan intense seismotectonic zone. Situated east of the longitudinal valley fault, it is a positive gravity anomaly zone. The crust is transitional between oceanic and continental types and attains a thickness of 20 km. A set of NNE-trending fractures has been developed. Earthquakes are distributed along the fractures. The zone shows high seismicity, and earthquakes occur frequently. Since 1900, two earthquakes of magnitude > 6 have occurred.

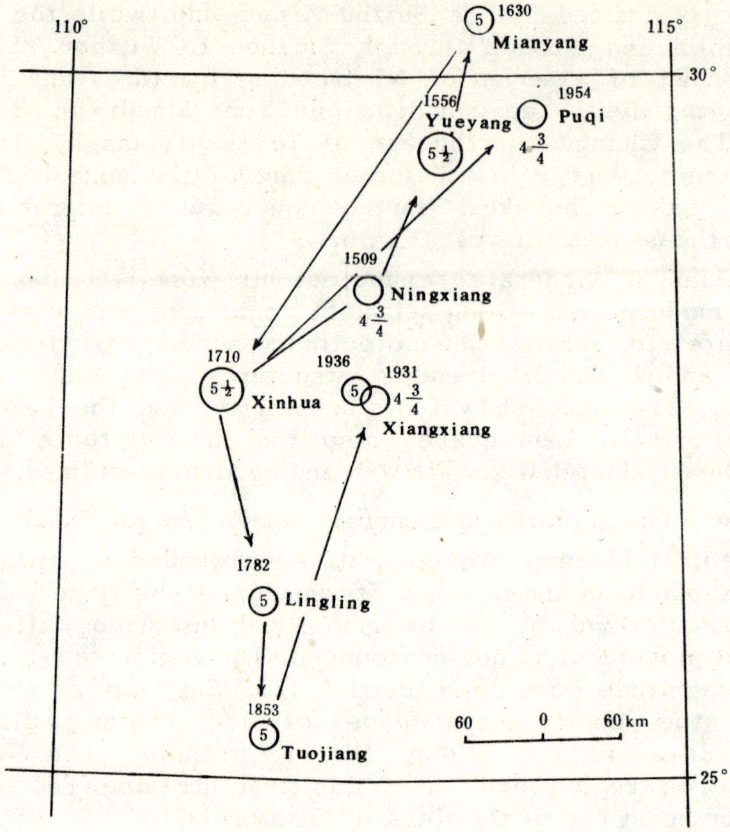

Fig.8. Reciprocating migration of earthquakes in the Yueyang-Wuzhou moderate seismotectonic zone.

The West Taiwan strong seismotectonic zone. Situated west of the longitudinal valley fault, it is cut across by two major NNE-trending fractures. Here the intensity and frequency of seismicity is lower than that in the eastern zone. The largest magnitude is 7 3/4. Since 1900, about forty earthquakes of magnitude > 6 have occurred.

The South China Sea Seismotectonic region (XIX)

Situated south of the SE diwa region and the Taiwan geosynclinal intense seismic region, it corresponds to the South China Sea diwa region in geographical limits. Seismicity and its relations to tectonics in this region need to be investigated.

REFERENCES

1. Alexandrov, E. A., Metallogeny of autonomous activation (A. D. Shcheglov). Econ. Geol., 71 (5), U. S. A., 1976.
2. Bhandari, A. K., et al., Rb-Sr geochronology of some granite/gneissic rocks of Northwestern Himalaya and its significance. Abstract, 26th IGC, Vol I, Paris, 1980.
3. Cai Bojun et al., Metallogenetic features of Deqin-Weixi siderite deposit, West Yunnan diwa region. Geotectonica et Metallogenia, 8(1), 1984.
4. Cai Jiayou, A discussion on diwa-type granitic rocks and metallogeny in Tonghua region, Jilin Province. Mineral Deposits, 1983(3).
5. Cao Renqi, Geotectonic types of gold mineralization in Shanxi [illegible]vince. Geotectonica et Metallogenia, 6(2), 1982.
6. Chang Chengfa, Geological outlines of the Altun Mount[illegible] eol. Soc. China, 38: 2 (1958).
7. Chang Chengfa, Geotectonic outlines of Qinghai-Tibet Plateau. [illegible]blems of Tectonics, 1965.
8. Chang Chengfa et al., Tectonic features of Mount Qomolangma, south Tibet, China.* Scientia Geologica Sinica, 16 : 1 (1973).
9. Chen Guoda, Examples of "activizing regions" in the Chinese platform, with special reference to the "Cathysia" problem. Acta Geologica Sinica, 36 : 3 (1955).
10. Chen Guoda, On the tectonics of China---Presentation of the Tectonic Map of China, 1 : 4,000,000. Global Tectonics & Metallogeny, 1(3), 276-285, Washington D.C., U. S. A., 1981.
11. Chen Guoda, On the geotectonic nature of SE coastal regions of China. Scientia Sinica, 7(1), 1958.
12. Chen Guoda, The third main tectonic element of the earth's crust---the diwa region.* Kexue Tongbao (Science Bulletin), 10 : 3 (1959).
13. Chen Guoda, Theory of progression with transformation between active and "stable" regions of the earth crust.* Acta Geologica Sinica, 39 : 3 (1959).
14. Chen Guoda, Philosophical problems in geotectonics.* Kexue Tongbao, 11 : 2 (1960).
15. Chen Guoda, Theory of platform activation and its ore-searching significance.* Geological Press, 1960.
16. Chen Guoda, On the subdivisions of the red beds of Southeastern China, Bull. Geol. Soc. China, 18(3-4), 1938.
17. Chen Guoda, Some philosophical view points about tectonic regionalization.* Kexue Tongbao (Science Bulletin), 11 : 17 (1960).
18. Chen Guoda, Viewing geomechanics from the angle of historic-analytical method and relationships between them.* Chinese Geology, 8-9 (1962).
19. Chen Guoda, Diwa region---a new type post-platform mobile region. Geotectonic problems of China.* Science Press, 1965.

20. Chen Guoda, An earthquake record---notes on the 1941 Xunwu, Jiangxi Province, earthquake.* Geological Review, 32 : 4 (1965).

21. Chen Guoda, Preliminary notes on the igneous rocks in the vicinity of Yongcheng (Jungcheng) City, Shandong (Shantung), Sci. Journ. Sun Yetsen Univ., 7(3), 1937.

22. Chen Guoda, The Lingshan earthquake of April, 1, 1936. Special Publ. Geol. Surv. Kwangtung & Kwangsi, 17, 1939.

23. Chen Guoda, On the occurrence of Carboniferous coal series in Kiangsi. Sci. Rec. Sinica, 1(1-2), 1942.

24. Chen Guoda, On the occurrence of marine Devonian strata in the vicinity of Canton.* Geol. Rev., 14(1-3), 1949.

25. Chen Guoda, The shoreline problem of China.* Sci. Sinica, 1(2-4), 1951.

26. Chen Guoda, An introduction to the oil shale of China.* Geol. Rev., 16(2), 1951.

27. Chen Guoda, The geotectonics of Nanling and its environs. Journ. Central-South Inst. Mining & Metallurgy, 2(2), 1957.

28. Chen Guoda, Construction---A new criterion for differentiating geotectonic regions.* Kexue Tangbao(Science Bulletin), 7, 1959.

29. Chen Guoda, Geology of China.* Central-South Inst. Mining & Metallurgy, 1959.

30. Chen Guoda, Characteristics and nature of the diwa region compared with the so-called "paraplatform". Acta Geologica Sinica, 40(2), 1960.

31. Chen Guoda & Chen Enmin, Geotectonics.* Central-South Inst. Mining & Metallurgy, 1962.

32. Chen Guoda, Huang Su et al. Preliminary conclusion on the main characteristics of the regional geotectonics of China.* Geotectonic Problems of China, pp. 96-116. Science Press, 1965.

33. Chen Guoda et al., A brief introduction to the tectonics of China.* Scientia Geologica Sinica, 1975(3).

34. Chen Guoda, The N-S Diwa Region in relation to the evolutionary differentiation of the crust in China and to the cause of uplift of the Qinghai-Xizang Plateau.* Geotectonica et Metallogenia, 3(1), pp.1-10, 1979.

35. Chen Guoda, Guan Yinwen et al., Geotectonic metallogenetic conditions of the Shilu-type iron ore deposits, Hainan Island. Geotectonica et Metallogenia, 2(2), 1978.

36. Chen Guoda et al., Outline of geotectonics of China.* (Explanation for the Tectonic Map of China, 1 : 4,000,000), pp. 1-122., Seismology Press, 1977.

37. Chen Guoda et al., Tectonic Map of China, 1 : 4,000,000 (based on diwa theory). Map Press, Beijing, 1977.

38. Chen Guoda, Diwa (geodepression)theory and its practical significance. Resume,26th IGC, Vol. I, Sect.5, pp.324, Paris, 1980.

39. Chen Guoda, Polygenetic compound ore deposits and their origin in the context of regularities in crustal evolution. Geotectonica et Metallogenia, 6(1) 1-55, 1982.

40. Chen Guoda, The diwa-type metallogensis in E. China. Resume, 27th IGC, Vol. 6, Sect. 12, p.62, Moscow. 1984.

41. Chen Guoda, Methods of Metallotectonic researches.* Geology Press, Beijing, 1978. 2nd edition, 1985.

42. Chen Guoda et al., On the geotectonic types of the oil and gas fields of China. Geotectonica et Metallogenia, 3(1), 1979.

43. Chen Guoda and Fei Baosheng, Oil fields of diwa (geodepression) type. Geol. Oil & Gas, 1(3), 1980.

44. Chen Guoda, Elevated wave-cut benches near Canton and the origin of the compound shoreline of Kwangtung. Sci. Journ. Sun Yetsen Univ., New Ser., 1(1), 1948.

45. Chen Guoda, Huang Su et al., Geotectonics and oil-gas, Northern Xinjiang. Monograph. Changsha Inst. Geot., Academia Sinica, 1983.

46. Chen Guoda et al., On the Geotectonic character of Tianshan in Meso-Cenozoic Time. 60 Anniversary Congress, Geol. Soc. China, 1983.

47. Chen Guoda, Southern Himalaya Diwa (Geodepression) Region. Geotectonica et Metallogenia 9(1), 1985. Resume, 27th IGC, Moscow, 1984.

48. Chen Guoda, Geotectonic Nature of the Fen-Wei rift system. Resume, 27th IGE, Moscow, 1984.

49. Chen Enmin et al., Preliminary discussion on the 1605 Qiongzhou earthquake and its seismogenetic structure. Seismology and Geology, 1(12), 1979.

50. Chen Liyu, On the geotectonic character of the Tongren area at the Qinghai-Gansu boundary.* Geotectonica et Metallogenia, 5(1), 1981.

51. Chen Qinwu, Transition from active to quiet period in diwa development is an important period of uranium mineralization. Geotectonica et Metallogenia, 6(1), 1982.

52. Chen Rung and Chen Yun-tai, Preliminary results of the investigation of the crustal structure of the South-Central part of Shansi Province. Acta Geophysica Sinica, 17(3), 1974.

53. Chen Xiangzhong, Tectonics of Southern Jiangxi and controlling factors of tungsten deposits.* Geotectonica et Metallogenia, 5(2), 1981.

54. Chen Xianying, Transformation of platform coal-bearing sediments by diwa activity. Geotectonica et Metallogenia, 6(2), 1982.

55. Cheng Yuqi et al., The Pre-sinian of north and northeast China.* Scientia Geologica Sinica, 1(1973).

56. Chinese Academy of Geological Sciences, Geological Atlas of the People's Republic of China.* China Geological Map Printshop, 1973.

57. Compilation Group of "Chronological table of Important Events in Natural Sciences", Chronological table of important events in natural sciences.* Shanghai People's Press, 1975.

58. Cong Bailin et al., Magmatic activities in Xichang Prefecture and their bearing on tectonics.* Scientia Geologica Sinica, 2(1973).

59. Ding Menglin et al., The Datong volcanic swarm.* Geographical Knowledge, 2(1976).

60. Deng Kangling, Geology of Gorge area.* Geological Review, 20 : 4 (1960).

61. Deng Qidong et al., Seismotectonic conditions of the fault depression earthquake zone, Shanxi uplift area, and trend in earthquake development.* Scientia Geologica Sinica, 1(1973).

62. Ding Peimin et al., Helan-Zhuozi Mountain activated platform.* Geological Review, 18 : 6 (1958).

63. Du Fangquan, A discussion of the tectonic conditions of formation of Daye Type Iron Ore Deposit, Hubei Province. Bull. Changsha Inst. Geot., Acad. Sinica, 1, 1984.

64. Fu Taizhi, A preliminary study of the beginning time of diwa (geodepression) stage in Tonghua, Jilin Province.* Geotectonica et Metallogenia, 4(2), 1980.

65. Gu Qingge e al., Stratigraphic division of Xizang.* Geological Bureau of Xizang. 1979.

66. Guan Yinwen, On the characteristics of the sedimentary formations of mobile regions. Geotectonica et Metallogenia, 8(2), 1985.

67. Guo Zengjian et al., Focus mechanism studies of China's great earthquakes from 1949 through 1960.* Acta Geophysica Sinica, 11 : 1 (1960).

68. Geological Institute, Academia Sinica, et al., Geology of Qilian Mountain* (vol.2). Science Press, 1960.

69. Geological Institute, Academia Sinica, et al., Geology of Qilian Mountain (vol.1).* Science Press, 1960.

70. He Chunsun, Tectonic history of Taiwan. Tectonophysics, 4 4-6 (1967).

71. He Yiewei, The geological characteristics of Sandu coal field of diwa type at Zixing in Hunan. Geotectonica et Metallogenia, 8(2), 1984.

72. Heilongjiang Geological Bureau, Geology and metallogeny of the Da Hinggan Mountains and adjoining regions.* Geological Press, 1959.

73. Hu Chaoyuan, Block crustal movement and Mesozoic-Cenozoic development in China.* Geological Review, 18 : 4 (1958).

74. Huang Peihua, On the problem of geomorphological development in Yunnan.* Chinese Quaternary Research, 3 : 1-2 (1960).

75. Huang Ruihua, Some geochemical features of the platform stage in East China. Geotectonica et Metallogenia, 8(3), 1984.

76. Huang Su and Wang Daojing, Tectonic features of North Xinjiang and formation of oil and gas pools. Geotectonica et Metallogenia, 8(4), 1984.

77. Huang Su, An approach to the basic geotectonic characteristics of North Xinjiang. Geotectonica et Metallogenia, 9(1), 1985.

78. Huang Su et al., Study on the geotectonic conditions of petroleum formation in the Eocene of Dongting Diwa. Bull. Changsha Inst. Geotect., Acad. Sinica, 1, 1984.

79. Huang Xian, Influence of diwa tectonism upon the coal field of Hunjiang River. Geotectonica et Metallogenia, 5(2), 1981.

80. Huang Yukun, Problem of upheaval and subsidence since Quaternary time on the coast of south China.* Acta Universitatis Sunyatseni, 2 (1974).

81. Igneous Rock Group, Nanling Regional Geological Survey Team, Guangdong Geological Bureau, Preliminary comprehensive report on Nanling intrusive rocks.* Geological Press, 1959.
82. Huang Ruihua, Some geochemical characteristics of the diwa regions of East China. Geochemistry, 1978 (3).
83. Han Kecong et al., Nappe structure in the Maoshan area and its geological significance. Geotectonica et Metallogenia, 9(1), 1985.
84. Jia Lanpo et al., Age determination of activity of the Datong volcanic swarm.* Geographical Knowledge, 1 (1976).
85. Jin Chengfo, Volcanic rocks of Tibet.* Magmatism and metamorphism of Tibet. Science Press, 1980.
86. Jiang Zhongti, Tectonic outlines of the Qinghai-Xizang Plateau and the problem on its origin.* The Fourth Geological Team of Xizang. 1979.
87. Jiao Shupei, On characteristics of the evolution of geological history of Tarim-Qaidam diwa region and the problem on its belongingness. Collected Works on Geology of Qinghai-Xizang Plateau, 15, Geology Press, 1984.
88. Kobayashi, Tectonic development of Taiwan (translated by Liu Xingyi).* Geological Publishing House, 1959.
89. Li Hongji et al., Statistical regionalization of the N-S earthquake zone in central China.* Scientia Geologica Sinica, 2 (1973).
90. Li Guopeng et al., Mesozoic-Cenozoic tectonic development and seismicity of east China.* Scientia Geologica Sinica, 2 (1973).
91. Li Peizheng, An analysis of the characteristics of metallogenetic feature and their relationship to mineralization in Yangjiazhangzi porphyry molybdenum ore field of diwa type. Journ. Central-South Inst. Mining & Metallurgy, 1982(4).
92. Li Shanbang, Earthquakes of China.* Earthquake Front, 1 (1974).
93. Li Zexin, Geology of Kangding and Tianquan, Sichuan Province.* Geological Review, 20 : 3 (1960).
94. Li Zucai and Huang Xian, Classification of geotectono-genetic types of coal-series in China, Kexue Tongbao (Science Bulletin), 27(4), 1982.
95. Li Zucai and Huang Xian, Geologic characteristics and genetic classification of coal-bearing formation in China. Bull. Changsha Inst. Geotectonics, Acad. Sinica, 1, 1984.
96. Li Zuwu, The relationship between reservoir earthquakes of China and geologic conditions.* Geotectonica et Metallogenia, 5(1), 1981.
97. Lin, C.C., Recommendation of the Tiwa (Diwa) Theory, and comment on the Tiwa (Diwa) in the western Taiwan (in Japanese). Jubilee Publication, Commemoration Prof. HAYASAKA, I, 60th Birthday, p. 379-395, 1967.
98. Liu Hongyun et al., Sinian of south China.* Scientia Sinica, 16 : 2 (1973).
99. Liu Hongyun et al., Tables of regional strata, China.* Science Press, 1956.
100. Liu Hongyun et al., Tables of regional strata, China (Continuation).* Science Press, 1958.

101. Liu Lijun and Wang Juncan, Uranium metallization of granite type and its bearing on the diwa tectonic movement. Geotectonica et Metallogenia, 7(2), 1983.

102. Liu Yixuan, Tectonic features of Guangdong.* Acta Geologica Sinica. 40 1 (1960).

103. Liu Yiyuan, A preliminary study on the metallogenetic zone of fault in the south China coastal region. Geotectonica et Metallogenia, 6(2), 1982.

104. Liu Yixuan, The geotectonics of South China Sea and continental margin activation. Geotectonica et Metallogenia, 8(3), 1984.

105. Liu Yixuan, A new type of diwa region---on the geotectonic evolution of the South China Sea. Geotectonica et Metallogenia, 8(2), 1984.

106. Liu Yuanlong et al., A study of the relation between the Mesozoic-Cenozoic oil fields and the crustal structure. Geotectonica et Metallogenia, 6(2), 1982.

107. Mo Hucen, Geological characteristics of diwa-type coal-bearing formation in the Datong coal field. Geotectonica et Metallogenia, 6(1), 1982.

108. Mo Zhusun, A discussion on the classification of Nanling granites according to geological environment. Geotectonica et Metallogenia, 9(1), 1985.

109. Mu Enzhi et al., Stratigraphy of Mount Qomolangma, south Tibet, China.* Scientia Geologica Sinica, 1 (1973).

110. Niu Guanglin et al., A preliminary study of the basic features of diwa development in Changbai-Xing'an region. Geotectonica et Metallogenia, 6(2), 1982.

111. Ouyang Shahuai, The Central Yunnan Diwa and sandstone copper deposit. Bull. Changsha Inst. Geot., Acad. Sinica, 1, 1984.

112. Peng Gelin, Geotectonic characteristics of Hunan, types of coal fields and the direction of search for coal.* Geotectonica et Metallogenia, 4(1), 1980.

113. Ruan Daoyuan, A preliminary research on geochemistry of REE mineralized district in Northeast Guangdong. Geotectonica et Metallogenia, 8(2), 1984.

114. Saklani, P. S., Tectonic geology of the Himalaya. Today and Tomorrow's Printers and Publishers, New Delhi, India, 1978.

115. Shi Zhenliang et al., Strong earthquake activity in China and plate tectonics.* Scientia Geologica Sinica, 2 (1973).

116. Sinha-Ray, S., Reworked Tibetan microplate in relation to Asian foldbelts. Abstract, 26th IGC, Vol. I, Paris, 1980.

117. Staritsky, Yu. G., Maymin Yu.S., Trofimov V. A., Tectonic development of North Vietnam. International Geology Review, 15 : 12 (1973).

118. Sui Jinxia, A preliminary study of the characteristics of diwa-type volcanic rock in Southeast Guangdong. Geotectonica et Metallogenia, 8(2), 1984.

119. Sun Jiafu, Features of Carboniferous polygenetic compound ore deposits in East Hubei. Geotectonica et Metallogenia, 8(1), 1984.

120. Sun Jiafu, On the genesis of Huangmei iron ore deposits in Hubei. Bull. Changsha Inst. Geotect., Academia Sinica, 1, 1984.

121. Sun Jiafu, Some features of porphyry copper deposits in Southern Hubei Province. Geotectonica et Metallogenia, 6(1), 1982.

122. Sun Dianqing, Tertiary strata and tectonic features of northern Qaidam Basin.* Geological Press, 1959.

123. Teng Chi-wen et al., Deep reflected waves and the structure of the earth crust of the eastern part of Chaidam Basin. Acta Geophysica Sinica, 17(2), 1974.

124. The Fourth Section, 106 Geological Team, Sichuan Geological Bureau, Pre-sinian geological characteristics of the central segment of the Kangdian Axis and its relationship with plate tectonics.* Scientia Geologica Sinica, 2(1975).

125. Tian Zaiyi et al., Collected papers on generation of oil in continental sediments and search for it in China.* Petroleum Industrial Press, 1960.

126. Tong Yuming, Characteristics of the Pingxiang-type coal-bearing formation of diwa aspect in Northwestern Jiangxi. Bull. Changsha Inst. Geotect., Acad. Sinica, 1, 1984.

127. Wang Mujing et al., On the diwa tectonism and the formation of uranium ore deposits. Geotectonica et Metallogenia, 7(1), 1983.

128. Wang Xuezheng et al., Some characteristics of uranium mineralization in the Hunan-Guangxi region. Geotectonica et Metallogenia, 9(1), 1985.

129. Wei Zhouling, On the tectonic character of Songliao Basin. Bull. Changsha Inst. Geot., Acad. Sinica, 1, 1984.

130. Wei Zhouling and Xie Jianhua, Formation types, oil potential and tectonic conditions of the diwa sediments in Dagang oil field. Geotectonica et Metallogenia, 7(2), 1983.

131. Wen Shanji, Geologic development of the N-E diwa region.* Geotectonica et Metallogenia, 3(1), 1979.

132. Wu Leibo et al., Analysis of regional structures in north Guangdong.* Collected Papers and Notes on Geomechanics, No. 2, 1960.

133. Xiang Dingqian, Geotectonic framework of Xinjiang.* Geological Review, 19 : 2 (1959).

134. Xiang Jixi et al., The formation and development of the Tancheng-Lujiang Rift Zone within Anhui Province and its control over the mineralization. Bull. Changsha Inst. Geotect., Acad. Sinica, 1, 1984.

135. Xiao Chuibin, Determination of the direction of prospecting for the Songshumao molybdenum deposits, Yangjiazhangzi, based on the conditions of its formation.* Geotectonica et Metallogenia, 5(1), 1981.

136. Xiao Xuchang et al., Tectonic features of chromium-bearing basic-ultrabasic rocks.* Chinese Industrial Press, 1965.

137. Xing Fengming, The nature of the Kangdian Axis in the context of the geology of Huili and Yanbian.* Scientia Geologica Sinica, 9 (1959).

138. Xu Ren et al., Quaternary paleobotanic studies and upheaval of the Himalayas in Mount Qomolangma.* Kexue Tongbao, 18 : 6 (1973).

139. Yang Hongzhi, The initial inquiry into the regional geochemical characteristics of tungsten in South China. Bull. Changsha Inst. Geotect., Acad. Sinica, 1, 1984.

140. Yang Hongzhi, An essay on diwa tectonics and plate movement in South China. Geotectonica et Metallogenia 8(1), 1984.

141. Yang Lihua, Neotectonics of Mount Qomolangma.* Scientia Geologica Sinica, 3 (1974).

142. Yao Chen-hing, Reflected waves from the M-discontinuity and the structure of the Lin Fen Basin. Acta Geophysica Sinica, 17(4), 1974.

143. Yao Zhenkai, Geology of strata-bound uranium deposits in diwa regions of China and regularities in their distribution.* Geotectonica et Metallogenia, 5(1), 1981.

144. Yao Zhenkai, The geotectonic types and their main character of uranium deposits in China. Geotectonica et Metallogenia, 7(2), 1983.

145. Yao Zhenkai et al., The metallotectonic features of a strata-bound uranium deposit formed by diwa metallogenesis. Bull. Changsha Inst. Geotect., Acad. Sinica, 1, 1984.

146. Ye Junying et al., Quaternary geology and geomorphology of the Dachaidan area, northeast Qaidam Basin.* Chinese Quaternary Research, 1 1(1958).

147. Yeh Hung et al., The analysis of the recent tectonic stress of the Himalaya arc-shaped mountain system and its vicinities.* Scientia Geologica Sinica, 1 (1975).

148. Yin Hezhong, A discussion on several tectonic problems of North Xinjiang. Geotectonica et Metallogenia, 9(1), 1985.

149. Yin Jixiang et al., Marine Jurassic of west Yunnan.* Scientia Geologica Sinica, 3 (1973).

150. Ying Sihuai, Magmatic, metamorphic and migmatite rocks in Mount Qomolangma, south Tibet, China.* Scientia Geologica Sinica, 2 (1973).

151. Zhang Bosheng, Precambrian of Zhongtiao Mountain and its geotectonic development.* Journal of Northwest University (Natural Sciences), 1958.

152. Zhang Qi et al., Metamorphic belts and metamorphism of Tibet.* Magmatism and metamorphism of Tibet. Science Press, 1981.

153. Zhou Shoulu, Characteristic features of granite in Tengchong County, Yunnan Province.* Geotectonica et Metallogenia, 5(2), 1981.

154. Zhou Wushi, Study of the metallogeny of pyrite deposits in China by applying diwa theory. Geotectonica et Metallogenia, 5(2), 1981.

155. Zhou Yunsheng et al., Granite belt of Himalaya.* Magmatism and metamorphism of Tibet, Science Press, 1981.

156. Белоусов,В.В., Явление тектонической активизации в развитии земной коры. Активизированные зоны земной коры. Наука,1964.

157. Казанский,В.И., Рудоносные тектонические структуры активизированных областей. Недра,1972.

158. Масайтис,В.Л., Старицкий,Ю.Г., Об особом типе структур Восточной Азии / третий основной структурный элемент континентальной земной коры -- дива/. Материалы по общей и региональной тектонике, том 85, 1963.

159. Масайтис,В.Л. Старицкий, Ю.Г., Структуры дива Восточной Азии. Строение и развитие земной коры. Наука,1964.

160. Николаев, Н.И., О некоторых теоретических и методических вопросах изучения новейшей тектоники Китая. Издательство"Наука", Пекин, 1959.

161. Чэнь Го-да, Область дива-послеплатформенная подвижная область нового типа. Китайская Наука, 14/10/,1965.

162. Щеглов, А.Д., Металлогения областей автономной активизации. Недра, 1968.

* Titles marked with asterisks are in Chinese.